Reassessing Epistemic Images in the Early Modern World

Reassessing Epistemic Images in the Early Modern World

Edited by
Ruth Sargent Noyes

Routledge
Taylor & Francis Group
LONDON AND NEW YORK

This book was generously made possible by the Mads Øvlisen Fellowship in Art History from the Novo Nordisk Foundation.

novo nordisk fonden

First published in 2023 by Amsterdam University Press Ltd.

Published 2025 by Routledge
4 Park Square, Milton Park, Abingdon, Oxon OX14 4RN
605 Third Avenue, New York, NY 10158

Routledge is an imprint of the Taylor & Francis Group, an informa business

ISBN: 9789463723350 (hbk)
ISBN: 9781041184911 (pbk)
ISBN: 9781003702306 (ebk)
NUR 685

Cover illustration: Abraham Bosse, Preparing the etching plate, from *Traicté des manières de graver en taille douce sur l'airin* (Paris: Chez Bosse, 1645), plate 1, etching. Duke University, Durham, NC, David M. Rubenstein Rare Book & Manuscript Library, NE1760 .B73 1645 c.1.

Cover design: Coördesign, Leiden

DOI: 10.5117/9789463723350

To Lembit
Lembitule

Table of Contents

Part 3 Imprint, Knowledge, and Affect

1. Prologue

For a Metaphorology of Engraving: From Epistemic Images to an Imaged Epistemology

Ralph Dekoninck

Abstract

The volume preface takes up the question of the reversibility between knowledge and image through the issue of engraving (and intaglio printmaking more broadly) as a metaphor for thinking knowledge in the early modern period. Reflecting on period epistemic-artistic metaphorology of engraving that enabled a thinking through of the actual plastic processes entailed in the reception and production of knowledge, this chapter interrogates how a new technique such as engraving generated or reactivated and thereby transformed rich metaphorical networks, enabling a re-thinking of certain issues at the intersection of knowledge, belief, and vision in early modernity.

Keywords: metaphorology, engraving, Claude Mellan, media studies, mnemonics, theology of images

In the guise of a preface to this volume, and as an overview and extension of the authors' reflections contributing to an exploration of prints' epistemic dimension, I would like to take up the question of the reversibility between knowledge and image through the issue of engraving (and intaglio printmaking more broadly) as a metaphor for thinking knowledge.[1] In his *Paradigms for a Metaphorology* of 1960, Hans Blumenberg interrogates the place of metaphors in Western thought. He thus underscores how metaphors

1 As it will be shown in what follows that period sources tend to gesture to burin engraving, acid etching, and sometimes even xylography interchangeably, this essay occasionally references specific techniques, while addressing intaglio techniques more generally.

Noyes, R.S. (ed.), *Reassessing Epistemic Images in the Early Modern World.*
Taylor & Francis Group, 2023
DOI 10.5117/9789463723350_CH01

(e.g. light as a metaphor for truth, or the legibility of the world, to which he dedicated a book),[2] far from simple ornaments of philosophical discourse or efficacious didactic means, allow for a consideration of that which resists conceptualization. Metaphors constitute *Grundbestände*, which is to say "fundamental elements" "from which and upon which philosophy operates, and do not allow themselves to be overtaken within a conceptuality."[3] That within this reflection on metaphorology artistic metaphors number among Western thought's fundamental elements can be confirmed, for example, in the metaphor of sculpture in Aristotle's theory of causality (doctrine of the four causes). They have enabled a thinking through of the actual plastic processes entailed in the reception and production of knowledge. Among such epistemic-artistic metaphors is that of engraving. This begs the question: how might a new technique such as engraving – a term that refers at once to procedure and product – generate or reactivate and thereby transform rich metaphorical networks, enabling a re-thinking of certain issues at the intersection of knowledge, belief, and vision in early modernity?[4]

In bringing an introductory reflection to this volume, I would thus propose an interrogation not so much of the ways that engraving specifically (and print-making more broadly) produced knowledge, but rather of how knowledge was thought of as engraving. I would likewise underscore that the latter, in return, by means of its own medialogical characteristics, preserves the material and imaginary memory of not only technical but also mental processes that presided over its genesis. If, according to McLuhan's formulation, the medium is the message, this is also in the sense that medium constitutes an extension of the individual (as McLuhan specified),[5] an extension that I understand in the sense of an analogy between the mental and material fabric of an image. From this follows the idea that the medium thinks and people think with the aid of the medium. Put differently: not only does the medium produce knowledge, but it also enters into an epistemic phase wherein the modes of knowledge production are conceived simultaneously to the production of the image.

On this point, it is important to relativize the modern rupture, as it is clear that engraving and the imaginary that conveys it are characterized by the survival and metamorphoses of a certain indexical imaginary inherited

2 Blumenberg, *Die Lesbarkeit*.
3 Blumenberg, *Paradigms*, p. 5.
4 I would permit a reference to Dekoninck, "*Formatur unicus una, non alter.*"
5 See *Understanding Media*.

from ancient philosophy and Christian theology.[6] Suffice it here to recall the importance of the metaphor of the imprint in ancient theories of the soul, since Plato's *Theaetetus* (191 c–d) and Aristotle's *De Anima* (427b, 18–22) in the field of philosophy, and Cicero's *De oratore* (II, LXXXVI, 354) and Quintilian's *Institutio oratoria* (XI, II, 17–22) in the domain of rhetoric, where the imprint is intimately connected to the *artes memoriae*. During antiquity, the soul was, in effect, conceived as a ductile material assimilated to wax, on or in which images conducted through the canal of the senses were imprinted in the manner of a seal. Such a noetic metaphor – or, more precisely, mnemonic metaphor since it essentially takes into account the *modus operandi* of memory – would remain among the principle epistemic analogies in the domain of theories of the soul through the early modern period. It is nonetheless interesting to note that this becomes progressively adapted to the new ascendant technologies of engraving and imprinting, although closer scrutiny reveals this to be much more than a simple transfer of meaning. Rather, we are witnessing the emergence of a new way of thinking about the close relationship between image and psychic functioning, whose effects on the very apprehension of engraving must be taken into account, particularly as regards its triple scientific, religious, and artistic vocation.[7]

Let us take two examples from the field of children's education: in a work dedicated to a defense of the religious image, French Jesuit Louis Richeome cites the example of a three-year-old child who, having seen and named birds in the work of naturalist Pierre Belon, can recognize and identify them.[8] Richeome's conclusion is particularly striking: the engravings, he claims, functioned as "*matrix and burin* to his little memory, so as to *engrave* so appropriately these *impressions* on the capacity of his age" (emphasis added).[9] Here, the perfect coincidence between the nature of perceived images and these images' effect on the mind, and particularly the memory, gives rise to engraved images that, in turn, engrave, according to the somewhat curious expression "matrix and burin" eliciting the sphere of agentive images

6 For a broader anthropological perspective, see Didi-Huberman, *La Ressemblance par contact*.

7 On this point, I would add William B. MacGregor's remarks: "how this conceptual system potentially structured the culture's experience of actual prints? If the mind was thought to be like an engraved copper plate or a sheet of paper imprinted with figures, what of the interface with its material referents? Put simply, what might be thought to happen when people with print- or plate-like minds looked at printed images?" In "The Authority of Prints," p. 411.

8 Richeome, *Trois discours*, p. 623.

9 *Ibid.*, p. 624.

(*imagines agentes*) that engraves the child's memory.[10] This physiological truth reappears in Fénelon's *Traité de l'éducation des filles* (Treatise on the education of girls), according to which the "softness of [children's] minds is such that everything is easily *imprinted* thereupon, and the images of all sensible objects are very vivid," while the author insists on that fact that "the images to be *engraved* there must be carefully chosen […] The first *engraved* images when the mind is still soft and nothing is yet written there are the deepest" (emphasis added).[11] One can see here the old hermeneutic of the imprint encountering that more recent of engraving, with the matrix remaining more or less similar to wax, whereas the imprinting mechanism is clearly taken from the realm of engraving.

Moving away from what appears tantamount to a psychological commonplace, it is clear that the notion of an almost instantaneous cerebral impression of the perceived image is progressively called into question, starting with Descartes. If the metaphor of the intaglio imprint (specifically acid etching) appears in the fourth discourse of his 1637 *Dioptrique* (Dioptrics), this is invoked not so much to explain the workings of memory, but rather to evoke the nature of the image produced by the imagination:

> As you see that etchings, made from nothing more than a little ink placed here and there on paper, represent to us forests, towns, people, and even battles and storms, and although they make us think of countless different qualities in these objects, it is only in respect of shape that there is any real resemblance. And even this resemblance is very imperfect, since [these prints] represent to us bodies of varying relief and depth on a surface which is entirely flat. […] Thus, it often happens that in order to be more perfect as an image and to represent an object better, etchings ought not to resemble it. Now, we must think of the images formed in our brain in just the same way, and note that the problem is to know simply how they can enable the soul to have sensory perceptions of all the various qualities of the objects to which they correspond, and not to know how they can resemble these objects.[12]

10 The deep mark left by the perceived image moreover imprints the imagination, which is largely dependent on it, as Malebranche will recall several years later: "[…] we imagine such things that much more strongly, if these marks are deeper and better *engraved*." Malebranche, *De la Recherche*, p. 207.

11 Fénelon, *De l'éducation des filles*, esp. pp. 55–69.

12 Descartes, *La Dioptrique*, p. 204. Adapted from a translation in Descartes, *Philosophical Writings*, I, pp. 165–166. See also MacGregor, "The Authority of Prints," pp. 404–408; Marion, *Sur la théologie*, pp. 249–255; Dumont, *Descartes et l'esthétique*, pp. 82–93; Leonhard and Felfe, *Lochmuster und Lienienspiel*, pp. 29–33.

The previously prevailing analogical link has, in a certain sense, been broken, hereafter there is no longer a relationship of contiguity between the referent and its image, no confusion between the model and its reproduction, which were suggested by the reference to the imprint. That transference gives way to translation or transposition signals the end of real presence and inaugurates the reign of representation.[13] In this new *episteme*, the intaglio print is divested of its indexical symbolism and supplanted by the underlying abstractive process of the model of the *camera obscura*. This allowed a great economy of means for an "infinity of different qualities," or, put differently, for an optimal result in terms of knowledge of reality.

In 1673, this idea reappears summarized and simplified in a completely different context, that of the "philosophy of images" as elaborated by Jesuit Claude-François Ménestrier. In one of the treatises that comprise his vast iconological project, he traces metaphors borrowed from the field of artistic techniques to explain the work of each of the six faculties of the soul. He wrote regarding intaglio printmaking (including xylography): "as it is in the nature [of sensible images] that are graven on copper or wood to be printed [...] [and] those that are printed and pulled from engraved images [...], imagination engraves images in the soul and on the body, [...] [and] memory prints and arranges them."[14] It is clear from this passage that the image of gravening as a creative process is preferable to that of the impression, as the imagination was no longer solely conceived of as a receiving faculty. Even the memory is no longer restricted to imprinting, but now arranges as well. As for the notion that a lively imagination engraves not only the soul but also the body, this constitutes another point of intersection with ancient physiology, which will reappear in seventeenth-century spiritual literature.

If the reign of the indexical paradigm thus seems to come to an end in the domain of the sciences, it effectively survives in the field of spirituality, where it tends toward a different referential horizon: no longer only that of ancient theories of the soul, but also that of Christian theology. To designate the imaginal condition of man, Augustine explicitly invokes the indexical paradigm. The image of God in man is of the order of the trace in search of its matrix: "The true honor of man," he writes in *De Trinitate*, "is being the image and resemblance of God, an image that can only be preserved by recourse the one by whom it is printed [*a quo imprimatur*]" (Trin., XII, 11, 16). This same anthropological principle resurfaces in Ménestrier. For him,

13 See Havelange, *De L'œil et du monde*, pp. 332–333.

14 Ménestrier, "Avertissement," in *Le veritable art*, n.p.

if man thinks and creates only through and in images, it is because he is an image-copy of the Creator:

> Man has such an inclination for his origin that he loves all of its copies, to the point of forgetting the original from which they were taken. This love seems right, since it is founded on resemblance, and being the Image of God he relates to pictures, and is sympathetic to portraits and painting.[15]

This passage has the merit of being quite explicit regarding the conceptualization of the divine creature as taken from its original: certainly a diminished copy, but one that nevertheless retains the trace of the original resemblance of an ontological nature, which makes him naturally appreciate any form of reproduction.

This genetic relationship is commonly thought of in the mode of *vestigium*, a concept widely exploited across Latin patristic and scholastic literature, as well as its Greek counterpart, *typos* (τύπος), which covers a vast semantic field, closely related to the lexicon of the Christian image.[16] That in the seventeenth century the calcographic metaphor largely inherits from the Christian *vestige* is witnessed, for example, by a passage in François de Sales' 1616 *Traité de l'amour de Dieu* (Treatise on the Love of God):

> Let us imagine, I pray you, on the one hand, a painter making a picture of Our Saviour's birth (and I write this in the days dedicated to this holy mystery). Doubtless he will give a thousand and a thousand touches with his brush, and will take not only days, but weeks and months, to perfect this picture, according to the variety of persons and other things to be represented. On the other hand, let us look at a printer of pictures, who having spread his sheet upon the plate which has the same mystery of the Nativity cut in it, gives but a single stroke of the press: in this one stroke, Theotimus, he will do all his work, and instantly he will pull [from the plate] a picture representing in a fine etching all that has been imagined, as sacred history records it. Though he performed the work with one movement, yet it contains a great number of personages, and other different things, each one well distinguished in its order, rank, place, distance and proportion: so that one not acquainted with the secret would be astonished to see proceed from one act so great a variety of effects.[17]

15 Ménestrier, *L'Art des emblèmes*, pp. 1–2.
16 See Goyet, "De la rhétorique," pp. 46–67.
17 de Sales, *Traicté*, 73. See Legros, *François de Sales*, pp. 159–163.

This rather unconventional *paragone* between painting and intaglio printmaking – or, to be more precise, between the work of a painter and that of a printer of images – thus rests on the following analogy: the time-consuming labor of the painter is contrasted against the almost instantaneous work of the printer. For the thousand brushstrokes, a single pull of the intaglio press seems to conflate the very realization of the image itself, thereby effacing the long and fastidious work of the engraver (or etcher). Here the image's printing is equivalent to its revelation. This technical act appears almost miraculous: a single movement begets infinite effects.

That this double analogy appears in a spiritual work, however, raises the question of the deep import of this parallelism between the ends and means of the respective media. This metaphor no longer occludes a scientific truth as in Descartes, but rather a spiritual meaning:

> ...like the painter, nature multiplies and diversifies her acts accordingly, as the works she has in hand are various, and it takes her a great time to finish great effects, but God, like the printer, has given being to all the diversity of creatures which have been, are, or shall be, by one only stroke of his omnipotent will. He draws from his idea as from a well-cut plate, this admirable difference of persons and of things, which succeed one another in seasons, in ages, and in times, each one in its order, as they were to be.[18]

The comparison can be summarized thus: God is to nature, what the printer of images is to the painter – a comparison whose full implications (including for example the ambivalence between the act of engraving and that of printing) are beyond the scope of this chapter, but whose essence for our present purposes can be limited to the immediacy or instantaneity superintending the act of creation, achieved through a single strike or pull from the divine press.

In the Christian tradition, the trace left by this process is nevertheless perceived as an obscure and indistinct mark, removed from and inferior to its matrix. Only the Incarnation could reveal a true imprint, as one pulled from a graven plate. The Son, perfect Icon of the Father, thus reconciled the image and its model, restoring the perfect identification between the seal and its imprint.[19] The image that incarnates in the strict sense this

18 *Traicté de l'amour de Dieu*, p. 74.

19 The sigillary metaphor that dominated in eleventh- and twelfth-century pre-scholastic theology, largely derived from Neoplatonism, assimilated the *imago*, understood as the seal's

identification, rendering it visible, is the Holy Face, in the form of three archetypes: the Mandylion, the Veronica, and the Turin Shroud. In the second half of the sixteenth century, Catholics exalted these three prestigious contact relics – miraculous imprints resulting from a real impression – as tangible proof of the legitimacy of the image, as they were held to perpetuate in some way the incarnational economy. Their striking kinship with the printed image engendered evocative connections, justifying most notably the multiplication of these *acheiropoieta*.[20]

Louis Richeome, speaking of the various *sudaria* in Turin, Besançon, and Spain, thus explicitly invoked the metaphor of pulling a print: the Shroud is composed of "two large sheets, where our Savior is printed [*tiré*] front and back life-size, as he was when entombed. This is preserved in Turin and Besançon; both have performed many miracles. The Veronica is likewise preserved, in a church in Spain. That these are found in different places is no cause for amazement: because it could be that many were printed [*tirées*] together, or that he who miraculously printed the first also miraculously multiplied them."[21]

The printing or "pulling" – a master engraver would pull proofs – of these *acheiropoieta* enabled engraving to furnish a model that, in turn, empowered a conceptualization of the miracle. In return, the model of the venerable icon profoundly imbued the ideation of engraving. The deliberate linking of the Veronica's impression to printing is evidenced by the frequency with which this *sudarium* was used as a printer's mark. The acheiropoietic miracle could thus be viewed as printing's origin myth, the ideal of a direct revelation of truth – not to be confused with a true imitation of reality, as *mimesis* here gives way to *genesis*, the latter divested of any presumption of mimetic identity.[22] Suffice it to think, too, of intaglio prints reproducing pilgrimage cult images and conserving their auratic traces, which the accompanying inscriptions never fail to emphasize. Thus, Israhel van Meckenem's engraving *Imago pietatis* is accompanied by the legend "This image was *counterfeited* [*contrafacta*] according to the manner and resemblance of the first *Imago Pietatis* in the church of Santa Croce [di Gerusalemme] in the city of Rome

imprint, to the human condition, whereas only Christ could be himself assimilated and co-extensive with the seal, i.e. consubstantial with the divine matrix. On this, see especially the work of Bedos-Rezak, e.g. "Replica."

20 The prototypical account was that of the legend of the Mandylion of Edessa, produced through auto-duplication by means of an impression on cloth (*Keramidion*). See Kessler and Wolf, pp. 95–108; Lecercle, "De la relique à l'image."

21 Richeome, *Trois discours*, pp. 611–612.

22 Lecercle, "Le signe et la relique," II: 487.

[...]" (emphasis added). Peter Parshall has thoroughly underscored the importance of the term "counterfeit" (*conterfeit, contrafactur*) in the domain of not only religious intaglio prints (with the attending notion of the contact relic), but also the scientific (with the idea of objective eyewitness).[23] Distinct from the realm of imitation, this concept suggests the idea of an identical reproduction, as if the world had come to rest on the image.

While the religious and scientific spheres began to separate in early modernity, they nonetheless shared a common indexical paradigm grounding a particular notion of truth. In order to set them into mutual dialog, I would like to close with an evocation of two engravings by Claude Mellan: one, his celebrated *Holy Face* of 1649; the other, his less well-known but equally fascinating series of three engravings of the moon's phases (commissioned by Nicolas-Claude Fabri de Peiresc and Pierre Gassendi), engraved after sketches the artist made in Aix-en-Provence in early 1636 based on telescopic observations.

It is worth recalling the technical and artistic tour de force that is Mellan's *Holy Face*[24] (Figure 1.1). This image is formed by the spiraling circonvolution, starting from the tip of the nose, of a single alternatively swelling and tapering line, a continuous and undulating line that winds or unwinds depending on the viewer's perspective, to the point of completely filling the void of the sheet of paper, whose limits coincide with the veil's edge. This creates a kind of identity between paper and veil, material support and figured support, as the latter at once conceals and reveals the former – or perhaps just the opposite, as this coincidence engenders a particular trouble between presence and representation, even a confusion between engraving and acheiropoietic imprint. This confusion is admittedly not total, since the veil's presence is revealed at the bottom of the image by a discreet curvature revealing a slight margin, a fictive stony background on which are engraved (or rather chiseled) both the date and artist's name, and the continuation of the legend printed on the *sudarium* itself. The fact remains that, except for this detail, as Irving Lavin writes, "the sheet of paper is and represents at the same time the single image which represents the single face."[25] As if this perceptive elision was not sufficiently explicit, Mellan added the legend: *Formatur unicus una, non alter* ("uniquely formed, [like] no other"), a species of *motto* that emblematizes the image, *motto* invented by Michel

23 Parshall, "*Imago contrafacta.*"

24 See Sgard, *La Sainte Face*; Préaud and De Lavergnée, *L'œil d'or*, 92–96; Macgregor and Bonfait, *Il Dio Nascosto*, 170–173.

25 Lavin, "Il Volto Santo de Claude Mellan."

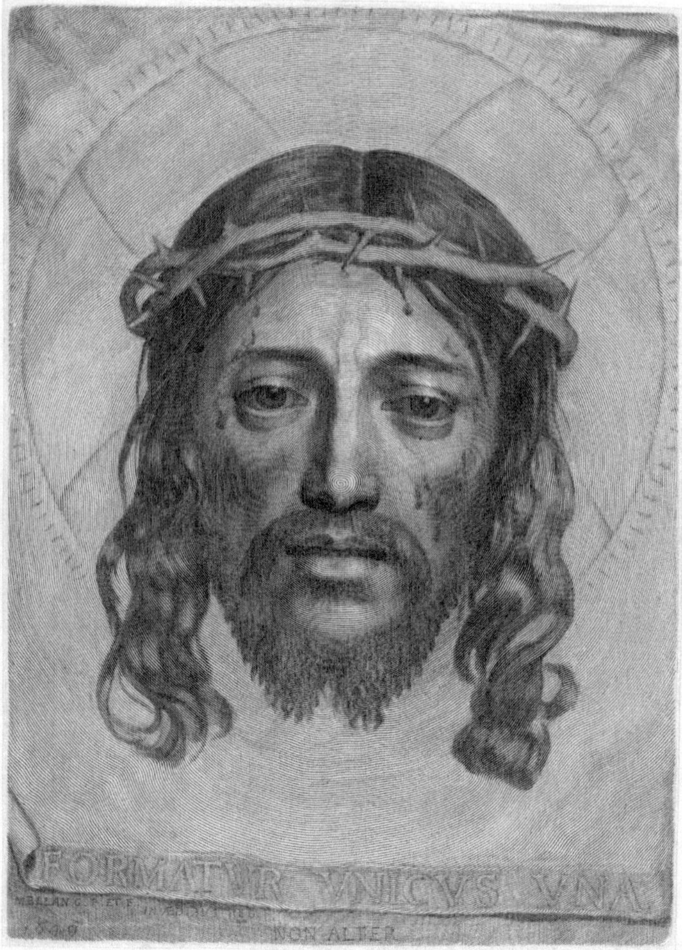

Figure 1.1 Claude Mellan, *Face of Christ on St. Veronica's Cloth* (*Holy Face*), 1649. Engraving, 46 × 34.8 cm. Rijksmuseum, Amsterdam, RP-P-OB-69.798.

de Marolles, friend of Mellan and avid collector of the artist's works. In his Memoirs, de Marolles gives the following explication:

> *Formatur unicus una* alludes to the beauty of the only Son of the Eternal Father, born of a virgin, and with a single spiral line with which the artist has so well drawn the portrait, with this other phrase written below, *Non alter*, because there is no one who resembles this First of

> the Predestined, and because the engraver of this image has made such a masterpiece that another would have difficulty imitating it and creating its equal.[26]

In a kind of cascade or procession of nested prototypes, *unicus* therefore refers not only to Christ himself, unique Image of the Eternal Word and prototype of every Christian, but also to the unique image of Christ, i.e. the Veronica, prototype of all Christological (and, indeed, Christian) images. The term *una* designates at once the Virgin, often called *unigenita*, and the artistic masterpiece, both conceived as corporal and visual mediation of the incarnation. The mystery of the universal line, creating an uncircumscribable portrait, meets the Christological mystery, representing in one stroke human and divine. The chiropoietic wonder merges the acheiropoietic miracle to the point of confusion, the union of technique and subject accomplished so as to become a figure of the incarnation.

If the first horizon is indeed of a theological nature because of the subtle play on the infinite line uniting the Christological prototype to its pictorial type – which is really only the image (the engraving) of an image (the Veronica veil) of an Image (Christ himself) – by means of a progressive shifting of thresholds toward the ultimate prototype, this game is imperceptibly transformed into a purely mental exercise, provoking an admiration that is no longer so tethered to the model in the image, but rather to the image's inventor. The prototype is now relocated to the brilliant mind of the artist; the masterpiece becomes no more than a portrait of his technical and artistic *maestria*. A zone of indecision remains, however, given that the artist's hand tends both to assert and occlude itself, as if to better pay homage to the prototype represented, were it not for his own talent.

The result of this dialectic is an eloquent effect in the rhetorical sense: in this case, it is the imprint that leaves a lasting mark in the mind. The figure of Christ becomes a veritable figure of speech that simultaneously diverts and attracts the gaze, intensifying the almost supernatural revelation by obscuring the technical process that makes it possible, and exposing the geometric underpinnings to the point of vanishing the model as the gaze approaches it – all ultimately a function of the beholder's distance from the image. These hypotheses are not mutually exclusive: this movement is, indeed, twofold, and the two artistic and theological dimensions are coextensive. To take up the distinction made by Thomas Aquinas concerning

26 *Les Mémoires de Michel de Marolles*, 266. Cited in Préaud and Brejon de Lavergnée, *L'œil d'or*, p. 121.

the just adoration due the image of Christ (*Summa theologiae*, pars 3, quaest. 25, art. 3), the movement of the gaze toward the image as an image and the movement toward the model in the image to some extent coincide.[27]

Due to this overlap of type and prototype with theological-artistic resonances, Mellan's engraving can be considered as representative of a visual culture still trying to achieve synthesis between two spheres that then tend to grow progressively and mutually distant. The image printed at the dawn of the era of mechanical reproduction clearly appears as the multiple type of a single prototype, where the latter refers not only to the engraved matrix, but also to the divinely created universe whose trace is preserved more or less distinctly in every image, with man living only in a world of "deferred transmission." The new technique takes hold of the symbolic universe of an indexical Christian paradigm that precedes it and gives it meaning.

That which is valid for the relationship between art and religion in the formation of knowledge of the divine is likewise pertinent for their relationship with the production of scientific knowledge. To close the loop opened in Ruth Noyes's introduction to the volume, regarding Galileo's *Macchie Solari* ("Sunspot Letters") and Greuter's *intaglio finissimo*, I will close by way of Mellan's representations of the lunar phases (First Quarter, Full, and Final Quarter), called *icons* in the accompanying legend[28] (Figure 2.2). Although the selenographies were not realized with a single unbroken burin line, they are still the result of a series of parallel continuous lines that evince no less of a technical tour de force than *Holy Face*. Apart from technique, what do the selenographic prints have in common with the latter? It could be said that the representation of the moon, as the face of Christ, is a challenge to representation. That Peiresc dedicated the work to Gassendi as "a memorable work for all time,"[29] and Mellan himself spoke of "a very new thing,"[30] can be compared to Cesi's remarks on the images illustrating Galilean telescopic solar observations that "delight in the wonder of the spectacle and the accuracy of the expression." Moreover, Peiresc maintained, writing to dal Pozzo, that the technical-representational challenge was such that Mellan initially could find neither "artisan nor a machine suited to the delicacy of the work, the proofs being stained and poorly printed, because the printers did not know how to work the ink."

27 Dekoninck, "Le double mouvement."

28 Préaud and Brejon de Lavergnée, *L'œil d'or*, pp. 115–119. Jaffé, "Mellan and Peiresc."

29 Cited in Préaud and Brejon de Lavergnée, *L'œil d'or*, p. 118.

30 Cited in *ibid.*, p. 118.

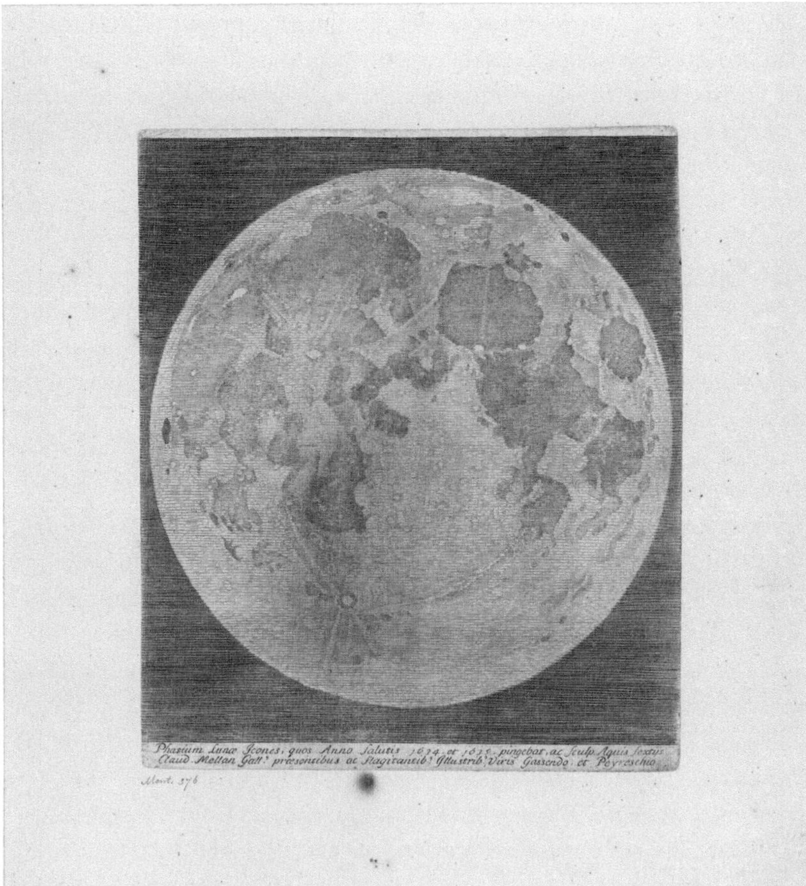

Figure 1.2 Claude Mellan, *Full Moon*, from *Three Representations of the Moon*, 1635. Engraving, 24.8 × 21.7 cm. Rijksmuseum, Amsterdam, RP-P-OB-71.150.

This might be construed as an implicit challenge to François de Sales' topos of the facile printer of images, because, in this case, the engraved matrix after nature exceeded the printer's abilities.

Beyond the well-established period analogies assimilating the Virgin with the moon and Christ with the sun,[31] and apart from the same striking effect, which can be likened to that of an image-eye (Mellan's self-proclaimed "eye of gold," a phrase he deployed to describe his geometrical science of representation) that fixes and hypnotizes the beholder, what emerges from this comparison between the Holy Face and the phases of the moon is again

31 On these tropes, see Ostrow, "Cigoli's Immacolata"; Booth and Van Helden, "The Virgin and the Telescope."

exegesis of the relationship between type and prototype.[32] When it comes to probing the invisible – be it natural or supernatural – such as aspects of the Christological mystery or that which is revealed through the telescope, engraving not only becomes the instrument of knowledge and its diffusion, but also continues to carry the idea of unmediation, of a deposit/transfer of the real, or revelation (in the photographic sense, so to speak) of the truth. This notion aligns with Christian anthropology and iconology, which persisted in conceiving of man and the pictorial productions of his mind according to the paradigm of the imprinted trace. The intaglio image, which is the result of the epistemic dynamic, and the hermeneutical dynamic that it generates, continue to be conceptualized and experienced within the framework of this congruence of *techne* and *episteme*.

Bibliography

Bedos-Rezak, Brigitte Miriam. "Replica: Images of Identity and the Identity of Images in Prescholastic France." In *The Mind's Eye: Art and Theological Argument in the Middle Ages*, edited by Jeffrey Hamburger and Anne-Marie Bouché, pp. 46–64. Princeton, NJ: Princeton University Press, 2006.

Blumenberg, Hans. *Paradigms for a Metaphorology.* transl. Robert Savage. Ithaca, NY: Cornell University Press, 2010.

Blumenberg, Hans. *Die Lesbarkeit der Welt.* Frankfurt am Main: Surhkamp, 1981.

Booth, Sara Elizabeth, and Albert Van Helden. "The Virgin and the Telescope: The Moons of Cigoli and Galileo." *Science in Context* 13, no. 3–4 (2000): 463–486.

Dekoninck, Ralph. "*Formatur unicus una, non alter.* Enjeux théologiques, Rhétoriques et Artistiques du Prototype dans le Monde de la Gravure Durant la Première Modernité." *Degrés: Revue de Synthèse à Orientation Sémiologique* 146 (2011): 1–16.

Dekoninck, Ralph. "Le Double Mouvement de l'Âme vers l'Image: Une Théorie Aristotélico-Thomiste au Cœur des Débats du Milieu du XVIe Siècle sur le Juste Rapport à l'Image Religieuse." *Revue de Théologie et de Philosophie*, 154 (2022), 277-296.

Descartes, René. *La Dioptrique* [1637], in *Œuvres et lettres*, edited by André Bridoux. Paris: Bibliothèque de la Pléiade, 1953.

32 According to Antoine Furetière's definition, "type" is the "Copy of a model, character engraved or imprinted by some other thing. Less used than its cognates *prototype* and *archetype*, which are originals made without a model. This word comes from the Greek *typos*, signifying *figure*." In *Dictionaire Universel*, p. 761.

Descartes, René. *The Philosophical Writings of Descartes*, transl. John Cottingham, Robert Stoothoff, and Dugald Murdoch. 3 vols. Cambridge: Cambridge University Press, 1985.

Didi-Huberman, Georges. *La Ressemblance par Contact. Archéologie, Anachronisme et Modernité de l'Empreinte*. Paris: Éditions de Minuit, 2008.

Dumont, Pascal. *Descartes et l'Esthétique. L'Art d'Émerveiller*. Paris: Presses Universitaires de France, 1997.

Fénelon, François de Pons de Salignac de La Mothe. *De l'Éducation des Filles* [1687], in *Œuvres*, edited by Jacques Le Brun. Paris: Bibliothèque de la Pléiade, 1983.

Furetière, Antoine. *Dictionaire Universel*. The Hague & Rotterdam: Arnout & Renier Leers, 1690.

Goyet, Francis. "De la Rhétorique à la Création. Hypotypose, Type, Pathos." In *La Rhétorique. Enjeux de ses Résurgences*, edited by Jean Gayon, Jacques Poirier and Jean-Claude Gens, pp. 46–67. Brussels: Ousia, 1998.

Havelange, Carl. *De L'œil et du Monde. Une Histoire du Regard au Seuil de la Modernité*. Paris: Fayard, 1998.

Jaffé, David. "Mellan and Peiresc." *Print Quarterly* 7 (1990), 168–175.

Kessler, Herbert L, and Gerhard Wolf, eds. *The Holy Face and the Paradox of Representation: Papers from a Colloquium Held at the Bibliotheca Hertziana, Rome and the Villa Spelman, Florence, 1996*. Bologna: Nuova Alfa Ed., 1998.

Lavin, Irving. "Il Volto Santo de Claude Mellan. *Ostendatque etiam quae occultet*." In *République des Lettres, République des Arts. Mélanges offerts à Marc Fumaroli*, edited by Christian Mouchel and Colette Nativel, pp. 385–414. Geneva: Librairie Droz, 2008.

Lecercle, François. "Le Signe et la Relique. Les Théologies de l'Image à la Renaissance." PhD Dissertation, 2 vols. Montpellier: Université de Montpellier, 1987.

Lecercle, François. "De la Relique à l'Image: La Promotion du Suaire de Turin." In *Symboles de la Renaissance III*, ed. Daniel Arasse, Maurice Brock, and Georges Didi-Huberman, pp. 95–112. Paris: Éditions Rue d'Ulm, 1990.

Legros, Philippe. *François de Sales, une Poétique de l'Imaginaire. Étude des Représentations Visuelles dans l'*Introduction à la Vie Dévote *et le* Traité de l'Amour de Dieu. Tübingen: Gunter Narr Verlag, 2004.

Macgregor, Neil, and Olivier Bonfait, eds. *Il Dio Nascosto. I Grandi Maestri del Seicento e l'Immagine di Dio*. Rome: Edizioni De Luca, 2000.

MacGregor, William B. "The Authority of Prints: An Early Modern Perspective." *Art History* 22 (1999): 389–420.

McLuhan, Marshall. *Understanding Media: The Extensions of Man*. New York: McGraw-Hill, 1964.

Malebranche, Nicolas de. *De la Recherche de la Vérité* [1674–1675], edited by Geneviève Rodis-Lewis. Paris: Bibliothèque de la Pléiade, 1979.

Marion, Jean-Luc. *Sur la Théologie Blanche de Descartes. Analogie, Création des Vérités Éternelles et Fondement*. Paris: Presses Universitaires de France, 1981.

Marolles, Michel de. *Les Mémoires de Michel de Marolles*. Paris: Antoine de Sommaville, 1656.

Ménestrier, Claude-François. *L'Art des Emblèmes où s'Enseigne la Morale par les Figures de la Fable, de l'Histoire, & de la Nature*. Lyon: Benoît Coral, 1662.

Ménestrier, Claude-François. *Le Veritable Art du Blason ou l'Usage des Armoiries*. Paris: Estienne Michallet, 1673.

Ostrow, Steven F. "Cigoli's Immacolata and Galileo's Moon: Astronomy and the Virgin in Early Seicento Rome." *Art Bulletin*, 78, no. 2 (1996): 218–235.

Parshall, Peter. "*Imago Contrafacta*: Images and Facts in the Northern Renaissance." *Art History* 16 (1993): 554–579.

Préaud, Maxime, and Barbara Brejon de Lavergnée, eds. *L'Œil d'Or: Claude Mellan, 1598–1688*. Paris: Bibliothèque nationale, 1988.

Richeome, Louis. *Trois Discours pour la Religion Catholique. Des Miracles, des Saincts et des Images*. Bordeaux: Simon Millange, 1598.

Sales, François de. *Traicté de l'Amour de Dieu*. Lyon: Pierre Rigaud, 1626.

Sgard, Jean. *La Sainte Face de Claude Mellan. Étude des Bases Géométriques du Dessin*. Abbeville: Société d'Émulation Historique et Littéraire d'Abbeville, 1957.

Trémolières, François. "L'Art Sacré au Prisme de Fénelon." *Dix-Septième Siècle* 230, no. 1 (2006): 49–69.

About the Author

Ralph Dekoninck is professor of early modern art history at the Université catholique de Louvain, co-director of the Centre for Early Modern Cultural Analysis (GEMCA) and member of the Royal Academy of Belgium. His research focuses on early modern image theories and practices, specifically in their relation to spirituality.

2. Introduction

Pittura filosofica: Etching Galileo's Sunspots and the Discursive Field of Early Modern Epistemic Images

Ruth Sargent Noyes

Abstract
Galileo Galilei's *History and Demonstrations on Solar Spots and Their Properties*, a collection of letters on observations of sunspots, came off the Roman presses in spring 1613 with 38 printed illustrations that minutely reproduced Galileo's telescopic drawings. Printed from etched copperplates by artist Mattheus Greuter, these prints' production consumed half of the book's entire budget. This introductory chapter explores the crucial rhetorical and epistemological role that Greuter's sunspot etchings and the artist's painstaking methods played in conveying and even constituting Galileo's controversial Copernican arguments against broader themes throughout essays in the book as whole, with special attention to re-examining the discursive field of early modern epistemic images in recent scholarship.

Keywords: Galileo, Lincean Academy, astronomy, etching, Mattheus Greuter, telescope, sunspots

I. Ink on Paper: Etching as Sensible Experience

As I mentioned, his Lordship the Marquis of Monticelli sent me your *Discourse* on the solar spots at the onset of my illness. Because I couldn't read it myself, I held onto it for a long time to think on it or have a copy made. Since His Holiness [Pope Paul V] needed to show it to certain Cardinals, I just read it by myself, and as much as was possible I considered it carefully, and looked together with the images of the [sun]spots. The first pleasure was this: to see that for many of the days when I myself had

Noyes, R.S. (ed.), *Reassessing Epistemic Images in the Early Modern World*.
Taylor & Francis Group, 2023
DOI 10.5117/9789463723350_CH02

> looked and noted [the sunspots], your notes compared to mine, allowing that my images did not have so many small mini-spots nor so distinct like your own, because I had not made use of such a perfect [telescope] in observing them [as you did]. The second was to hear the arguments, five it seemed to me, with which Your Lordship sufficiently proved, as far as I am concerned, that the spots are contiguous with the solar body. [...] And I believe that they will appear the same to whomever may have even some small understanding of [linear] perspective, or natural judgement.[1]

This passage was written by Giovanni Battista Agucchi, a Bolognese art theorist and collector, prelate and papal courtier, amateur astronomer and Galilean acolyte. He was among the first recipients of a partial manuscript copy of a book on solar telescopic astronomy in preparation in Rome in the autumn of 1612.[2] *Istoria e dimostrazioni intorno alle macchie solari e loro accidenti* (*History and Demonstrations on Solar Spots and their Properties*, hereafter *Macchie Solari*) collected letters on observations of sunspots made by Galileo Galilei and a pseudonymous rival astronomer (who was, in fact, German Jesuit astronomer-mathematician Christoph Scheiner).[3] The book that came off the Roman presses in spring 1613 in a first edition of about 1400 copies coupled an epistolary dialog between Galileo and Scheiner with printed illustrations of their respective observations of the solar spots.[4] Even before Pope Paul V (Camillo Borghese) himself, in October 1612, at the onset of a long illness and around the time Galileo's text was under revision by Vatican censors, Agucchi received a partial manuscript copy that contained Galileo's first two of three letters on the spots (the third letter was only finished in December). Having studied *Macchie Solari* while convalescing, on 1 December 1612, Agucchi dispatched a triumphant review, cited in part above, in a letter from Rome to his astronomer friend in Florence, linking his recovery from disease and reclamation of his faculty of reason to the victory of Galileo's arguments over "natural judgment" and the book's passage from manuscript to print in overcoming the censors.[5] His evocation of physical

1 Galilei, *Le Opere di Galileo* (hereafter: *OG* followed by volume number), XI: 441–442.

2 On Agucchi see Toesca and Zapperi, "AGUCCHI."

3 Galilei, *Istoria e dimostrazioni*. For an excellent recent translation with commentary, see Galilei and Scheiner, *Galileo and Scheiner on Sunspots*. See also Reeves and Van Helden, "Idiom and Image." For an overview of surviving copies of the book, see Mayer, "An Interim Report." On Scheiner, see Daxecker, *The Physicist and Astronomer Christoph Scheiner*.

4 On the printed illustrations, see Biagioli, "Picturing Objects in the Making." See also Noyes, "Mattheus Greuter's Sunspot Etchings," and further bibliography below.

5 On Agucchi's relations with Galileo, see Favaro, "Giovanni Battista Agucchi."

suffering paralleled the censorial travail inflicted upon Galileo's text by Church revisers before it was conferred the *imprimatur* on 4 November, a trial Agucchi himself doubtless had some knowledge of, despite his professed diminished mental state.[6]

Agucchi had been temporarily entrusted with an advance copy from the publication's chief editor and sponsor, Umbrian nobleman Federico Cesi, then known as the Marquis of Monticelli.[7] Cesi also presided as founder and patron of the *Accademia dei Lincei* (Academy of the Lynxes or Lincean Academy), an international society based in Rome and dedicated to the generation of knowledge across all domains of the natural sciences, through observation, experiment, and inductive methods, of which Galileo was a recent member.[8] Despite operating during the Counter-Reformation when the censorship of the Inquisition dominated the papal city, thanks to Cesi's stature and diplomatic skills the *Lincei* managed to pursue an intellectual agenda sometimes at the fringes of philosophical and theological orthodoxy, winning elite members and powerful allies like Agucchi, whose benefaction and protection were essential to continue prosecuting Lincean knowledge-making pursuits.[9] In the case of *Macchie Solari*, for example, Cesi pre-circulated manuscript copies amongst high-ranking cardinals and their family members.

That he also sent one for the convalescing Agucchi to examine and forward to the Vatican was a shrewd move: formerly in service to the cardinal-nephew of Pope Clement VIII, the Bolognese ecclesiast had excellent connections and knowledge of the papal court, and his interests in art and astronomy intersected with those of the Tuscan scientist and his Lincean confrères, as well as the dual missive of the provocative book, which Agucchi clearly recognized despite his professed impairments.[10] An avid art collector, c. 1610–1615 he authored a *Trattato della pittura* (Treatise on painting), published posthumously.[11] Following the art theorizing of Bolognese archbishop Gabriele Paleotti, Agucchi may have sympathized with

6 On the censorship of the book, see Mayer, "The Censoring of Galileo's Sunspot Letters."

7 On Cesi, see Baldriga, *L'occhio della lince*; Freedberg, *The Eye of the Lynx*; Graniti, *Federico Cesi*; Galluzzi, *The Lynx and the Telescope*.

8 For a recent overview of the extensive bibliography on the Accademia dei Lincei, see Brevaglieri, *Natural desiderio di sapere* and "Science, Books and Censorship in the Academy of the Lincei."

9 See Freedberg, *The Eye of the Lynx*.

10 On Agucchi's astronomical activities in connection to Galileo and the Lincei, see Favaro, "Giovanni Battista Agucchi," and Bucciantini, "Teologia e nuova filosofia."

11 On Agucchi's art theoretical writings, see Mahon, *Studies in Seicento Art and Theory*, pp. 109–154, 231–275.

Cesi's own conceptualization of *pittura filosofica* (philosophical picturing), image-making with the primary end of effectively conveying knowledge: "not just aimed at pure enjoyment, which is just a vane abuse, but meant to be a vivid and efficacious discipline and fruit of deep utility."[12]

Agucchi's invocation of his experience of viewing the text of *Macchie Solari* in tandem with images of the sunspots raises the question of which images, if any, he had in hand besides his own comparatively poorly detailed telescopic drawings. Perhaps he was given copies of the painstaking detailed series of telescopically projected drawings made by Galileo and assistant(s) of the spots' daily movements across the face of the sun over several months.[13] Galileo maintained the methods he employed were "absolutely precise" and figured the spots' changing form and position "without a hairsbreadth of error" (*senza error d'un minimo capello*).[14] Or, perhaps he was given proofs (or partial proofs) of the 38 illustrations that minutely reproduced the Galilean drawings, printed from etched copperplates and delivered as loose unbound sheets fresh from the presses to be viewed side by side with the manuscript text[15] (Figure 2.1). These prints' production consumed half of the book's entire budget, a fact that underscores the importance of especially copperplate engraving and etching to the Lincean agenda more broadly – "whence we will advance [our order] and will be able to represent each and every observation," according to Cesi.[16]

That they were crucial to *Macchie Solari* specifically is seconded by Agucchi's claim that the illustrations in tandem with the text's arguments would convince anyone with "some small understanding of [linear] perspective, or natural judgement." More forcefully, Cesi declared that Galileo's telescopic images "delight both by the wonder of the spectacle and by the accuracy of expression. It is now your adversaries' turn to be horsewhipped by this sensible experience, because in arguing with it they abuse reason."[17] Cesi's gesture to the *sensate esperienze* ("sensible experiences") – which Galileo

12 On Cesi, the Lincei, and this concept, see Baldriga, *L'occhio della lince*, p. 14 n21. See also discussion in *Idem*, "'La fatiga di pigliar i disegni dalle piante'" and "Reading The Universal Book of Nature."

13 On Galileo's methods, see Reeves, "Mere projections," p. 53. See also van Helden, "Galileo and Scheiner on Sunspots."

14 Galilei and Scheiner, *Galileo and Scheiner on Sunspots*, pp. 104–105.

15 *OG* XIX: 265–266. See Biagioli, "Picturing Objects in the Making." On Gabriele Paleotti's *Discorso intorno alle imagini sacre e profane* (Discourse on Sacred and Profane Images), see Noyes, "*Aut Numquid Post Annos Mille Quingentos.*"

16 On the costs of producing *Macchie Solari*, see Gabrieli, *Il Carteggio Linceo*, pp. 67–68. On the crucial role of visual pursuits in the Lincean projects, see Freedberg, *The Eye of the Lynx*.

17 *OG* XI: 393.

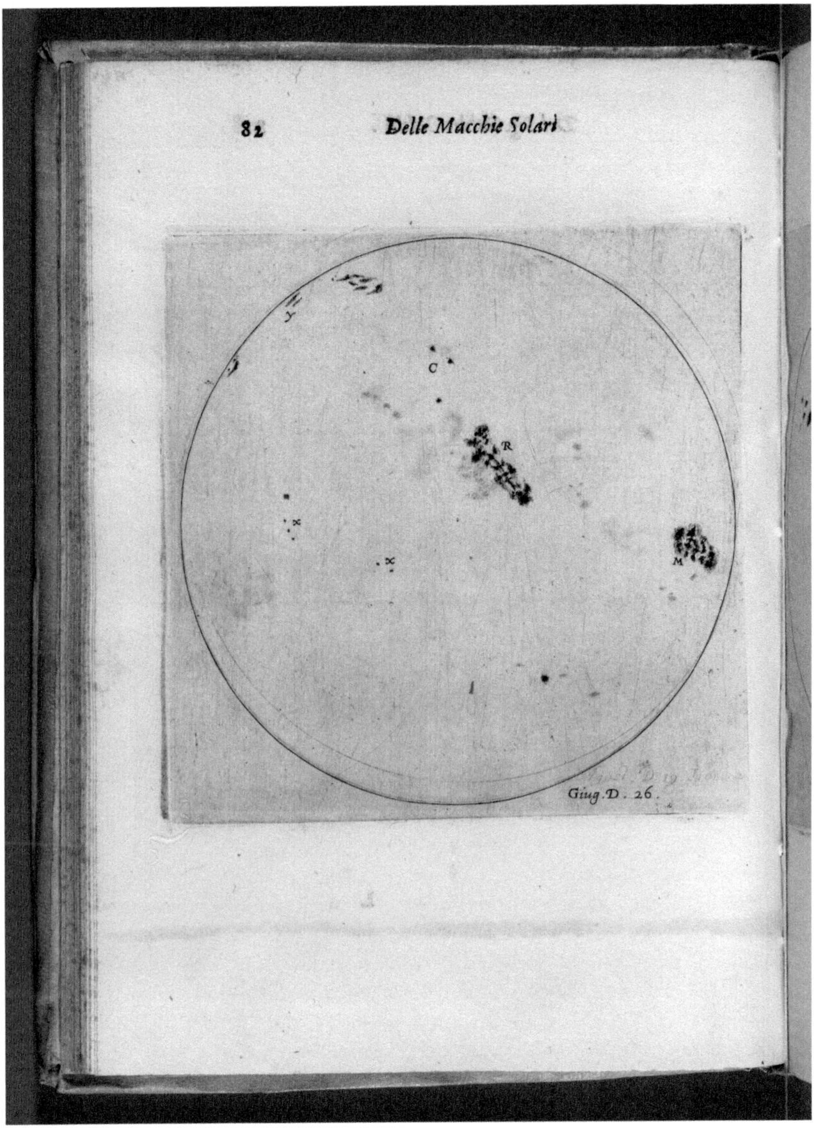

Figure 2.1 Mattheus Greuter, Sunspots, from Galileo Galilei, *Istoria e dimostrazioni intorno alle macchie solari e loro accidenti*, Rome: Appresso Giacomo Mascardi, 1613, page 82, etching, diam. of circle 4 7/8 in. (12.5 cm), showing multiple offsets and ink bleed-through from verso. Harvard University, Cambridge, MA, Houghton Library, hyde ic6 g1333 613ia.

would mention as crucial in his scientific method[18] – here in connection to imprinted images, raises the issue of the possible role that printed pictures

18 De Caro, "On Galileo's Platonism, Again," esp. p. 99.

and their production may have played in this method, particularly in connection to Galileo's use of experiments both real and imagined.[19]

While the ailing Agucchi was scrutinizing *Macchie Solari* against his own observational notes and professedly inferior telescopic drawings, he may also have had in hand loose unbound proof sheets freshly printed from the Roman workshop of artist Mattheus Greuter, a German Catholic convert refugee recruited for this project through Cesi and Galileo's mutual friend, Tuscan painter Lodovico Cardi (or *il Cigoli*), who had worked with Greuter during the latter's sojourn in Florence a few years before.[20] Between late September and early November, in close collaboration with Cesi and Cigoli, Greuter quickly translated his signature *intaglio finissimo* (ultrafine manner), hitherto used for burin-engraving miniature sacred devotional prints into an acid-etching technique that inimitably captured the stain-like sunspots' vaporous appearance, blurred contours, and elusive luminescence.[21] He then personally oversaw the printing of his 38 copperplates, likely in his own well-equipped workshop, adjusting various inks and papers to "make darker those [spots] in the center [of the solar sphere]" for a heightened perspectival effect, and best preserve the plates' ultrafine scratches through close to 2,000 runs under the intaglio press's massive pressure, which forced the paper into the faint abrasions in the copper.[22] The vast amount of negative space on each plate relative to the small etched patches that receive ink entailed painstakingly wiping excess ink from the soft metal surface for

19 On the question of if and to what extent Galileo engaged in thought experiments versus real experiments, particularly in relation to the legacy of Alexandre Koyré (who claimed Galileo did not perform real experiments, but only thought experiments), see Settle, "An Experiment in the History of Science"; Drake, "Galileo's Discovery of the Law of Free Fall"; *Idem*, "Galileo's Experimental Confirmation of Horizontal Inertia"; MacLachlan, "Galileo's Experiments with Pendulums"; *Idem*, "Experimenting in the History of Science." On the concept more broadly in the context of early modern science, see Dear, *Discipline and Experience*; *Idem*, "The Meanings of Experience."

20 On Greuter, see Baglione, *Le vite de' Pittori*, pp. 398–400; Diefenbacher and Leuschner, *The Greuter Family*. For helpful overviews, see also Guerrieri Borsoi, "GREUTER, Matthäus" and Bell, "Johann Friedrich Greuter" and "Matthaus Greuter." On Cigoli's collaboration with Galileo, see Galilei and Scheiner, *Galileo and Scheiner on Sunspots*, pp. 48–49, 75, 129, 172, 233, 246, 257, 267. On his work with Greuter, see Chappell, "Lodovico Cigoli" and Diefenbacher and Leuschner, *The Greuter Family*, I: xli. For his work with Greuter on the sunspots, see e.g. *OG* XI: 405, 422, 428 and Noyes, "Mattheus Greuter's Sunspot Etchings."

21 For a helpful introduction to the etching process with video demonstrations, see also Spira *et al.*, "Etching." On Greuter's manner and its confessional implications in the Roman context, see Noyes, "'One of Those Lutherans'."

22 *OG* XI: 422, 474–475. On Greuter's workshop in this period, see Bilancia, "Le lastre e le stampe di Matthäus Greuter."

each printing without leaving scratches, which would mar the uniformity of unmarred cream paper that stood in for the sun's disk, or even be misread as sunspots. The estimated print run for the book of around 1400 copies, each of which required 38 sunspot copperplates run through the presses, would have thus necessitated a total of well over 50,000 tedious wipings of the plates. This process, together with the plates' delicate etching, surely made them a technical nightmare for Greuter and his workshop to print. Each inking risked marring the copper and each run through the press degraded the subtle incised grooves and the raised burr, which was likely made with a drypoint needle after the etching stage. The burr characteristically captured a small amount of ink and contributed to the impression of the spots' paradoxical darkly luminescent quality.[23]

The etchings of the spots, whose precision and correlation with his own observational drawings delighted Agucchi, reified Galileo's theories that the spots were cloud-like vapors issued from the sun's body by force of extreme pressure induced by heat, thus intrinsic to the sun (rendering the hitherto immaculate solar body corruptible), and that their movement implied the sun's rotation around a central axis (resituating it in the cosmos).[24] With his own artistic sensitivities, Agucchi immediately grasped the importance of the crucial effect of stereographic projection and the foreshortened depiction of the sunspots: "To those who understand, because of [the laws of] perspective, what the receding of the spherical surface near the extremities of the visible hemisphere means, this will be a clear argument for both the sphericity of the Sun and the proximity of the spots to the solar surface."[25] Despite the artist's great care in printing, unintentional scrapes were inevitably incurred upon the copper in the course of wiping the plates and transferred some ink to paper, particularly in the earliest prints, when any residual burr was more likely to leave such marks, which the press subsequently gradually effaced with successive printings.[26] In this way, across copies of the book the fading inadvertent plate abrasions and the gradual degradation of the etched spots together rehearsed the transmutations and evaporations of the sunspots themselves argued for in Galileo's text, such that aspects of the print medium's inherent non-fixity and instability resonated with

23 On this effect, see Ivins, *How Prints Look*, 16; Reeves, "Mere projections," 57.

24 On these two points, see *OG* IV, 64; V, 140; XI: 354–355, 376, 428, 437–439, 446, 453, 460, 465. See also Galluzzi, *The Lynx and the Telescope*, pp. 106–107.

25 Galilei and Scheiner, *Galileo and Scheiner on Sunspots*, pp. 111–112. For the contrast between Galileo's treatment of his illustrations and those of his rival, Jesuit astronomer Christoph Scheiner, as a function of their differing views on the sunspots, see Reeves, "Mere projections," p. 51.

26 On this effect, see Angela Campbell's essay in this volume.

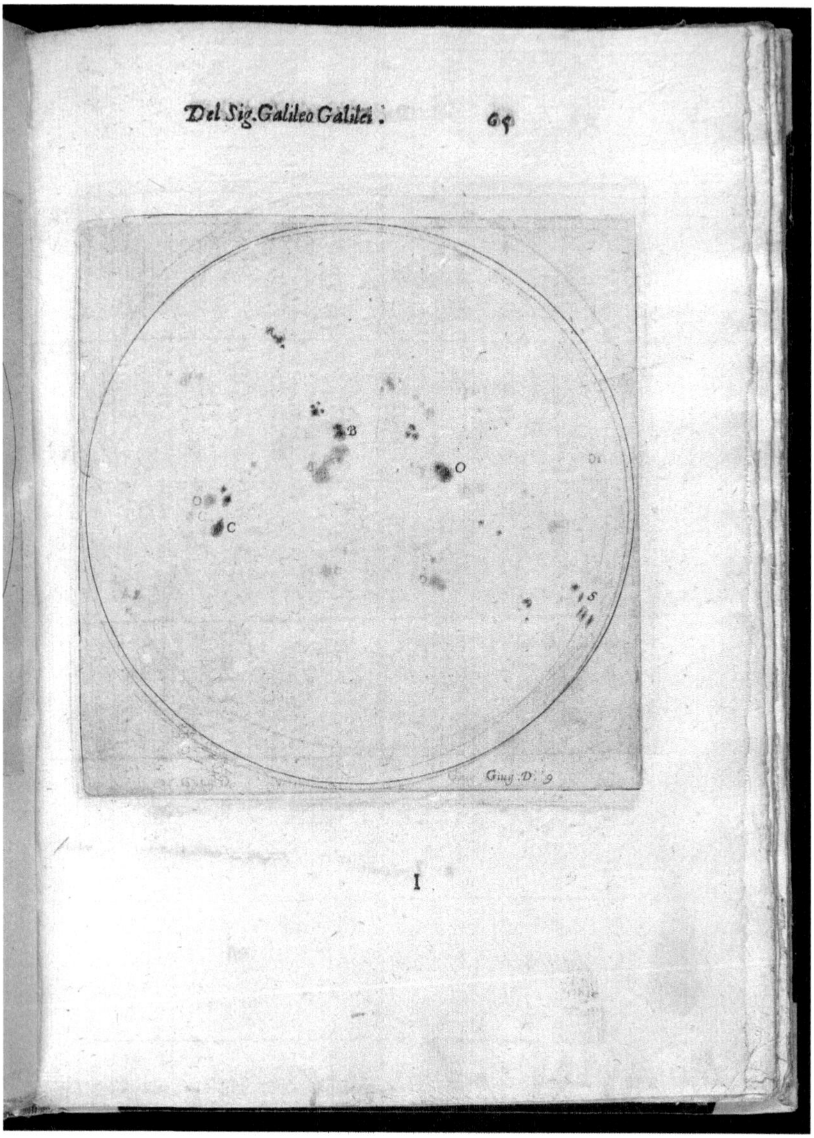

Figure 2.2 Mattheus Greuter, Sunspots, from Galilei, *Macchie Solari*, page 65, etching, showing foul biting from etching mordant (lower left corner). Yale University Library, New Haven, CT, Beinecke Library, QB41 G325 1613B.

the theoretical content of the scientific arguments in *Macchie Solari*. The intrinsic volatility of Greuter's etching technique registered too: distinct traces of foul biting left on certain plates by the etching mordant solution (mineral acid or solution of various salts in strong vinegar) can be found across prints in the series, appearing and disappearing as faint blackish stains (another meaning for *macchie*) and whorls against the light-colored void of the paper support[27] (Figure 2.2).

II. Bitumen on Iron: Printing Images as Galilean Experimental Method

Galileo's analogy in the text of the contrast of black "ink compared to this paper" rhetorically gestured to the sunspots' paradoxical brightness and translucence.[28] Viewed against Galileo's own artistic sensibilities, his bootstrapping of the discursive potential of Greuter's etchings merits reconsideration.[29] That Galileo was thinking more broadly about printmaking when theorizing on the spots can be inferred from his adaptation of a transfer technique from engraving and etching, where it was used to translate design to plate without inverting the original, to rectify the orientation of his original projected sunspot sketches.[30] From September 1612 through February 1613, as the text of *Macchie Solari* rapidly evolved under the intense pressure of Lincean and Vatican scrutiny, the Tuscan remained closely engaged with the German artist's own quickly developing working process, as Cesi regularly forwarded changing iterations of the print series.[31]

In keeping with his larger project to relocate from the mathematics of pure astronomy to a new philosophical astronomy that prosecuted "the true constitution of the universe,"[32] Galileo shifted the heuristic emphasis of his Copernican arguments about sunspots from the realm of mathematical abstraction to philosophical reality, using the rhetorical elaboration of analogies (between sun and earth, sunspots and clouds, celestial and terrestrial matter) "in order to make the unobservable in some sense observable, and

27 On this process and its constituent materials, see Stijnman, *Engraving and Etching*, pp. 199–202, 204–205.

28 Galilei, *Macchie Solari*, 13. For discussion of this passage, see Reeves, "From Dante's Moonspots to Galileo's Sunspots," p. 203. See also *Idem*, "Mere projections."

29 For various approaches to the issue of Galileo's artistic sensibilities, see Panofsky, *Galileo as a Critic of the Arts*; *Idem*, "Galileo as a Critic of the Arts"; Edgerton, "Galileo, Florentine 'Disegno,' and the 'Strange Spottednesse' of the Moon"; Reeves, *Painting the Heavens*; Booth and van Helden, "The Virgin and the Telescope"; Bredekamp, "Gazing Hands and Blind Spots"; *Idem.*, *Galilei der Künstler.*

30 See Bury, *The Print in Italy*, pp. 15–16; Reeves, "Mere projections," p. 53.

31 Cesi to Galileo on 6 October 1612, in OG XI: 409.

32 Mayer, "The Censoring of Galileo's Sunspot Letters," p. 7. On Galileo's belief in the reality of his vision of the universe, see also Stabile, "Linguaggio della natura e linguaggio della Scrittura in Galilei," p. 48.

the immutable in some sense manipulable."[33] This move would be the cause of much trouble in the ensuing Galileo Affair, beginning with the First Trial two years after the publication of *Macchie Solari*.[34] Galilei crucially invoked an experiment "proceeding on the basis of analogy with materials known and familiar to us" in his third letter on the sunspots that involved the sense experience of manipulating mundane materials to recreate, inferentially and in miniature, the sunspots: "Imitation of the spots. I would liken [the sunspots] to our clouds or smokes. Surely whomever might wish to imitate [the sunspots] by means of our [earthly] materials, I believe they could very easily find the most fitting imitation by putting on a scorching-hot iron plate some drops of incombustible bitumen, which would imprint on the iron a black spot, whence there will arise a black smoke that will disperse in strange and changing shapes."[35]

He clearly possessed an understanding of the material properties of bitumen or asphaltum, a naturally occurring petroleum residue (e.g. coal tar), which, in admixture with the principal ingredients of intaglio printers' ink (turpentine, pitch, and various boiled oils), enjoyed a near-ubiquitous presence in premodern industry, medicine (to treat wounds), and art, as it was used in dark-brown paint.[36] He likely knew, thanks to exchanges with Cigoli or another friend who worked in oils, Peter-Paul Rubens, that, despite bitumen's rich depth of hue and sheen, as an artist's pigment it had drawbacks: asphaltum prohibitively slowed the drying process and developed craquelure, corrupting bitumen paint into dark patches with an uneven cracked surface, so dramatically in some paintings that the process of "bitumization" deformed painted areas where asphaltum oxidized with wound-like eruptions resembling burn blisters.[37] Bituminous paint's para-

33 Feldhay, "Producing Sunspots on an Iron Pan."

34 The bibliographic references on the so-called Galileo Affair are staggeringly voluminous. On the censorial difficulty triggered by *Macchie Solari*, however, see Mayer, "Sunspot Letters." An excellent overview with bibliography is McMullin, "The Galileo Affair." For a helpful historiographical overview, see Finocchiaro, *Retrying Galileo*. For translations and discussion of relevant primary documents, see Finocchiaro, *The Galileo Affair* and *The Trial of Galileo*. For an overview of the circumstances of the First Trial, see Shea and Artigas, *Galileo in Rome*.

35 Galilei, *Macchie Solari*, p. 142. On this passage, see Feldhay, "Producing Sunspots on an Iron Pan."

36 For a premodern history of bitumen/asphaltum, see Forbes, *Bitumen and Petroleum in Antiquity*. For premodern medicinal applications, see Ambrose, "The Conquest of Pus." For artistic applications, see Featherstone, "Bitumen," and White and Kirby, "Rembrandt and his Circle," p. 71. See also further bibliography on bitumen in printmaking below.

37 On Galileo and Rubens, see Huemer, "Rubens's Portrait of Galileo" and Reeves, *Painting the Heavens*, pp. 68–76.

doxically glossy darkness and its tendency toward corruption counterfeiting effects of scorching heat conjured an evocative imaginary corollary for the sunspots familiar to art connoisseurs or, indeed, artists.[38]

This mention of burning bitumen redounded upon that of ink on paper, the latter made in the context of arguments regarding the relative brightness of solar spots compared to lunar spots.[39] Such ostensibly paradoxical discourse inflecting the luminescence of tar-black artists' pigments through analogical arguments about the sunspots' greater relative brilliance was not exclusive to Galileo, but pervaded Lincean rhetoric too, most pointedly in relation to the manufacturing of the etchings. In reporting on Greuter's choice of different weight papers to print his plates, Cesi relayed how the artist had selected thicker paper (so-called *carta doppia* or "double paper") for the donation copies, so that the printed spots on one side of the sheet would not "shine through" to the other (*acciò non traspaiano*).[40] In using the Italian verb *trasparire* – normally used for the action of light penetrating through another substance, most typically to describe the solar corona during an eclipse – Cesi ascribed a seemingly illogical luminosity to the dark patches of ink, as Galileo did in his letters, underscoring the normative utility of references to knowledge of the various properties of the constituent material elements of print for explicating otherwise imperceptible celestial properties in realist terminology.[41]

Asphaltum also appeared in some period recipes for intaglio printing ink, though it posed the challenge of drying time for printers.[42] Composed in late autumn 1612, during an intensive period of writing and revision of the text of *Macchie Solari*, which ran concurrent to Greuter's hasty manipulations of the etchings, Galileo's passage on experimenting with bitumen on a hot metal plate seemingly inflected knowledge of the material and visual properties of the dark pigmenting substance, its counterintuitive brilliance and its temporal corruption, even as Cesi rushed Galileo to complete the book so it could be hurriedly printed early in the new year, in time for a

38 I have argued elsewhere that another of the visual referents for Greuter's etched spots was naturally occurring stains and discolorations on metals (e.g. copper and brass) familiar to any period collector or maker of fine instruments. See Noyes, "Mattheus Greuter's Sunspot Etchings," pp. 474–475.

39 See OG XI: 463; Reeves, "Mere projections," p. 55.

40 OG XI: 475.

41 See also Biagioli, "Picturing Objects in the Making."

42 See Stijnman, *Engraving and Etching*, pp. 267–270 and Campbell's essay in this volume. For a study taking up issues of early modern printer's ink to explore the instability and un-fixity of print, see Adrian Johns, "Ink."

delegation departing Rome to deliver copies to Germany.[43] In worrying about the deadline to transmit his completed book to Germany, Galileo was surely aware of etching's early evolution in turn-of-the-sixteenth-century Germany from methods for decorating armor and other metal objects to iron plates to produce printed images.[44] He was also no doubt mindful of Giorgio Vasari's account of the rise of the "fine manner" in intaglio engraving – of which Greuter's *intaglio finissimo* represented an iteration – that attributed the method to fifteenth-century Florentine goldsmiths who transferred niello metalworking techniques to graven metal plates and thence to printed pictures on paper.[45] The issue of printing *Macchie Solari* and its images in and against time was such a concern that Cesi would suggest leaving the third letter containing this analogical experiment in manuscript form for delivery to Germany; for Greuter and his workshop, the proverbial ink drying adumbrated in Galilei's parable of the iron pan posed such a problem that "in no two of [copies of the book] are [Greuter's plates] in exactly the same position on the page and there is often offsetting between plates," indicating "the printing was done rapidly and sometimes sloppily," a point discussed further below[46] (Figure 2.1).

Greuter's *intaglio finissimo* was familiar to many in Rome – and certainly those in the papal ambit like Agucchi with whom Cesi shared advance copies of *Macchie Solari* –from small-scale prints of Christ, the Virgin, and saints that constituted a well-known specialty of Greuter's oeuvre amongst contemporaries, which the artist even advertised when applying for copyright privileges from the Vatican's Sacred Palace.[47] His ultrafine manner cloaked beneath a pious veil Galileo's provocative hypotheses about the spots pointing toward the reality of a Copernican heliocentric cosmos filled with celestial bodies susceptible to corruption that troubled

43 Mayer, "Censorship," p. 7.

44 For a recent collection of cross-cultural studies in the early history of etching in Europe, see Jenkins, Orenstein, and Spira, *The Renaissance of Etching*.

45 For an account contextualizing the advent of the fine manner in Florence, see Landau, "Printmaking in the Age of Lorenzo." For a recent study further evaluating technical and intercultural aspects, see Roberts, "Inventing Engraving." For the cross-cultural inflection of early fine manner engraving, see *Idem*, "Francesco Rosselli." Vasari's history ran counter to other period accounts attributing the fine manner to northern (Netherlandish) artists, for which, see Noyes, "Mattheus Gretuer's Sunspot Etchings," pp. 477–480 and Melion, "Prayerful Artifice."

46 Mayer, "The Censoring of Galileo's Sunspot Letters," p. 5 n62, who wrote having examined thirty copies of the book.

47 On Greuter's specialty in small-scale devotional engravings, see Baglione, *Le vite de' Pittori*, p. 398; Bilancia, "Le lastre e le stampe di Matthäus Greuter," p. 347. On his *intaglio finissimo*, see Noyes, "One of Those Lutherans." See also Leuschner, "The Papal Printing Privilege."

not only Vatican censors, but even Lincean colleagues like Cesi.[48] On multiple occasions, after conferring with Greuter in his workshop and examining the etchings as the artist was working on them, Cesi admonished Galileo of censure on these points or cautioned alternate theories on the spots (that they were "jumbled stars" casting shadows on the sun, an argument very close to that of rival Scheiner), seemingly perturbed by the implications simultaneously disclosed and disguised by the subtle images and their *techne*.[49] I have previously connected the fine manner with a particular Counter-Reformation hermeneutics of spiritual perception that claimed to disclose Catholic truths otherwise indiscernible to earthly vision and focused on incarnational mysteries revealed by meditative engagement with subtle renderings of Christ's glorified wounded body.[50] Coupling Greuter's etched sunspot splotches with the oxidized sores endemic to bitumen oil paint reinforced another coupling, that of a comparison between sunspots and moonspots, by means of a reference to the ink on the book's page and contemporary analogies of the moon with the body of the Virgin and the sun with Christ.[51]

If Galileo had not visited a workshop like Greuter's in Rome himself, he had probably learned second hand of their workings from Cigoli and understood the processes of producing intaglio prints; moreover, he had witnessed (thanks to Cesi's intervention) second hand the rapid transmutations wrought upon the sunspot etchings as the autumn months progressed. This raises another even more crucial period use for asphaltum: as a key ingredient of etching ground, commonly recorded in seventeenth-century recipes calling for mixtures of asphaltum, one or more resins, and beeswax. Greuter carefully applied a layer of ground (or resist) to his copperplates, then, with a stylus, scratched this coating away according to the design to be printed, revealing the metal that was then exposed to the mordant solution and corroded away in a process comparable to alchemy, whereby the actual incising of the metal was enacted by the chemical reaction of acids dissolving or "biting" metal.[52] Furthermore, the method Greuter most likely used to apply an even layer of ground to his plates entailed heating

48 Mayer, "The Censoring of Galileo's Sunspot Letters."

49 *OG* XI: 422, 428. See also Galluzzi, *The Lynx and the Telescope*, pp. 113–120.

50 See Noyes, "One of Those Lutherans" and Melion, "Prayerful Artifice," pp. 628–629.

51 See Booth and Van Helden, "The Virgin and the Telescope," and Ostrow, "Cigoli's Immacolata."

52 For a period account of etching, see Bosse, *Traicté des manières de graver*, pp. 9–43. For an overview of the materials and techniques discussed here, see Orenstein and Stijnman, "Bitten with Spirit." Overlap with alchemy was largely due to shared materials (the acid) and techniques (enacting acid on metals); see Karpenko, "Some Notes."

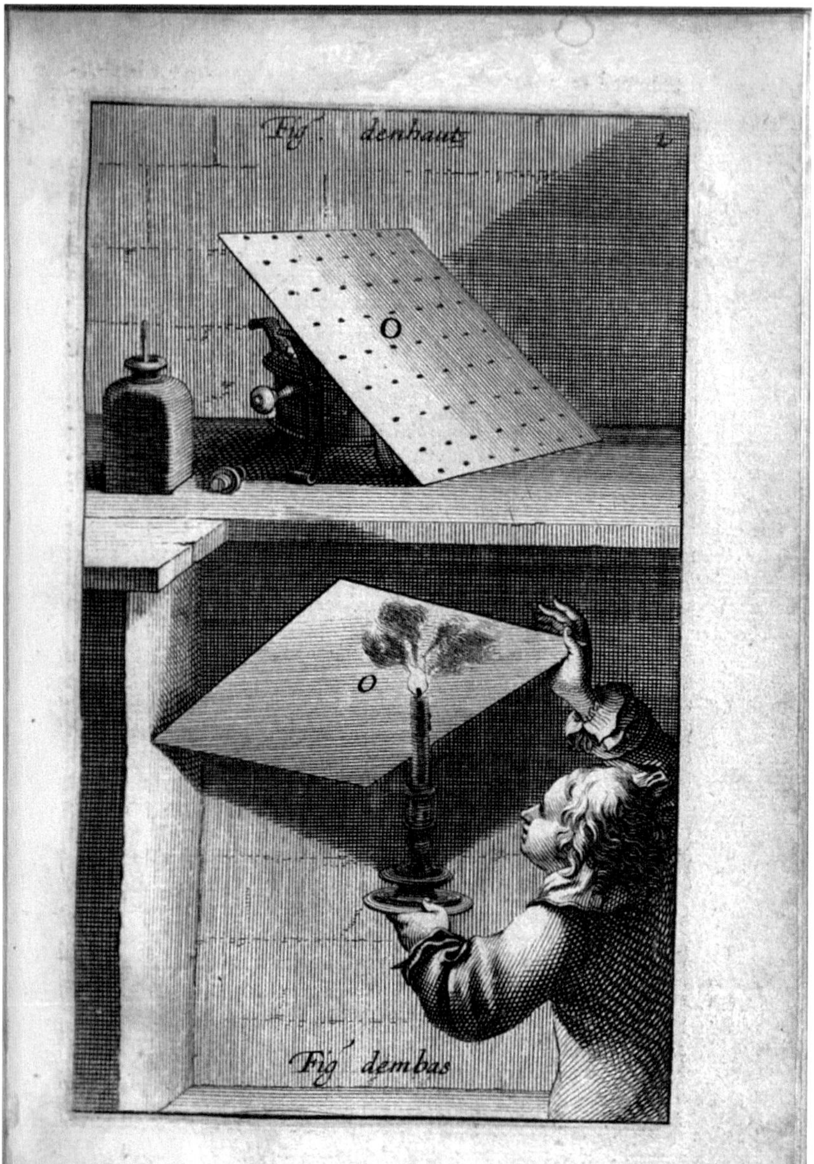

Figure 2.3 Abraham Bosse, Coating the etching plate with ground (above) and smoking the plate (below), from *Traicté des manières de graver en taille douce sur l'airin* (Paris: Chez Bosse, 1645), plate 1, etching. Duke University, Durham, NC, David M. Rubenstein Rare Book & Manuscript Library, NE1760 .B73 1645 c.1.

the copper and applying small drops of the bituminous oil mixture to the hot metal and spreading the amalgam over the plate's surface, then searing the plate to raise smoke from the ground, which, after cooling, hardened into a strong and resistant layer. This two-stage process was illustrated in Bosse's mid-century manual on etching[53] (Figure 2.3).

Agucchi, for his part, received from Cesi in late Spring 1613 a printed copy of *Macchie Solari* with Galileo's full text – including his third letter, not included with the materials of the previous Fall – and Greuter's whole cycle of etchings. He then wrote his Tuscan astronomer friend, first with congratulations commending Galileo's victory and reaffirming support for his interpretation of the spots, then, a month later, with a longer caution against Galileo's adherence to the reality of Copernicanism in contravention of Holy Scripture.[54] In suddenly shifting course, Agucchi ironically parroted Cesi's earlier prediction that those who might argue against the overwhelming "sense experience" generated by the book's images would "abuse reason." Having fully recovered from the illness that had impaired him six months earlier when he had before him the partial proofs (and possibly no etchings), Agucchi now had the third letter, which featured the iron pan experiment, and all 38 plates. Confronted with the full force of text and images, whose "wonder" and "accuracy" Cesi had claimed would stop naysayers dead in their tracks, the Bolognese prelate outlined scriptural and philosophical reasons against the heliocentrism forcefully unveiled in the final version of *Macchie Solari*, then declared: "my third reason [for objecting to your arguments] is finally reason itself."[55] He never wrote to Galileo again.

III. Prints as Drawings: Epistemic Image as Discursive Field

While Greuter's sunspot etchings fall within the remit of the epistemic image, the present-day analogue for Cesi's *pittura filosofica*, like many of the case studies in this volume, that of his prints for *Macchie Solari* does not dovetail so neatly with a nascent dichotomy in scholarship taking up issues of early modern epistemic images as a self-sufficient interdisciplinary

53 For the preparation and application of bituminous etching ground in this period, see e.g. Bosse, *Traicté des manières de graver*, pp. 41–43; Salmon, *Polygraphice*, pp. 76–77 and 80–81. See also Stijnman, *Engraving and Etching*, pp. 199–202, 204–205.

54 See letters of 8 June 1613 and 13 July 1613, in *OG* XI: 520–521 and 532–535, respectively.

55 See the discussion in Torrini, "Galileo copernicano," pp. 26–27 and Mayer, "The Censoring of Galileo's Sunspot Letters," p. 4.

field of research.[56] An emergent juxtaposition contrasts two accounts of the admittedly capacious terminology: on one hand, "epistemic image" is made to refer "to any image that was made with the intention of expressing, demonstrating or illustrating a theory."[57] On the other, it is "made with the intent not only of depicting the object of scientific inquiry but also of replacing it" and "can be shared by a dispersed community."[58] Amongst Galileo, Cesi, Agucchi, and their contemporaries, the diverse ends served by Greuter's print series included those of expressing, demonstrating, and illustrating Galileo's theories about the sunspots – including some too controversial for words – and exceeding the depiction of their object to replace it, proliferating within and between dispersed communities. This case, and those presented in essays in this volume, suggest that, rather than contrasting these definitions as respectively underdetermined and overdetermined poles on a notional spectrum, in some instances they might be suspended in dynamic equilibrium within a coextensive discursive field.

Galilei's nonchalant directing of readers to the ink-saturated pages of the book, as if readers could easily compare their scrutiny of the imprint with that of the sunspots, imagined a collective of observers who might virtually witness the astronomer's findings, since the very few who (like Agucchi) actually possessed the right equipment lacked the best instruments and know-how to see what Galileo professed to have seen, such that "observational simultaneity in the sunspot debate should be understood as the product of the [book's] images, rather than the other way around."[59] Greuter's efforts and this sleight of hand proved successful: the images invoked a claim to supplanting their prototype insofar as the book's nomenclature somewhat disingenuously referred to the etchings not as prints but *disegni* (drawings), avatars for Galileo's drawings, his telescopic observational method, and, indeed, the spots themselves, which were constituted as new targets[60] (Figures 2.4–2.6). Agucchi was not alone amongst the community of elite readers with and without telescopes who accepted this elision: he may have been amongst the group of "many important prelates in the Quirinal Gardens" to whom Galileo claimed to have demonstrated the spots with

56 See e.g. Marr, "Knowing Images," p. 1001; Fransen and Reinhart, "The Practice of Copying"; Marr and Heuer, "Introduction."

57 Lüthy and Smets, "Words, Lines, Diagrams, Images," p. 399 n2.

58 Daston, "Epistemic Images," pp. 17–18.

59 Wilding, "Galileo and the Stain of Time."

60 Biagioli, "Picturing Objects in the Making"; Noyes, "Mattheus Greuter's Sunspot Etchings." This differs from the interpretation in Reeves, "Mere projections," p. 51.

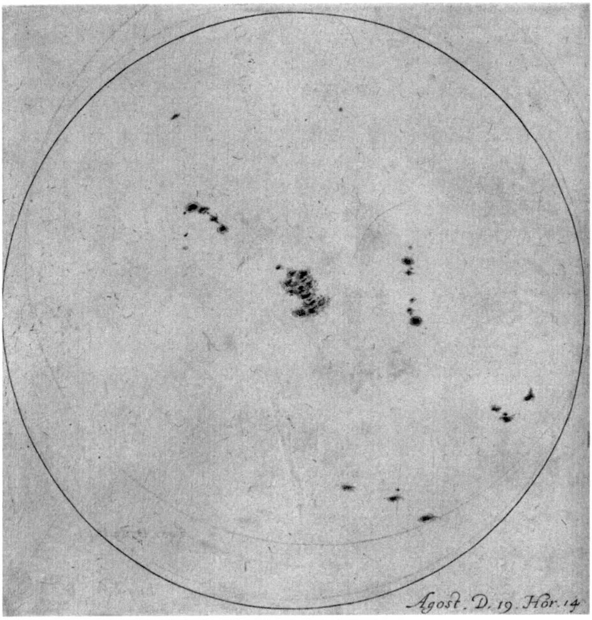

Figure 2.4 Mattheus Greuter, Sunspots, from Galilei, *Macchie Solari* (first edition, second issue), page 94, etching. Detail: spot from first variant. Yale University Library, New Haven, CT, Beinecke Library, QB41 G325 1613.

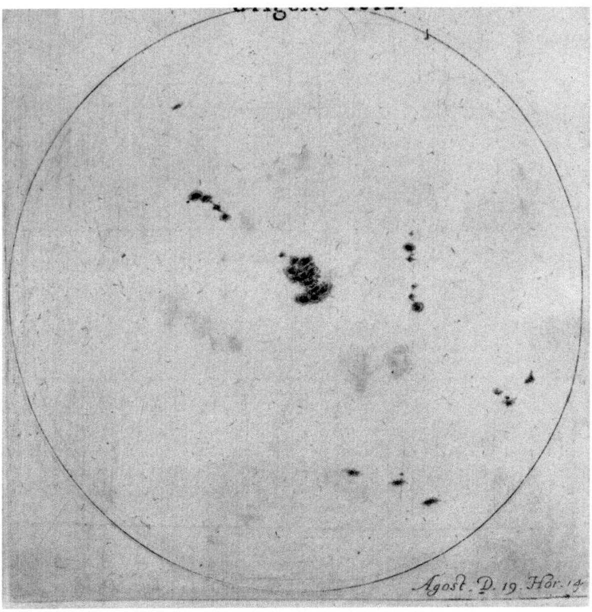

Figure 2.5 Mattheus Greuter, Sunspots, from Galilei, *Macchie Solari* (first edition, first issue), page 94, etching. Detail: spot from second variant. Yale University Library, New Haven, CT, Beinecke Library, QB41 G325 1613B.

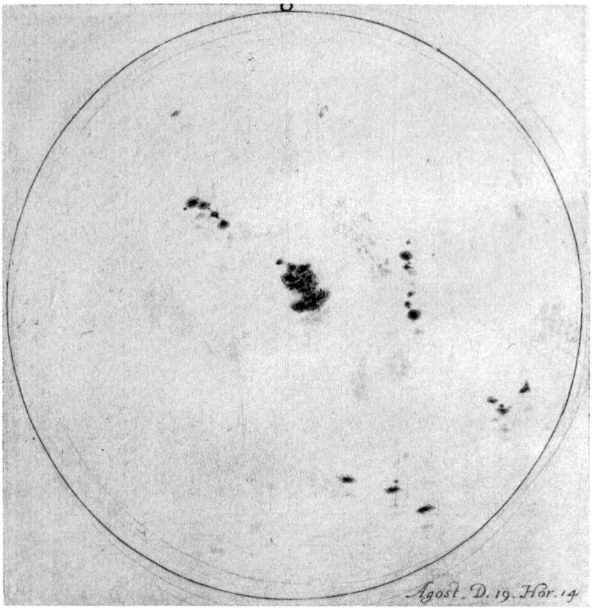

Figure 2.6 Mattheus Greuter, Sunspots, from Galilei, *Macchie Solari* (first edition, second issue), page 94, etching. Detail: spot from third variant. Yale University Library, New Haven, CT, Beinecke Library, QB41 G325 1613B.

his own instrument months before his rival Scheiner began observations in Germany.[61]

The first sunspot etchings and other early modern epistemic imprints that serve as the investigatory focus of this book, their constituent materials and manufacture, and their relations and interactions with period manuscript and textual culture, complicate "epistemologically and socio-historically grounded model[s] of how pictures helped to manufacture science."[62] In describing the thought experiment performed by a third-person "whomever" with bitumen on a searing iron plate, Galileo gestured to a transalpine history of intaglio etching that was still being codified and whose proverbial endpoint was Greuter's own Roman workshop, casting the artist in the role of experimenter. By virtue of their technical facture as acid-etched imprints transferred from sun to paper by telescopic projection, paper to metal plate by heat-induced chemistry, and thence to paper again by the intaglio presses' force, Greuter's prints and the *techne* they subsumed not only expressed and

61 *OG*, XVII: 297. See similar claims amongst contemporaries in Galluzzi, *The Lynx and the Telescope*, p. 106.

62 Swan, "Illustrated Natural History," p. 188.

illustrated Galileo's theories on the spots and supplanted the astronomer's observations, but also performatively articulated how the fiery vaporous spots in the celestial realm might be reproduced in miniature in the artist's workshop near San Marcello al Corso, midway between the Pantheon and Quirinal gardens and around the corner from the Jesuit *Collegio Romano*.[63] Thus, the artist's activities attending the production of *Macchie Solari*, and the ways in which these activities may have actively informed the book's arguments, methods, and reception, might be reframed in terms of the embodied artisanal reality of the experimental approach that crucially informed what has come be known as the Galilean scientific method.

A re-examination of Greuter's sunspot etchings, together with studies in this book on both lesser-known and canonical prints and related objects, materials and media – ranging from the anatomical woodcuts for Andreas Vesalius's *Fabrica* and the copperplate matrices for Albrecht Dürer's *meisterstiche*, to botanical woodblock matrices, engraved metal astronomical instruments, and manuscripts collating hand-drawn copies of pamphlet prints – contributes to recent scholarship recalibrating the foundational theories of William Ivins, Jan van der Stock, Elizabeth Eisenstein, and Bruno Latour, who have variously argued from different angles for the constitutive role of the purported identicality, fixity, imitability, profligacy, and longevity specific to prints in knowledge-making practices in the early modern period.[64] In a recent critical reassessment of Ivins's notion of printed images' exact repeatability, Dániel Margócsy points out that the genesis of Ivins's concept can be found in his study of early modern anatomical prints.[65] Ivins's arguments and those of Van der Stock on print's profligacy in this period are revised in Caroline O. Fowler's recent study exploring the creative and expressive potential of inherently impermanent and precious physical aspects of intaglio print's constituent materials, especially the copper matrix, which also redounds against ideas of printed images' fixity and permanency.[66] Eisenstein's theorizing on the fixity and authority of print with special emphasis on texts and books, since extended to printed images,[67] has been reassessed by Adrian Johns' account of early modern printing practices, Sean Roberts' investigation of

63 On the locations of Greuter's workshop in Rome, see Guerrieri Borsoi, "GREUTER, Matthäus."
64 Ivins, *Prints and Visual Communication*; Eisenstein, *The Printing Press as an Agent of Change*; Van der Stock, *Printing Images in Antwerp*; Latour, *Science in Action*; *Idem*, "Visualisation and Cognition."
65 Margócsy, "From Vesalius through Ivins to Latour."
66 Fowler, "Res Papirea." See also the essay by Angela Campbell in this volume.
67 See e.g. Reeves, "Mere Projections."

the advent of burin engraving, and Matthew C. Hunter's destabilization of paper more broadly as a reliable bearer of knowledge.[68] On Eisenstein and on Bruno Latour's related concept of "immutable mobiles," Michael John Gorman's critical reassessment argues "against the inevitable stability of knowledge propagated via print" in the context of the circulation of early modern astronomical knowledge in print.[69]

In the case of *Macchie Solari*, two separate issues and three variants of the printed text of the first 1613 edition have already been identified by scholars, though Mayer's census notes relatively few anomalies in the text.[70] In contrast, Greuter's surviving etchings reveal that while under the presses the images underwent numerous incremental corruptions and mutations that exceeded the textual changes: profligate offsets from other plates, ink smudges including fingerprints, misalignments on the page, variable foul biting incurred by the etching mordant, and scratches made while wiping ink from the plates. In addition to these fluctuations, no fewer than eighteen discrete variants of the sunspots series statistically iterate according to three principle factors: paper weight and type, sequencing and typographical errors and emendations, ink consistency and viscosity, and wear to the incised copperplates over successive printings, resulting in incrementally blotchy, corrupted, and muddled impressions. The delicacy of the plates and the array of inks and papers used to print them over the entire edition meant that recipients of an early print (like Agucchi) and those of later prints actually beheld very different looking spots, as did those who received presentation copies on heavy paper versus those with nearly translucent paper (Figures 2.4–2.6). Despite Greuter's best efforts at mitigating the printing challenges and inherently short lifespan of his plates, the precise iteration of the etched spots seen by Agucchi on his advanced proofs would be seen by no one else. Thus, comparing any two copies, which many surely did given that texts of the first and second issue of the first edition differed considerably, contemporaries would have been struck by the prints' variability, preciosity, fragility, and instability. These were all qualities that Greuter and the production process he oversaw apparently tried to minimize and obscure, but that, paradoxically, enhanced and revealed aspects of Galileo's provocative Copernican arguments. Essays

68 Johns, *The Nature of the Book*; Roberts, "Tricks of the Trade"; Hunter, *Wicked Intelligence*, pp. 68–97. See also Fowler, *The Art of Paper* and the chapter by Margócsy, Somos, and Joffe in this volume.

69 Gorman, "The Elusive Origins." See also Margócsy, "From Vesalius through Ivins to Latour."

70 Mayer, "An Interim Report."

in this book similarly revisit prints to re-evaluate questions surrounding the symbiosis between knowledge claims of truth and priority and print media in the early age of mechanical reproduction, within and beyond the European cultural sphere, including colonial contexts.

That the same text hailing Greuter's etchings in *Macchie Solari* as "drawings of the spots of the sun" also specified that they were "seen and observed by" Galileo and shown "to many others" at a specific time and place claimed for them the status of so-called *ad vivum* (from life) images, thereby advertising the images' claim "to be a faithful likeness or a bearer of reliable data," in this case about the sun.[71] A number of the images explored in the chapters that follow likewise could be qualified as *ad vivum*. However, following Greuter's evolution of the measured hatch marks of his *intaglio finissimo*, from burin engraving to acid etching and from icons of holy figures to icons of the solar body, highlights issues at the triangulation of scholarly accounts maintaining the reliance of early modern knowledge production on artistic techniques of and interest in naturalism:[72] those stressing tensions between early modern "emphases on verisimilitude" and concurrent "claims for the artistry of images" in the context of knowledge production,[73] and those exploring the ramifications of prints' highly conventional syntax in this same context.[74] The stylistic evolution in the case of *Macchie Solari* veiled Galileo's truth and priority claims against his rival (a Jesuit priest), as well as provocative Copernican arguments and perhaps more problematic use of the Bible, beneath an aesthetically normalized cloak of ostentatious piety whose paradoxical insistence on mimetic accuracy enhanced the credibility of its motivating ideologies, by grounding them in a hybrid visual language of scientific and religious truthiness.[75]

The case studies in this book also pinpoint intersections, for example exploring ways in which emulative imitations of earlier *ad vivum* prints critically problematized the concept's growing normativity across and between competing early modern scientific experts like anatomists, surgeons,

71 Balfe and Woodall, "Introduction," p. 2.

72 See e.g. Panofsky, *Artist, Scientist, Genius*; Ackerman, *Distance Points*, pp. 185–207; Ogilvie, *The Science of Describing*.

73 Quoting Swan, "Illustrated Natural History," p. 189. For recent explorations of these questions with special attention to prints, see Egmond, *Eye for Detail*; Idem, "The Ad Vivum Conundrum"; Kusukawa, "The Role of Images"; Idem, *Picturing the Book of Nature*.

74 Swan, "Illustrated Natural History." See also Jorink, "Beyond the Lines of Apelles"; Doherty, "Discovering the 'True Form'"; Lawson, "Crafting the Microworld."

75 On this, see Spiegel, "Forging the Past," p. 277.

and optical theorists.[76] A recent survey of early modern *ad vivum* images eschews for them such present-day qualifications as "informational" images (per James Elkins),[77] "technical" or "scientific" images (given, above all, to Horst Bredekamp),[78] and even "epistemic" images (retraced to Hans-Jörg Rheinberger),[79] none of which "adequately capture the character of early modern images of this kind, which operated between the modern art-science divide."[80] Cesi's own concept of *pittura filosofica* gestures at once to the interstitial nature of these images, to the inadequacies of current-day constructs and language to address such interstitially, and to a nascent awareness at the cusp of the Scientific Revolution of the beginning of such a constructed divide. To negotiate new approaches to analyzing images at the interstices of nascent early modern fields, studies in this book engage with a range of established and emerging methodologies and investigative tools and tactics, ranging from (re)learning to see early modern prints through the "eyes" of artificial intelligence, to establishing new iconographical classifications for printed images and their matrices and enhancing scrutiny of prints through digital instruments and applications.

Investigating the ways in which technical, historical, and aesthetic truth claims were transmitted with and encoded in the facticity of Greuter's telescopic imprints likewise reminds us of the historical contingency of teleological narratives. Albeit proleptically, Greuter's technical evolution of his *intaglio finissimo* gestured to not-so-distant controversies with the Roman Inquisition and the Papacy that, within a decade, would prove calamitous for Galileo. That the technique of acid etching entailed a temporal process of deterioration and entropy only compounded the exegetical audacity of Galileo and Greuter's undertaking, amplifying the effect of the book's title that extended "history" and "demonstrations" to the sun, and thus temporalizing the hitherto timeless superlunary cosmos and subjecting it to performative spectacle, mundane experimentation, artisanal manufacture, and interpretation. The artisanal process that engendered Greuter's etchings visibly performed the course of corruption, mutation, and generation enacted upon Aristotelian philosophy itself in Galileo's text, which attempted to efface not only the traces of its own making, but also stage the Galilean episteme's generation from and replacement of Aristotelian episteme, in

76 See also Leitch, "Dürer's Rhinoceros Underway."
77 Elkins, "Art History and Images That Are Not Art."
78 E.g. Bredekamp, Dünkel, and Schneider, *The Technical Image*.
79 E.g. Rheinberger, *An Epistemology of the Concrete*.
80 Balfe and Woodall, "Introduction," p. 2.

the very act of rethinking – or, in the print's case, representing – corruption as generation.[81] Greuter's plates and their production offer a retrospective glimpse of both a surprisingly fragile artistic medium and a vulnerable system of knowledge production on the brink of collapse even at its apogee.

To generate new knowledge about early modern epistemic imprints, their constitutive materials, and making, as well as adjacent media, and the knowledge they helped to make in turn, *Reassessing Epistemic Images in the Early Modern World* brings together case studies from museum professionals and scholars from the History of Art and History of Science that investigate along an unfolding thematic axis, from essays taking up diverse issues of materiality, to the mind's eye and how images played in the imagination, to those plotting new methodologies against historical print networks. The Prologue raises issues of historical models of vision, perception, and knowledge reception through case studies in certain printed pictures, which then inflect through and across the subsequent essays. Part 1, "Approaches to Print Matrices," features studies attending to the material facture of woodblock and copperplate print matrices. Chapters variously explore aspects of their materiality, using such methods as digital imaging to exploit new understandings about the functioning of early modern printing workshops, the chronology of individual impressions of within a print series, the embodied knowledge of early modern printshop artisans, and even what kinds of historical information can be extracted from matrices themselves. In Part 2, "Imprints as Instruments," contributions undertake to expand the notional imprint to include incised metal objects and manuscript and non-mechanical forms and technologies that coexisted, engaged, and interacted with early modern prints and print culture. The essays investigate the crucial role of such print-adjacent media in educational methods and institutions of learning, as well as how embodied interaction with such objects by early modern consumers of diverse ages and socio-cultural backgrounds vitally contributed to notions of communal, patronal, and epistemological identity. Affective and performative dimensions are further explored in Part 3, "Imprint, Knowledge and Affect." Chapters engage with questions of how the invention of a single, complex engraving instantitated Galenic theory, and reconstruct reception histories nuancing how instances of what has been conventionally branded "plagiarism" might be re-understood as emulative heuristic practices whereby canonical visual constructs might be reframed, adapted, and critiqued. The Epilogue broadens the volume's discursive, temporal, and geographical scope, marshaling present-day questions issuing

81 Paraphrasing Wilding, "Galileo and the Stain of Time."

from the evolving ontology of computer vision to revisit issues attending the reception of European prints in early modern colonial contexts, and probe print's ontological operations and epistemic limits.

Bibliography

Early modern published sources

Baglione, Giovanni. *Le Vite de' Pittori Scultori et Architetti.* Rome: Nella Stamperia d'Andrea Fei, 1642.

Bosse, Abraham. *Traicté des Manières de Graver en Taille Douce sur l'Airin.* Paris: Chez Bosse, 1645.

Galilei, Galileo. *Istoria e dimostrazioni intorno alle macchie solari e loro accidenti. Comprese in tre lettere scritte all'illustrissimo Signor Marco Velseri.* Rome: Appresso Giacomo Mascardi, 1613.

Galilei, Galileo. *Le Opere di Galileo Galilei* (hereafter: *OG* and volume number), 21 vols, ed. Antonio Favaro. Florence: G. Barbèra, 1929–1939.

Galilei, Galileo, and Christoph Scheiner. *Galileo and Scheiner on Sunspots 1611–1613*, ed. and transl. Eileen Reeves and Albert Van Helden. Chicago, IL: University of Chicago Press, 2010.

Salmon, William. *Polygraphice, or, The Arts of Drawing, Engraving, Etching, Limning, Painting, Washing, Varnishing, Gilding, Colouring, Dying, Beautifying and Perfuming.* London: R. Jones, 1685.

Secondary sources

Ackerman, James. *Distance Points: Essays in Theory and Renaissance Art and Architecture.* Cambridge, MA: MIT Press, 1991.

Ambrose, Charles T. "The Conquest of Pus: A History of Bitumen, Creosote and Carbolic Acid." *Journal of Infectious Disease and Preventive Medicine* 6, no. 2 (2018): 1–8.

Baldriga, Irène. *L'occhio della lince. I primi Lincei tra arte, scienza e collezionismo; 1603–1630.* Rome: Accademia Nazionale dei Lincei, 2002.

Baldriga, Irène. "'La fatiga di pigliar i disegni dalle piante'. Federico Cesi, la pittura filosofica e la riproduzione del mondo vegetale." In *Federico Cesi. Un principe naturalista. Acquasparta, 29 e 30 settembre 2003*, edited by Antonio Graniti, pp. 503–523. Rome: Bardi, 2006.

Baldriga, Irène. "Reading The Universal Book of Nature: The Accademia dei Lincei in Rome (1603–1630)." In *The Reach of the Republic of Letters: Literary and Learned Societies in Late Medieval and Early Modern Europe*, edited by Arjan van Dixhoorn and Susie Speakman Sutch, pp. 353–388. Leiden: Brill, 2008.

Balfe, Thomas, and Joanna Woodall. "Introduction: From Living Presence to Lively Likeness – the Lives of Ad Vivum." In *Ad vivum? Visual Materials and the Vocabulary of Life-Likeness in Europe before 1800*, edited by Thomas Balfe, Joanna Woodall, and Claus Zittel, pp. 1–43. Leiden: Brill, 2019.

Bell, Peter Jonathan. "Johann Friedrich Greuter" and "Matthaus Greuter," *Allgemeines Kunstlerlexikon: Die bildenden Kunstler aller Zeiten* 61: 500–505. Munich: De Gruyter, 2009.

Biagioli, Mario. "Picturing Objects in the Making: Scheiner, Galileo and the Discovery of Sunspots." In *Wissensideale und Wissenskulturen in der Frühen Neuzeit: Ideals and Cultures of Knowledge in Early Modern Europe*, edited by Wolfgang Detel and Claus Zittel, pp. 39–96. Berlin: Akademie Verlag, 2002.

Bilancia, Fernando. "Le lastre e le stampe di Matthäus Greuter in una lista di beni assegnati in dote alla figlia." In *Roma nel primo Seicento. Una città moderna nella veduta di Matthäus Greuter*, edited by Augusto Roca De Amicis, pp. 347–353. Rome: Editoriale Artemide, 2018.

Booth, Sara Elizabeth, and Albert van Helden. "The Virgin and the Telescope: The Moons of Cigoli and Galileo." *Science in Context* 13, no. 3–4 (2000): 463–486.

Bredekamp, Horst. "Gazing Hands and Blind Spots: Galileo as Draftsman." *Science in Context* 13, no. 3–4 (2000): 423–462.

Bredekamp, Horst. *Galilei der Künstler. Der Mond, die Sonne, die Hand*. Berlin: Akademie Verlag, 2009.

Bredekamp, Horst, Vera Dünkel, and Birgit Schneider, eds. *The Technical Image: A History of Styles in Scientific Imagery*. Chicago, IL: University of Chicago Press, 2015.

Brevaglieri, Sabina. "Science, Books and Censorship in the Academy of the Lincei: Johannes Faber as Cultural Mediator." In *Conflicting Duties. Science, Medicine and Religion in Rome (1550–1750)*, edited by Maria Pia Donato and Jill Kraye, pp. 109–133. London: Warburg Institute Colloquia 15, 2009.

Brevaglieri, Sabina. *Natural Desiderio di Sapere. Roma Barocca fra Vecchi e Nuovi Mondi*. Rome: Viella, 2019.

Bucciantini, Massimo. "Teologia e Nuova Filosofia. Galileo, Federico Cesi, Giovambattista Agucchi e la Discussione sulla Fluidità e Corruttibilità del Cielo." In *Sciences et Religions de Copernic à Galilée*, pp. 411–441. Rome: École Française de Rome, 1999.

Bury, Michael. *The Print in Italy, 1550–1620*. London: British Museum Press, 2001.

Chappell, Miles. "Lodovico Cigoli, Matthäus Greuter and Santa Maria del Fiore in Florence." *Paragone/Arte* 3, 58.75/76 (2007): 96–112.

Daston, Lorraine. "Epistemic Images." In *Vision and Its Instruments. Art, Science, and Technology in Early Modern Europe*, edited by Alina Payne, pp. 13–35. University Park, PA: Pennsylvania State University Press, 2015.

Daxecker, Franz. *The Physicist and Astronomer Christoph Scheiner: Biography, Letters, Works.* Innsbruck: Leopold-Franzens-University of Innsbruck, 2004.

De Caro, Mario. "On Galileo's Platonism, Again." In *Hypotheses and Perspectives in the History and Philosophy of Science: Homage to Alexandre Koyré 1892–1964*, edited by Raffaele Pisano, Joseph Agassi, and Daria Drozdova, 85–104. Dordrecht: Springer, 2018.

Dear, Peter. *Discipline and Experience: The Mathematical Way in the Scientific Revolution*. Chicago, IL: University of Chicago Press, 1995.

Dear, Peter. "The Meanings of Experience." In *The Cambridge History of Science*, edited by Katharine Park and Lorraine Daston, pp. 106–131. Cambridge: Cambridge University Press, 2006.

Drake, Stillman. "Galileo's Discovery of the Law of Free Fall." *Scientific American* 228 (1973): 84–92.

Drake, Stillman. "Galileo's Experimental Confirmation of Horizontal Inertia: Unpublished Manuscripts." *Isis* 64, no. 3 (1973): 291–305.

Diefenbacher, Jörg, and Ekhard Leuschner, eds. *The New Hollstein German Engravings, Etchings and Woodcuts 1400–1700, vol. 10, The Greuter Family, pts. 1–2, Matthäus Greuter.* 2 vols. Ouderkerk aan den IJssel: Sound & Vision Interactive, 2016.

Doherty, Meghan. "Discovering the 'True Form': Hooke's *Micrographia* and the Visual Vocabulary of Engraved Portraits." *Notes and Records of the Royal Society* 66 (2012): 211–234.

Edgerton, Samuel Y. "Galileo, Florentine 'Disegno,' and the 'Strange Spottednesse' of the Moon." *Art Journal* 44, no. 3 (1984): 225–232.

Egmond, Florike. *Eye for Detail: Images of Plants and Animals in Art and Science, 1500–1630.* London: Reaktion, 2017.

Egmond, Florike. "The Ad Vivum Conundrum: Eyewitnessing and the Artful Representation of Naturalia in Sixteenth-Century Natural Science." In *Zeigen – Überzeugen – Beweisen: Methoden der Wissensproduktion in Kunstliteratur, Kennerschaft und Sammlungspraxis der Frühen Neuzeit*, edited by Elisabeth Oy-Marra and Irina Schmiedel, pp. 33–62. Heidelberg: arthistoricum.net, 2020.

Eisenstein, Elizabeth L. *The Printing Press as an Agent of Change: Communications and Cultural Transformations in Early-Modern Europe*, 2 vols. Cambridge: Cambridge University Press, 1979.

Elkins, James. "Art History and Images That Are Not Art." *The Art Bulletin* 77, no. 4 (1995): 553–571.

Favaro, Antonio. "Giovanni Battista Agucchi." In *Amici e Corrispondenti di Galileo Galilei,* 3 vols. Florence: Libr. Editrice Salimbeni, 1983, II: 373–395.

Featherstone, Rupert. "Bitumen." *Grove Art Online* 2003, https://www.oxfordartonline.com/groveart/view/10.1093/gao/9781884446054.001.0001/oao-9781884446054-e-7000009066 (accessed 5 March 2021).

Feldhay, Riva. "Producing Sunspots on an Iron Pan: Galileo's Scientific Discourse." In *Science, Reason, and Rhetoric*, edited by Henry Krips *et al.*, pp. 119–144. Pittsburgh, PA: University of Pittsburgh Press, 1995.

Finocchiaro, Maurice A. *The Galileo Affair: A Documentary History*. Berkeley, CA: University of California Press, 1989.

Finocchiaro, Maurice A. *Retrying Galileo, 1633–1992*. Berkeley, CA: University of California Press, 2005.

Finocchiaro, Maurice A. *The Trial of Galileo: Essential Documents*. Indianapolis, IN: Hackett Publishing, 2014.

Forbes, Robert James. *Bitumen and Petroleum in Antiquity*. Leiden: Brill, 1936.

Fowler, Caroline. "Res Papirea: Mantegna's Paper Things." *The Art Bulletin* 99, no. 1 (2017): 8–35.

Fransen, Sietske, and Katherine M. Reinhart, "The Practice of Copying in Making Knowledge in Early Modern Europe: An Introduction." *Word & Image* 35, no. 3 (2019): 211–222.

Freedberg, David. *The Eye of the Lynx: Art, Science and Nature in the Age of Galileo*. Chicago, IL: University of Chicago Press, 2002.

Gabrieli, Giuseppe, ed. *Il Carteggio Linceo della Vecchia Accademia di Federico Cesi (1603–1630)*. "Atti della Reale Accademia Nazionale dei Lincei," ser. 6, vol. 7. Rome: Dott. Giovanni Bardi, Tipografo Della R. Accademia Nazionale Dei Lincei, 1938.

Galluzzi, Paolo. *The Lynx and the Telescope: The Parallel Worlds of Cesi and Galileo*. Leiden: Brill, 2017.

Gorman, Michael John. "The Elusive Origins of the Immutable Mobile." 2001, https://web.archive.org/web/20120628201904/http://www.stanford.edu/group/STS/immutablemobile.htm_(accessed 5 March 2021).

Graniti, Antonio, ed. *Federico Cesi. Un Principe Naturalista*. Rome: Bardi, 2006.

Guerrieri Borsoi, Maria Barbara. "GREUTER, Matthäus." *Dizionario Biografico degli Italiani* 59 (2002), www.treccani.it/enciclopedia/matthaus-greuter_(Dizionario-Biografico) (accessed 4 March 2021).

Hunter, Matthew C. *Wicked Intelligence: Visual Art and the Science of Experiment in Restoration London*. Chicago, IL: University of Chicago Press, 2013.

Ivins, William M. *Prints and Visual Communication*. Cambridge, MA: MIT Press, 1969.

Ivins, William M. *How Prints Look: Photographs with Commentary*, rev. and exp. by Marjorie B. Cohn. Boston, MA: Beacon, 1987.

Jenkins, Catherine, Nadine Orenstein, and Freyda Spira, eds. *The Renaissance of Etching*. New York: The Metropolitan Museum of Art, 2020.

Johns, Adrian. *The Nature of the Book: Print and Knowledge in the Making*. Chicago, IL: University of Chicago Press, 1998.

Johns, Adrian. "Ink." In *Materials and Expertise in Early Modern Europe: Between Market and Laboratory*, edited by Ursula Klein and E. C. Spary, pp. 101–124. Chicago, IL: University of Chicago Press, 2010.

Jorink, Eric. "Beyond the Lines of Apelles: Johannes Swammerdam, Dutch Scientific Culture, and the Representation of Insect Anatomy." In *Art and Science in the Early Modern Netherlands*, edited by Eric Jorink and Bart Ramakers, pp. 148–184. Zwolle: WBooks, 2011.

Karpenko, Vladimir. "Some Notes on the Early History of Nitric Acid: 1300–1700." *Bulletin of the History of Chemistry* 34, no. 2 (2009): 105–116.

Kusukawa, Sachiko. "The Role of Images in the Development of Renaissance Natural History." *Archives of Natural History* 38 (2011): 189–213.

Kusukawa, Sachiko. *Picturing the Book of Nature: Image, Text and Argument in Sixteenth-Century Human Anatomy and Medical Botany*. Chicago, IL: University of Chicago Press, 2012.

Landau, David. "Printmaking in the Age of Lorenzo." In *Florentine Drawing at the Time of Lorenzo the Magnificent: Papers from a Colloquium Held at the Villa Spelman, Florence, 1992*, edited by Elizabeth Cropper, pp. 175–180. Baltimore, MD: Johns Hopkins University Press, 1992.

Latour, Bruno, ed., *Science in Action*. Cambridge, MA: Harvard University Press, 1987.

Latour, Bruno. "Visualisation and Cognition: Drawing Things Together." In *Knowledge and Society: Studies in the Sociology of Culture Past and Present* 6, edited by H. Kuklick, pp. 1–40. Greenwich: Jai, 1990.

Lawson, Ian. "Crafting the Microworld: How Robert Hooke Constructed Knowledge About Small Things." *Notes and Records of the Royal Society of London*, 70 (2016): 23–44.

Leitch, Stephanie. "Dürer's Rhinoceros Underway: the Epistemology of the Copy in the Early Modern Print." In *The Primacy of the Image in Northern European Art, 1400–1700: Essays in Honor of Larry Silver*, edited by D. Cashion *et al.*, pp. 241–255. Boston, MA: Brill, 2017.

Leuschner, Eckhard. "The Papal Printing Privilege." *Print Quarterly* 15, no. 4 (1998): 359–370.

Lüthy, Christoph, and Alexis Smets, "Words, Lines, Diagrams, Images. Towards a History of Scientific Imagery." *Early Science and Medicine* 14 (2009): 398–439.

MacLachlan, James. "Galileo's Experiments with Pendulums: Real and Imaginary." *Annals of Science* 33, no. 2 (1976): 173–185

MacLachlan, James. "Experimenting in the History of Science." *Isis* 89, no. 1 (1998): 90–92.

Mahon, Denis. *Studies in Seicento Art and Theory*. London: The Warburg Institute University of London, 1947.

Margócsy, Dániel. "From Vesalius through Ivins to Latour: Imitation, Emulation and Exactly Repeatable Pictorial Statements in the Fabrica." *Word & Image*, 35, no. 3 (2019): 315–333.

Marr, Alexander. "Knowing Images." *Renaissance Quarterly* 69, no. 3 (2016): 1000–1013.

Marr, Alexander, and Christopher P. Heuer. "Introduction: The Uncertainty of Epistemic Images." *21: Inquiries into Art, History, and the Visual* 2 (2020): DOI: https://doi.org/10.11588/xxi.2020.2.76226.

Mayer, Thomas F. "The Censoring of Galileo's Sunspot Letters and the First Phase of His Trial." *Studies in History and Philosophy of Science* 42 (2011): 1–10.

Mayer, Thomas F. "An Interim Report on a Census of Galileo's Sunspot Letters." *History of Science* 50, no. 2 (2012): 155–196.

Mayer, Thomas F. *The Trial of Galileo, 1612–1633*. Toronto: University of Toronto Press, 2012.

McMullin, Ernan. "The Galileo Affair: Two Decisions." *Journal for the History of Astronomy* 40, no. 2 (2009): 191–212.

Melion, Walter S. "Prayerful Artifice: The Fine Style as Marian Devotion in Hieronymus Wierix's Maria of ca. 1611." In *The Authority of the Word: Reflecting on Image and Text in Northern Europe, 1400–1700*, edited by Celeste Brusati, Karl A. E. Enenkel, and Walter S. Melion, pp. 589–637. Leiden: Brill, 2011.

Noyes, Ruth Sargent. "*Aut Numquid Post Annos Mille Quingentos Docenda est Ecclesia Catholica quomodo Sacrae Imagines Pingantur?* Post-Tridentine Image Reform and the Myth of Paleotti." *Catholic Historical Review* 99, no. 2 (2013): 239–261.

Noyes, Ruth Sargent. "Mattheus Greuter's Sunspot Etchings for Galileo Galilei's *Macchie Solari* (1613)." *The Art Bulletin* 98, no. 4 (2016): 464–485.

Noyes, Ruth Sargent. "'One of Those Lutherans We Used to Burn in Campo de' Fiori': Engraving Sublimated Suffering in Counter-Reformation Rome." In *Visualizing Sensuous Suffering and Affective Pain in Early Modern Europe and the Spanish Americas*, edited by Heather Graham and Lauren Kilroy-Ewbank, pp. 116–165. Leiden: Brill, 2018.

Ogilvie, Brian W. *The Science of Describing: Natural History in Renaissance Europe*. Chicago, IL: University of Chicago Press, 2006.

Ostrow, Steven F. "Cigoli's Immacolata and Galileo's Moon: Astronomy and the Virgin in Early Seicento Rome." *The Art Bulletin* 78, no. 2 (1996): 218–235.

Panofsky, Erwin. *Artist, Scientist, Genius: Notes on the "Renaissance-Dämmerung"*. New York: Metropolitan Museum of Art, 1953.

Panofsky, Erwin. *Galileo as a Critic of the Arts*. The Hague: Martinus Nijhoff, 1954.

Panofsky, Erwin. "Galileo as a Critic of the Arts: Aesthetic Attitude and Scientific Thought." *Isis* 47, no. 1 (1956): 3–15.

Reeves, Eileen. *Painting the Heavens: Art and Astronomy in the Age of Galileo.* Princeton, NJ: Princeton University Press, 1997.

Reeves, Eileen. "Mere Projections: Sunspots and the Camera Obscura." *Galilaeana* 4 (2007): 47–77.

Reeves, Eileen. "From Dante's Moonspots to Galileo's Sunspots." *MLN* 124, no. 5 (2009): 190–209.

Reeves, Eileen, and Albert Van Helden. "Idiom and Image: Translating the Letters on Sunspots." *Isis* 109, no. 4 (2018): 767–773.

Rheinberger, Hans-Jörg. *An Epistemology of the Concrete: Twentieth-Century Histories of Life.* Durham, NC: Duke University Press, 2010.

Roberts, Sean. "Francesco Rosselli and Berlinghieri's Geographia Re-Examined," *Print Quarterly* 28 (2011): 4–17.

Roberts, Sean. "Tricks of the Trade: The Technical Secrets of Early Engraving." In *Visual Cultures of Secrecy in Early Modern Europe*, edited by Timothy McCall, Sean Roberts, and Giancarlo Fiorenza, pp. 182–207. Kirksville, MO: Truman State University Press, 2013.

Roberts, Sean. "Inventing Engraving in Vasari's Florence." *Intellectual History Review* 24, no. 3 (2014): 367–388.

Settle, Thomas B. "An Experiment in the History of Science." *Science* 133, no. 3445 (1961): 19–23.

Shea, William R., and Mariano Artigas. *Galileo in Rome: The Rise and Fall of a Troublesome Genius.* Oxford: Oxford University Press, 2003.

Spiegel, Gabrielle. "Forging the Past: The Language of Historical Truth in the Middle Ages." *History Teacher* 17 (1984): 267–288.

Stabile, Giorgio. "Linguaggio della Natura e Linguaggio della Scrittura in Galilei. Dalla 'Istoria' Sulle Macchie Solari alle Lettere Copernicane." *Nuncius* 9 (1994): 37–64.

Stijnman, Ad. *Engraving and Etching, 1400–2000: A History of the Development of Manual Intaglio Printmaking Processes.* London: Archetype, 2012.

Swan, Claudia. "Illustrated Natural History." In *Prints and the Pursuit of Knowledge in Early Modern Europe*, edited by Susan Dackerman, pp. 186–191. Cambridge, MA: Harvard Art Museums, 2011.

Toesca, Ilaria, and Roberto Zapperi. "AGUCCHI, Giovanni Battista." *Dizionario Biografico degli Italiani* 1 (1960), https://www.treccani.it/enciclopedia/giovanni-battista-agucchi_(Dizionario-Biografico)/ (accessed 27 February 2021).

Torrini, Maurizio. "Galileo Copernicano." *Giornale Critico della Filosofia Italiana* 72, no. 1 (1993): 26–42.

van Helden, Albert. "Galileo and Scheiner on Sunspots: A Case Study in the Visual Language of Astronomy." *Proceedings of the American Philosophical Society* 140, no. 3 (1996): 358–396.

van der Stock, Jan. *Printing Images in Antwerp: The Introduction of Printmaking in a City: Fifteenth Century to 1585*. Rotterdam: Sound and Vision Publishing, 1998.

White, Raymond, and Jo Kirby. "Rembrandt and his Circle: Seventeenth-Century Dutch Paint Media Re-Examined." *National Gallery Technical Bulletin* 15 (1994): 64–78.

Wilding, Nick. "Galileo and the Stain of Time." *California Italian Studies* 2, no. 1 (2011): https://escholarship.org/uc/item/7rh9r96p (accessed 5 March 2021).

About the Author

Ruth Sargent Noyes is Marie Skłodowska-Curie EU Senior Research Fellow in Art History at the National Museum of Denmark. Author of a number of books and articles, she is a 2014 Fellow of the American Academy in Rome and recipient of a number of research grants and awards.

Part 1

Approaches to Print Matrices

3. Sequencing Vesalius's *De Humani Corporis Fabrica*

Dániel Margócsy, Mark Somos, and Stephen N. Joffe

Abstract

This chapter examines the deterioration of the woodblocks in several editions of Andreas Vesalius' *De humani corporis fabrica,* the first fully illustrated anatomy atlas. Comparison of different woodcuts in the *Fabrica* reveals a consistent pattern of degradation across copies: perfect impressions of the woodblocks were selected to be bound to form a first-state book; impressions of the damaged woodblocks were bound together to form a later-state volume. This finding suggests the early modern print shop was not as disorganised as Adrian Johns and other historians have claimed in recent years.

Keywords: printing history, history of the book, history of anatomy, visual studies of science, Andreas Vesalius

I. Introduction

A plethora of scholarship from recent decades examines the complex processes whereby an early modern book came into existence.[1] While printing obviously speeded up the production of books compared to the manuscript era, a print run of hundreds (let alone thousands) still required many months.[2] During these months, printers could still fiddle with the format of the book, adding accidental and intended changes to text and

1 Gerritsen, "Printing at Froben's."

2 Eisenstein, *The Printing Press as an Agent of Change*; Grafton, "The Importance of Being Printed." For questions of relationships between manuscript and print culture, see the essay by Stephanie Leitch and Britta-Juliane Kruse in this volume.

Noyes, R.S. (ed.), *Reassessing Epistemic Images in the Early Modern World.*
Taylor & Francis Group, 2023
DOI 10.5117/9789463723350_CH03

images at the behest of correctors, authors, and typesetters.[3] Sometimes, as the first sheets of a volume were being printed at full speed, the author was still finishing the text's concluding section, writing up a fulsome epistle dedicatory, assembling the index, and composing an *errata* list of typographical errors.[4] Arguably, the complexities of these processes made each exemplar of a printed book unique. They can give the impression that the production of books was a messy and unstable enterprise in the early modern world, and not the orderly business of standardized twentieth-century production.[5] Our chapter contributes to this historiographical debate by examining the temporal sequence of printing, asking whether exemplars produced at the beginning and end of the process differed from each other.

A well-established distinction between the different states of woodcuts, engravings, and etchings exists in the field of art history. The hundreds of sheets that printmakers pulled from the same copperplate or woodblock can often be grouped into separate and distinct states based either on the matrices' deterioration, or on the artist's intervention, as in the case of Rembrandt, who obsessively reworked his etchings throughout his career.[6] For printed books, the scholarship certainly makes a distinction between the different editions of a volume, but there has been little discussion regarding how to differentiate between copies made early or late during the printing process of one edition.[7] Indeed, one could argue that such a differentiation makes no sense in the production of books that were made from several sheets of paper, unlike a single-sheet print.[8] If printing was, indeed, a messy process in early modernity, it would follow that printers might randomly bind together sheets printed early in the process with late-printed sheets, especially if those sheets were in different quires. It seems plausible that a printer decided to bind the very first state of the first sheets of a book with the very last state of the last sheets of the same book, thereby producing a hybrid.[9]

3 McKenzie, "Printers of the Mind."

4 Grafton, *The Culture of Correction in Renaissance Europe.*

5 Johns, *The Nature of the Book*; McKitterick, *Print, Manuscript and the Search for Order.*

6 On related questions of copperplate and woodblock degradation as index of print sequencing, see the essays in this volume by Angela Campbell and Jolien Vanden Bossche, respectively.

7 See, however, Schwenke and Preisendanz, "Zwei neue Exemplare," for the earliest Flugblatt in existence.

8 Needham, "Copy-Specifics in the Printing Shop."

9 Copies of the Gutenberg Bible appear to contain many such hybrids, see Agata, "Stop-Press Variants."

However, if the early modern bookshop was a reasonably well-organized workspace, one could expect that printers carefully laid one printed sheet on another. As they assembled books from their stacks of printed sheets, they paired early sheet with early sheet to make up quires, and then grouped early quire with early quire. Since newly printed sheets had to be individually hung to dry and then assembled in stacks, there is some reason to assume that the temporal order of printing was preserved during the process of assembly in orderly print shops.[10] In such a case, a distinction could be made between early or later states of a printed book even within one edition, and the temporal sequence of exemplars within a print run could be determined. This question can be resolved through empirical study. This chapter studies the deterioration of several woodcuts placed at different points in Andreas Vesalius's *De humani corporis*, the first major illustrated atlas of anatomy in the Western world, to test the viability of creating a simple chronological sequence to establish different states during the production of a multiple-sheet object. We use the findings of this study to argue for the early modern printing shop as a moderately organized entity, at least in this respect.

II. Printing the *Fabrica* and Its Variants

The printing history of Vesalius's *Fabrica* is reasonably well documented, although the lack of archival documentation from the side of the printer, Johannes Oporinus, means that some crucial details will remain forever obscure. As Vivian Nutton has shown, the *Fabrica* emerged from Vesalius's project to produce a revised version of his pirated edition of Guinther von Andernach's *Institutiones* in the years following 1538.[11] A professor at the University of Padova, Vesalius completed the text of the *Fabrica* in 1542.[12] He commissioned one or more artists to design over 200 woodcuts, probably including Jan Steven van Calcar.[13] He then had the manuscript and the woodblocks sent to Basel, to the print shop of Oporinus. As the printing proceeded, Vesalius went to Basel to supervise production of the book and introduced some minor changes to the text. The book was

10 Gaskell, *A New Introduction to Bibliography*, pp. 143–145; Hedges, "A Method for Dating Early Books and Prints."

11 Nutton, "More Vesalian Second Thoughts."

12 On Vesalius and the printing of the *Fabrica*, see O'Malley, *Andreas Vesalius of Brussels*; Margócsy, Somos and Joffe. *The* Fabrica *of Andreas Vesalius*; and Nutton, "Historical Introduction."

13 On the illustrators, see Kornell, "Jan Steven van Calcar."

completed in the summer of 1543. Vesalius personally delivered a dedication copy to Emperor Charles V in Speyer in August, although this does not necessarily mean that all exemplars had been printed by that time. Even with Vesalius in Basel, the finished book contained a number of errors, most notably when it came to pagination, with over 100 pages bearing the wrong page number in the completed product. The *errata* at the end of the volume lists only a fraction of the typographical errors in the text. Vesalius took twelve years to produce a second updated edition that added over a 100 pages of text, a few new woodcuts, and a re-cut version of the frontispiece. However, the majority of illustrations remained the same, with minor adjustments to the original woodblocks, focusing especially on the positioning of the characters identifying the organs depicted.[14] From these two editions, one can thus retrace the woodblocks' sequential deterioration both within and between editions. No authorized third edition of the *Fabrica* ever appeared, though Vesalius's notes for such a plan have recently been found.[15] The woodblocks survived, however, until World War II, having been used in the eighteenth and twentieth centuries for pulling new impressions.[16]

The scholarship has already identified some important variations within editions in Vesalius's publications.[17] In 1984, Horowitz and Collins discovered two variant editions of the 1543 *Fabrica*, differing in small but significant typographical details from the standard edition and printed in very small numbers. In Variant A (known in five exemplars), all of the changes recorded were in the prefatory matter, including the frontispiece. In Variant B (known in a single copy), further changes were made to a very few pages in the middle of the text (pp. 299–302). Horowitz and Collins were very cautious about determining when these two variants were printed. The variant editions were not published in high numbers and they are less pleasing aesthetically. This raises the possibility that these variants were advance copies printed in proof. At the same time, both variants contain typographical corrections that improve upon the text. They use a new font first documented in Swiss areas in 1549 and first used in France in 1543. Moreover, these variants also reveal damage to the woodblock of Vesalius's portrait not present

14 We eagerly await Monique Kornell's study of these changes.

15 Nutton, "Vesalius Revised."

16 Vesalius, *Icones anatomicae*. On the afterlife of Vesalian images, see Margócsy, "From Vesalius through Ivins to Latour."

17 Horowitz and Collins, "A Census of Copies."

in the regular edition, but present in Vesalius's *China Root Epistle* from 1546. On the basis of these arguments, Horowitz and Collins dated these variant exemplars to the second half of the 1540s, and raised the possibility that Oporinus was running out of copies of the 1543 edition at this point, with stocks of two quires depleted, and therefore needed to print the prefatory matter again.

Horowitz and Collins did not note all the typographic variations in Vesalius's pioneering atlas. The recent French census of the 1543 *Fabrica* has discovered a variant concerning the signature of sheet H3, which was erroneously marked as G3 in some copies in France. Again, in some other copies, the erroneous G3 was corrected, potentially by hand, to H3.[18] At the same time, the *Fabrica* appears to have less typographic variation on its other pages than otherwise presumed with early modern books. The *Gutenberg Bible* exhibits a large number of variations on its pages, as one would expect from a book produced at the dawn of printing.[19] Over ten per cent of the pages of Shakespeare's *First Folio* exist in at least two variants, and Andrew Walkling has recently noted an astonishing number of typographic variations in the libretto of Henry Purcell's *The Fairy-Queen*.[20] In contrast, the variants noted above comprise roughly two per cent of the *Fabrica*'s pages. While it is likely that further scrutiny and machine-aided collation techniques will reveal further variation in the text of the *Fabrica*, it is also possible that Vesalius's presence ensured that most changes were effected during a proof run and, potentially, that the working methods of Oporinus did not encourage the introduction of numerous typographic changes to a text already set.[21]

While most of the research on printed books focuses on changes to the text, we explore how the deterioration of the numerous woodcut illustrations of the *Fabrica* can help us understand the sequence of printing exemplars. Horowitz and Collins already documented the chipping of the woodblock in Vesalius's portrait, but they did not examine the rest of the illustrations in the text from this aspect.[22] Scholars working on other publications have explored the deterioration of images to establish the chronological order of different editions using the same image, but

18 Charreaux and Wijland, "Recensement et description."

19 See Agata, "Stop-Press Variants." See also Rangel and Alabert, "Two Corrections to the Recent Census."

20 Higgins, "Printing the First Folio"; Walkling, "Hinman, Redux."

21 For a review and potential of pre-digital collators, see Smith, "'Armadillos of Invention'."

22 For further study of the deterioration of the woodblocks 1543–1934, see Goree, "The Woodblocks of Vesalius."

they have rarely applied the technology to document variation within an edition.[23]

Our methods for determining the temporal sequence of books is based on a simple model of the orderly print shop. By identifying patterns of deterioration in two different images, we can first establish states 1 and 2 for woodcut A, and states 1 and 2 for woodcut B. Plotted against the timeline of the printing process, this means that the two groups may not be identical, but they must be complementarily exclusive. Pristine exemplars with first-state impressions of both woodcuts and damaged exemplars with second-state impressions for both images should exist. Importantly, it is likely that the two woodcuts will become damaged at different points of time. If woodcut A becomes damaged first, there should also be a number of exemplars that contain state 2 for woodcut A and state 1 for woodcut B. In a perfectly ordered workshop, there should therefore be only three states of the whole book: a) pristine exemplars; b) exemplars with damaged woodcut A but intact woodcut B; and c) exemplars with damage to both woodcuts. If one also finds exemplars with a damaged woodcut B, but an intact woodcut A, this is either the result of the print shop proceeding in a disorderly manner, or that the anomalous exemplar was tampered with after it left the printing shop (e.g. it is a hybrid copy assembled by a nineteenth-century book dealer from two different exemplars).

Imperfections in the impression of a woodcut are not always the result of the chipping of woodblocks. They may be the consequence of some loose material getting stuck between the woodblock and the paper in the process of printing. Alternatively, a pressman may fail to ink all areas of the woodblock with sufficient care. In these cases, the imperfections cannot be used to document temporal change, as perfect impressions can again be pulled once the loose material is removed and the pressman is disciplined. Fortunately, we have one set of controls that prove that the breaks in the 1543 and 1555 *Fabrica* illustrations are the result of woodblock chipping. The 1934 Bremer Presse edition of the *Icones anatomicae* used the original woodblocks that had been rediscovered, and applied extreme care toward their printing as their purpose was to create a collector's edition printed on special paper.[24]

23 Rasmussen, "The Date of Q4 'Hamlet'"; Bowen and Imhof, "18,257 Impressions from a Plate"; Hedges, "A Method for Dating Early Books and Prints." See also Angela Campbell's essay in this volume.

24 Joffe and Buchanan, "The Andreas Vesalius Woodblocks" ; Chevallier and Neidhardt, "La destinée des bois de la *Fabrique* de 1543."

Despite the Bremer Presse's care, the imperfections we have documented for the 1543 and 1555 impressions also manifest in this latter-day edition, strongly suggesting prior chipping or cracking.

Our chapter offers an exploration of the potential of this technique by examining four woodcuts in the 1543 and 1555 editions and one only in the 1555 edition. These are not the only woodcuts that display damage: for example, every known exemplar of the *Fabrica* shows a massive crack at the bottom left of the third muscleman, which must have occurred before printing started and therefore cannot be used for the purposes of sequencing.[25] In addition, while over 300 copies of the 1543 and more than 400 exemplars of the 1555 *Fabrica* are known to survive today, our exploratory survey is limited to a smaller sample of twenty 1543 and twenty 1555 editions, whose bibliographical details are listed in the appendix.[26] Consequently, our results should be considered with caution and will need to be confirmed across a larger sample of sources.

In the first edition we focus on woodcuts on pages 2, 101, 108, and 200. In the 1555 edition, these woodcuts are republished on pages 2, 124, 133, and 235; and a new woodcut is inserted on page 560. These woodcuts display the following signs of damage:

1. In the 1543 edition, the woodcut on page 2 shows a variety of bones, most prominently the *humerus* bone of the arm. At the bottom of the woodcut, Vesalius also displays a sesamoid bone of the toe cut into two halves, with the character *N* placed in between. This woodcut is intact in all known exemplars of the 1543 edition, but displays damage in some of the 1555 editions, at the top of the bone to the right of the character *N* (Figure 3.1A–C). 1555 editions in Kiel and two copies at the Countway Library in Boston exhibit such damage, amongst other exemplars.

2. The second illustration of the clavicle on page 101 in the first edition, marked "SECUNDA," shows its interior face together with its superior aspect. Here, almost all copies of the 1543 edition exhibit a sustained crack line to the cross-hatched shading to the left of the bone, near character *A*. Several exemplars, including those at Dartmouth College (I/224) and at Angers (I/25), reveal the original, intact state, but most copies feature a damaged impression. Importantly, Oporinus was able to retouch this damage to some extent for the 1555 edition (for page 124), either by correcting the hatching or by better positioning the woodblock (Figure 3.2A–C). The 1934 reprinting of the woodblocks also presents the corrected state, although

25 We thank Sachiko Kusukawa for alerting us to this.

26 Margócsy, Somos and Joffe, *The* Fabrica *of Andreas Vesalius*.

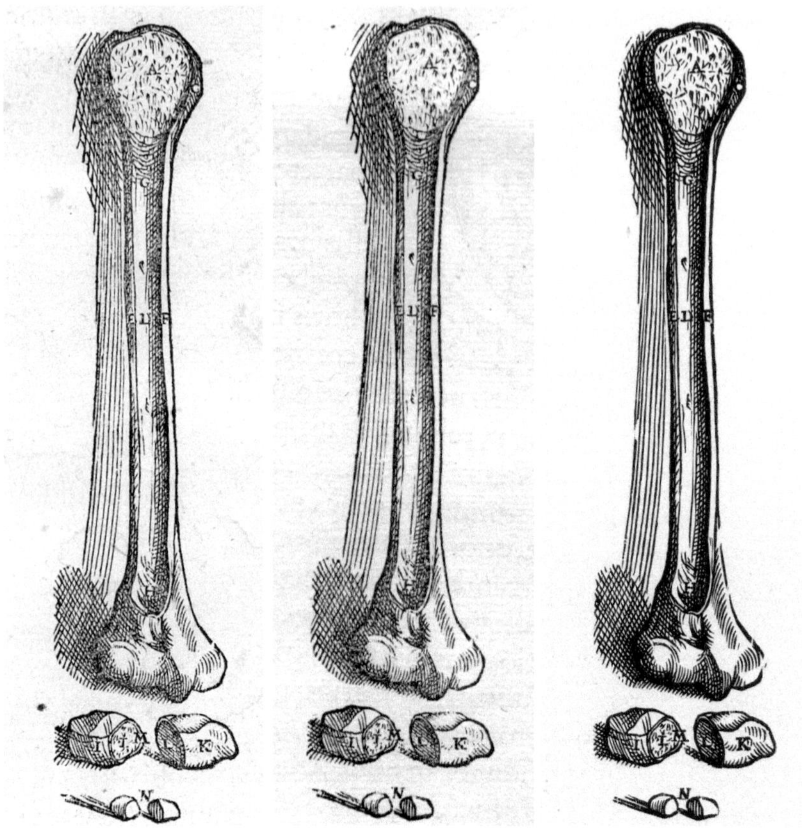

Figure 3.1A–C Deterioration in the woodcut of the sesamoid bones of the toe, to the right of character N at the bottom of the woodcut. Left-Right: A, Intact woodcut from Neuchâtel (II/238), BPU Neuchâtel; B, damaged woodcut from Kiel (II/88), credit Kiel University Library; C, damaged woodcut from the Icones edition of 1934, courtesy of the New York Academy of Medicine.

some new minor damage is visible right at the top of character *A*, affecting two hatched lines.

3. In the 1543 edition, the lower half of page 108 presents six images of the bones of the forearm. Proceeding from left to right, the first two images, marked "SECUNDA," show the radius and the ulna in their outer aspect. The third image, marked "TERTIA," shows the inner side of the radius, while the fourth, marked "QUARTA," shows its outer side. The last two images, marked "QUINTA" and "SEXTA," similarly reveal the two aspects of the ulna. Already in the 1543 edition deterioration manifests in some exemplars in the third image, "TERTIA." At the bottom of this image, to the left below character *M*, the contours of the radius become broken; the damage remains present in all copies of the 1555 edition on page 133. The 1934 edition also retains this damage and shows some further deterioration to the same

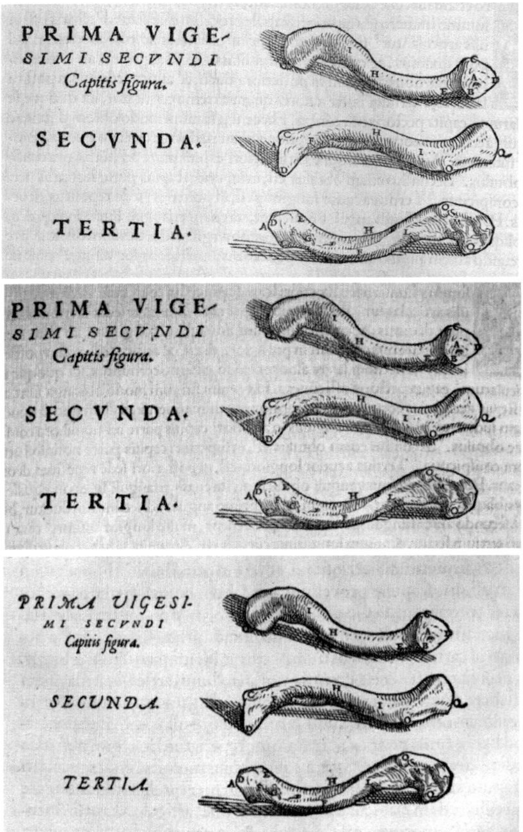

Figure 3.2A–C Deterioration and Correction of the Clavicle Woodcut, where the damage is shown to the shading to the left of the middle bone, next to characters A and C. Left-Right: A, Intact woodcut from Antwerp (I/4), cc-by 4.0 Erfgoedbibliotheek Hendrick Conscience; B, damaged woodcut from Augsburg (I/50); C, retouched woodcut from Augsburg (II/63).

image near the character *q* (Figure 3.3A–B). Only a few exemplars feature the intact woodcut, including those at the Getty in Los Angeles (I/231) and one in Angers (I/25).

4. The second book of the 1543 edition, devoted to the muscles, features on page 237[235] an illustration of anatomist's instruments arrayed on a table.[27] The table top has two layers, and the right vertical edge of the lower table top becomes increasingly worn until in disappears. That the deterioration clearly happens gradually and not as the result of one sudden chip may

27 As mentioned, the pagination of the 1543 edition is often wrong. This illustration is on one of two pages that are both marked "237." The correct pagination should be page 235.

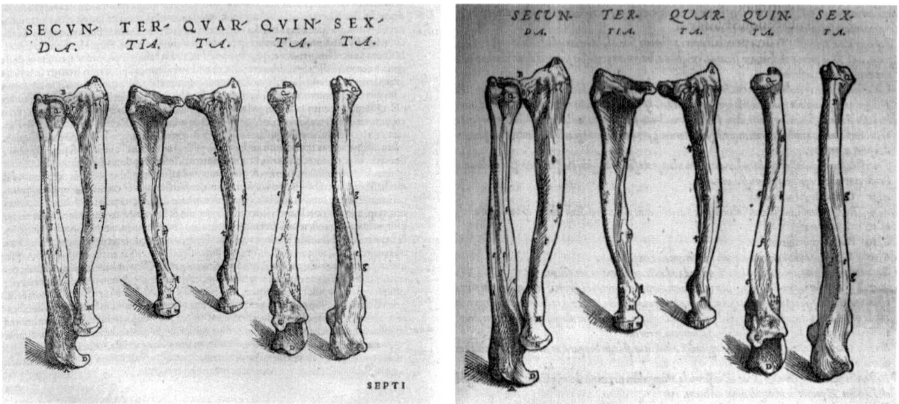

Figure 3.3A–B Deterioration of the Radius Woodcut, damage affecting the bone marked "TERTIA," at the bottom, to the shading to the left of the area marked with M. A (left): Intact woodcut from Madrid (I/144), cc-by 4.0 Biblioteca Nacional de España; B (right): damaged woodcut from Augsburg (II/63).

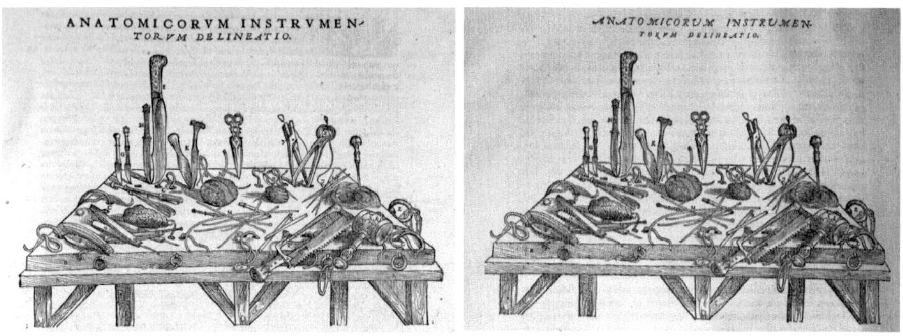

Figure 3.4A–B Deterioration of the Anatomist's Instruments woodcut, damage affecting the bottom vertical edge of the table on the right. A (left): Intact woodcut from Tours (I/49), cc-by-nc-sa 4.0 CESR Tours; B (right): damaged woodcut from Augsburg (II/63).

allow for a refined analysis of the extent of the damage in further studies (Figure 3.4A–B). Vesalius decided to print the illustration at a different location for the 1555 edition, on page 200, at the end of the first book devoted to bones. Every examined exemplar of this later edition consistently exhibits the deterioration at this spot. The 1934 printing of the woodblocks confirms the damage and reveals further, marked deterioration to the other edges of the table tops in the same area. A copy in Tours (I/49) and the variant edition of the Countway Library (I/209) are amongst those that feature an intact image.

5. The 1555 edition introduces a new woodcut on page 560, a view of the omentum. At the two sides of the woodcut, near characters *E* and *F*, two bows tie off the intestinal tract; the one on the left, near character *E*, shows increasing deterioration in a number of copies. Just as in the case of the table,

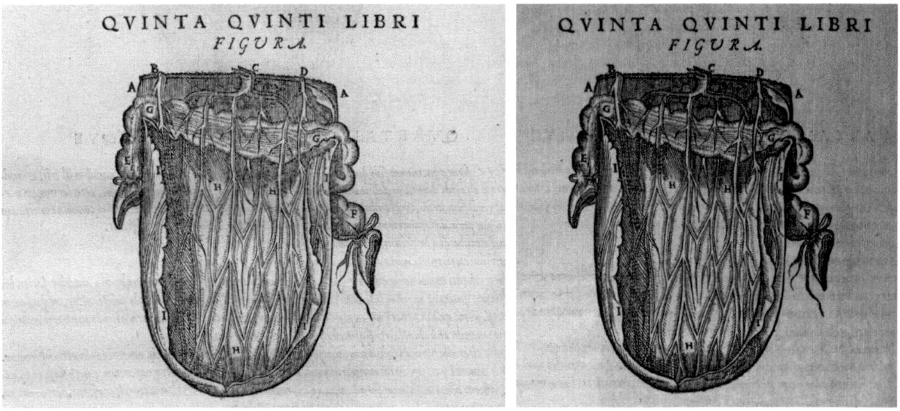

Figure 3.5A–B Deterioration of the Omentum Woodcut (no impression available from 1934), damage affecting the ribbon on the left below character E. A (left): Intact woodcut from Kiel (II/88), credit Kiel University Library; B (right): damaged woodcut from Augsburg (II/63).

a more nuanced method of analysis tracing the amount of deterioration step by step may be possible (Figure 3.5A–B). Unfortunately, this woodblock was lost before the 1934 re-printing, which instead features an ersatz image reproduced photomechanically from a copy of the 1555 edition that therefore cannot be used for the purposes of confirmation. Most exemplars feature an intact woodcut, but both exemplars at Dartmouth College (II/316, II/317) feature a damaged state.

To sum up, the 1543 edition exhibits variation at three locations, on pages 101, 108, and 237[235], to images of the clavicle, the radius, and the anatomist's instrument table, respectively. In each of these cases, both the intact and the damaged states are present in a good number of exemplars for this edition. Importantly, while the image of the clavicle was corrected for the 1555 edition, sustained damage remains in the images of the radius and the table in all exemplars of this second printing. The 1555 edition also exhibits two additional variants for the sesamoid bones of the toe on page 2 and the new woodcut of the omentum on page 560, for which both intact and damaged states are documented in several exemplars. Whenever possible, the 1934 *Icones* confirms that the damage to the woodblocks was also present when they were last used in the twentieth century.

III. Findings

Our analysis reveals that it is not possible to establish a straightforward chronology of exemplars that presumes a completely consistent printing pattern in Oporinus's shop. The emerging patterns strongly suggest that, while Oporinus's working methods were rather orderly, there was plenty of opportunity for mixing and matching sheets printed at different times in the printing process. It is not the case that early states of one woodcut are always paired with early states of another woodcut in other part of the volume. Late states of the woodcuts are not always paired together, either Tables 1 and 2 show our findings for both editions. The results for the 1543 edition reveal that there are very few exemplars containing all woodcuts in pristine condition. Only two out of 28 copies of the 1543 *Fabrica* contain first-state woodcuts of the clavicle, the radius, and the surgical instruments. The plurality of *Fabrica* exemplars (around 40 per cent) contain the second, damaged state of all three of these woodcuts, strongly suggesting that the woodblocks became worn relatively quickly. This finding qualifies S. Blair Hedges's claim that woodcut deterioration was correlated primarily with the passage of time, not use.[28] In the case of the *Fabrica*, the woodblocks clearly became damaged soon after they were put under the printing press.

Most *Fabrica* exemplars represent hybrids containing both first- and second-state woodcuts. The blocks for the clavicle and the instruments were probably damaged first, as most exemplars feature the second state of these images with an intact image of the radius on page 108. Just under half feature an intact woodcut of the radius, around 20 per cent of the clavicle woodcuts are intact, and around 15 per cent of the images of the anatomist's table. Given the high number of damaged woodcuts, the two pristine exemplars of the *Fabrica* with all intact woodcuts suggest that the gathering of sheets into complete volumes did proceed according to some temporal order in Oporinus's print shop. Half of the intact woodcuts of the anatomist's table were bound with sheets displaying intact woodcuts on other pages.

Establishing a more nuanced temporal sequence encounters difficulties. Based on our sample, it is not obvious which woodcut became damaged first. Our sample contains two exemplars that contain only one second-state woodcut: the copy in Angers (I/25) features a damaged woodcut of the table, while a copy at the Countway Library (I/209) in Boston features a damaged woodcut of the clavicle. At least one of these two exemplars demonstrates that printers did not follow a strict temporal order when pairing sheets to bind their volumes. The order in which sheets were pulled off the press corresponds to the order in which they were gathered and bound together, but by no means strictly. To some degree, one could explain this finding by the fact that both woodcuts were damaged relatively early in the process.

28 Hedges, "A Method for Dating Early Books and Prints."

Nonetheless, the analysis of those exemplars containing two second-state woodcuts further complicates the situation. Given that the radius woodcut survives in a larger number of intact impressions than the other two, it was probably damaged last. Yet, in four exemplars (almost 30 per cent of the second-state impressions of this woodcut), the damaged radius woodcut is bound together with an intact woodcut of either the clavicle or the table. This clearly suggests that Oporinus's print shop workers freely combined sheets printed at various times in the publication process. The absence of exemplars where a damaged radius woodcut is bound with intact impressions of both other woodcuts strongly suggests that there was some limit to mixing and matching. That sheets printed early on tended to be paired with other sheets printed early in the process, and sheets printed late tended to be paired with other sheets printed late, represents a pattern, though not a rule.

A similarly complicated picture emerges from comparison with other studies of variants in the 1543 edition. The French census's findings clearly suggest that those exemplars with an uncorrected version of the H3 signature contain at least one sheet printed early on. In our sample, two exemplars (Angers 1, I/25 and Angers 2, I/26) contain the erroneous H3 signature without correction. One of them (I/25) is paired with two sheets of intact woodcuts, and one damaged with damage; another (I/26) is paired with two damaged woodcuts. Of the four corrected exemplars, two are paired with the same two damaged woodcuts as Angers 2 (I/26), one in Tours is paired with damaged states of the clavicle and the radius, and one is paired with damaged states of all woodcuts. Again, despite a tendency for the uncorrected H3 copies to be paired with earlier states of the woodcuts, the pattern is not a very strong one.

Curiously, our analysis also included a variant 1543 edition (I/209) that features the irregular elements noted by Horowitz and Collins; moreover, Collins argued that these variant editions were printed either before or after the main batch of printing. Yet, a close study of the Countway variant reveals that it contains a damaged woodcut of the clavicle, but intact woodcuts of the two other images. This finding clearly suggests that the Horowitz and Collins variant editions were either not proofs, or if their prefatory matter was printed as a proof, those sheets were later bound with other sheets printed during the regular printing process. However, the presence of intact sheets of two woodcuts also indicates that the variant editions of Horowitz and Collins were not printed at a completely new stage, years after the original edition. Oporinus likely used the same batch of sheets for both the regular and variant editions, and the changes to the prefatory matter may simply reflect that these sheets were printed completely separately from the rest of the volume, or that Oporinus's pressmen experimented with a new font for a few exemplars of the *Fabrica* in the middle of the process of printing, and then rejected this newer version. In any case, the chronology proposed by Horowitz and Collins needs to be treated with extreme caution.

In the case of the 1555 edition, our findings are less conclusive because our data are based on variation in only two woodcuts, not three. The woodcut of the omentum appears intact in most copies, in twenty out of our sample of 26 exemplars. In contrast, the woodcut of the sesamoid bone was probably damaged early in the printing process, with the majority of the exemplars (c. 70 per cent) containing a second-state image on page 2. Five exemplars feature both woodcuts in an intact state, e.g. in Augsburg (I/63) or Neuchâtel (II/238). The image of the sesamoid bone probably became damaged well before the omentum woodcut, and fifteen exemplars feature the damaged woodcut on page 2 paired with an intact illustration of the omentum. At the same time, two exemplars in Paris (II/45) and Dartmouth College (II/316) show the opposite pattern: the damaged omentum is paired with an intact woodcut of the sesamoid bones. The remaining four damaged woodcuts of the omentum are paired with a damaged woodcut of the sesamoid bones. Oporinus's workshop was not completely orderly, but most exemplars of the 1555 edition do seem to follow a reasonably clear chronological sequence. Given the highly limited number of damaged woodcuts of the omentum, our findings for the second edition remain somewhat inconclusive and encourage further research on a larger sample. It is nonetheless noteworthy that we did not observe any deterioration to the woodcuts between the 1543 and 1555 editions and, in fact, the woodcut of the clavicle appears to have been repaired. This seems to confirm the traditional argument that woodblocks primarily become damaged while in use, not while they are sitting idle.

IV. Concluding Thoughts

The early modern print shop was a complex workplace, and Oporinus's firm was no exception. While the *Fabrica* was a landmark volume in the history of printing with its lavish outlay and sumptuous imagery, it was not a book without blemishes. As we have seen, even the close oversight by the author could not prevent the proliferation of typographic errors, and the pagination of the 1543 edition is riddled with mistakes. Our chapter has offered a new perspective on how to assess the relative state of more or less orderly and well-organized early modern print shops. We examined the correlation between the temporal sequence of how printed sheets were pulled off the press, and how they ended up bound together in volumes once the printing was finished. Our findings suggest that Oporinus's print shop was reasonably well-ordered. Sheets printed early in the process were more likely to end up bound with other sheets printed early in the process. In some cases, it is reasonably clear that a volume, with all its sheets, was printed early on, and in other cases, one can assume that most sheets of an exemplar were printed at the end of the print run.

Yet, the process of printing cannot be reduced to an idealized model. For a variety of reasons, not all of them clear, workers in the print shop could mix and match sheets printed early in the process with sheets printed later on. As a result, it is not possible to delineate a neat and perfect timeline of first, second, third, and subsequent state exemplars of the *Fabrica* at the level of a whole book. In a significant number of cases, an exemplar can contain a variety of first and second state woodcuts that go against our best estimate of how and when the woodblocks became damaged. No easy method exists to determine whether a particular copy of the *Fabrica* is a pristine volume with intact woodcuts throughout: reaching a conclusion requires examination of as many of the images as possible, on as many pages as possible. The further study of more woodcuts, on a larger sample, may shed greater light on how the *Fabrica* was produced. It may also prove beneficial to examine typographical variation in the *Fabrica* beyond those errors noted by Horowitz and Collins, and by the authors of the French census. If other early modern books are anything to go by, there must be much more variation hidden in the pages of Vesalius's atlas. Early modern printers devoted special efforts to print their books sheet by sheet, and scholars of bibliography can truly understand their working methods only if they examine the printed output with the utmost care, page by page.

Appendix

Details of *Fabricas* under consideration, using the numbering system of the census introduced in Margócsy, Somos and Joffe, op. cit.

1543 Editions:

I/4 Antwerp
Erfgoedbibliotheek Hendrick Conscience
Call No.: J 5833 [C2-572A]

I/7 Brussels
Royal Library of Belgium
Call No.: VH 7.413 C (RP)

I/8 Ghent
University Library Ghent
Call No.: BHSL. RES. 1950

I/215 Prague
National Library of the Czech Republic
Call No.: 65 B 230

I/25 Angers 1
Bibliothèque municipale d'Angers
Call No.: Rés SA 2570

I/26 Angers 2
Bibliothèque municipale d'Angers
Call No.: Rés SA 2570 bis

I/28 Boulogne-sur-Mer
Bibliothèque des Annonciades
Call No.: S1 3341

I/32 Montpellier
Médiathèques Montpellier
Call No.: C 769

I/34 Paris
Bibliothèque interuniversitaire de Santé
Call No.: 302

I/49 Tours
Centre d'études supérieures de la Renaissance
Call No: SR 57 C-Bibgen. Res-C. EPD

I/50 Augsburg
Staats- und Stadtbibliothek Augsburg
Call No.: 2 Med 193

I/63 Halle (Saale)
Universitäts- und Landesbibliothek Sachsen-Anhalt
Call No.: UB 213 2°

I/65 Jena
Thüringer Universitäts- und Landesbibliothek
Call No.Ö 2 Med. XIV, 4

I/130 Utrecht
Utrecht University Library
Call No.: M fol 92 (Rariora)

I/135 Coimbra
Universidade de Coimbra Biblioteca Geral
Call No.: 4 A-21-14-1

I/144 Madrid
Biblioteca Nacional de España
Call No.: R/34024

I/153 Basel
Universitätsbibliothek Basel
Call No.: AN I 15

I/207 Boston, MA 1
Francis A. Countway Library of Medicine
Call No.: ff QM21.V63 copy 1

I/208 Boston, MA 2
Francis A. Countway Library of Medicine
Call No.: ff QM21.V63 copy 2

I/209 Boston, MA 3
Francis A. Countway Library of Medicine
Call No.: ff QM21.V63 copy 3

I/210 Boston, MA 4
Francis A. Countway Library of Medicine
Call No.: ff QM21.V63 copy 1

I/215 Chicago, IL
Newberry Library
Call No.: Case 6A 156

I/224 Hanover, NH 1
Dartmouth College
Rauner Library
Call No.: QM25.V4

I/225 Hanover, NH 2
Dartmouth College
Rauner Library
Call No.: QM25.V4 c.2

I/230 Kansas City, MO
Linda Hall Library
Call No.: QM21 V395 1543

I/231 Los Angeles
Getty Research Institute
Call No.: ID 84-B27611

1555 Editions:

II/7 Vienna 1
Universitätsbibliothek der Medizinischen Univeristät
Josephinische Bibliothek
Call No.: JB 855a

II/9 Vienna 2
Universitätsbibliothek der Medizinischen Univeristät
Obersteiner Bibliothek
Call No.: HOB-A-00002

II/36 Lille
Bibliothèque Universitaire Centrale de Lille-3
Call No.: Réserve patrimoniale 177

II/45 Paris
Bibliothèque interuniversitaire de Santé
Bibliothèque Médicine
Call No.: 302A

II/59 Strasbourg
Université de Strasbourg
Médecine et Odontologie
Call No.: J R 277 [extant]

II/63 Augsburg
Staats- und Stadtbibliothek Augsburg
Call No.: 2 Med 195

II/72 Erlangen
Universitätsbibliothek
Erlangen-Nürnberg
Call No.: H61/2 TREW.C 224

II/73 Frankfurt am Main 1
Städel Museum
Call No.: Tresor 2/390

II/74 Frankfurt am Main 2
Zentralbibliothek der Universitätsbibliothek Johann Christian Senckenberg
Call No.: 2 R.175.622

II/79 Halle (Saale)
Universitäts- und Landesbibliothek
Sachsen-Anhalt
Call No.: UB 213 a 2°

II/88 Kiel
Universitätsbibliothek Kiel
Call No.: Arch2 136

II/90 Leipzig
Universitätsbibliothek Leipzig
Call No.: Anat. 66.

II/98 Munich
Bayerische Staatsbibliothek
Call No.: Rar. 2252

II/103 Nuremberg
Stadtbibliothek Nürnberg
Call No.: Med. 2. 155

II/160 Riga
Paul Stradins Museum of History of Medicine

II/190 Toruń
Nicolaus Copernicus University Library
Call No.: Ob.6.IV.117

II/215 Mahón
Biblioteca pública del estade de Mahón
Call No.: IN-79

II/230 Basel 1
Anatomisches Institut

II/231 Basel 2
Pharmazie-Historisches Museum der Universität Basel

II/238 Neuchâtel
Bibliothèque publique et universitaire de Neuchâtel
Call No.: ZU 65

II/294 Boston, MA 4
Boston University Library
Howard Gottlieb Archival Research Center
Call No.: YQM25. V4.1555

II/295 Boston, MA 1
Francis A. Countway Library of Medicine
Call No.: Rare Books 1.Mw.1555.V (Cab. 11, Shelf 5)

II/296 Boston, MA 2
Francis A. Countway Library of Medicine
Call No.: Rare Books ff 1.Mv.1

II/297 Boston, MA 3
Francis A. Countway Library of Medicine
Call No.: Rare Books QM21. V63 1555

II/316 Hanover, NH 1
Dartmouth College
Rauner Library
Call No.: QM25. V467

II/317 Hanover, NH 2
Dartmouth College
Rauner Library
Call No.: QM25. V467 c. 2

Table 1. The Different States of the Woodcuts in Exemplars of the 1543 Edition

	All Woodcuts Intact	**Intact Clavicle & Radius Good, Damaged Instruments**	**Intact Clavicle & Instruments Good, Damaged Radius**	**Intact Radius & Instruments, Damaged Clavicle**
	Hanover, NH 1	Angers 1	0	Boston 3
	Los Angeles			
Total	*2*	*1*	*0*	*1*
	Intact Instruments, Damaged Clavicle & Radius	**Intact Radius, Damaged Clavicle & Instruments**	**Intact Clavicle, Damaged Radius & Instruments**	**All Woodcuts Damaged**
	Tours	Angers 2	Antwerp	Brussels
		Boulogne	Halle	Ghent
		Montpellier	Kansas City	Paris
		Coimbra		Augsburg
		Boston 1		Jena
		Boston 2		Utrecht
		Boston 4		Basel
		Madrid		Newberry
				Prague
				Dartmouth 2
Total	*1*	*8*	*3*	*10*

Table 2. The Different States of the Woodcuts in Exemplars of the 1555 Edition

	Both Woodcuts Intact	Intact Sesamoid, Damaged Omentum	Intact Omentum, Damaged Sesamoid	Both Woodcuts Damaged
	Augsburg	Paris	Frankfurt 1	Toruń
	Neuchâtel	Hanover, NH 1	Frankfurt 2	Hanover, NH 2
	Boston 1		Kiel	Erlangen
	Basel 2		Boston 2	Leipzig
	Riga		Boston 3	
			Boston 4	
			Basel 1	
			Munich	
			Mahon	
			Halle	
			Strasbourg	
			Lille	
			Vienna 1	
			Vienna 2	
			Nuremberg	
Total	*5*	*2*	*15*	*4*

Bibliography

Early modern published sources

Vesalius, Andreas. *De Humani Corporis Fabrica Libri Septem.* Basel: Ex Officina Ioannis Oporini, 1543.

Vesalius, Andreas. *Andreae Vesalii Bruxellensis, scholae medicorum Patauinae professoris, suorum de humani corporis fabrica librorum epitome.* Basel: ex officina Ioannis Oporini, anno 1543 mense iunio.

Vesalius, Andreas. *De Humani Corporis Fabrica Libri Septem.* Basel: Ex Officina Ioannis Oporini, 1555.

Vesalius, Andreas. *Icones anatomicae.* Munich: Bremer Presse, 1934.

Secondary sources

Agata, Mari. "Stop-Press Variants in the Gutenberg Bible: The First Report of the Collation." *Papers of the Bibliographical Society of America* 97 (2003): 139–165.

Bowen, Karen L., and Dirk Imhof. "18,257 Impressions from a Plate." *Print Quarterly* 22 (2005): 265–279.

Charreaux, Stéphanie, and Jérôme van Wijland. "Recensement et Description des Exemplaires de la Première Édition du De Fabrica 1543 Conservés en France dans les Bibliothèques Publiques." In *La Fabrique du Vésale. La Mémoire d'un Livre*, edited by Jacqueline Vons, pp. 253–312. Paris: BIUS, 2014.

Chevallier, Jacques, and Jean-Christophe Neidhardt. "La Destinée des Bois de la *Fabrique* de 1543." *Histoire des Sciences Médicales* 48 (2014): 485–494.

Eisenstein, Elizabeth L. *The Printing Press as an Agent of Change: Communications and Cultural Transformations in Early-Modern Europe*, 2 vols. Cambridge: Cambridge University Press, 1979.

Gaskell, Philip. *A New Introduction to Bibliography.* London: Oak Knoll Press, 1972.

Gerritsen, Johann. "Printing at Froben's: An Eye–Witness Account." *Studies in Bibliography* 44 (1991): 144–163.

Goree, John. "The Woodblocks of Vesalius and the Printings: From the Renaissance to the Modern Era." *The Free Library* (2014): https://www.thefreelibrary.com/The+woodblocks+of+Vesalius+and+the+printings%3a+from+the+renaissance+to...-a0407670283 (accessed 27 February 2021).

Grafton, Anthony. "The Importance of Being Printed." *Journal of Interdisciplinary History* 11 (1980): 265–286.

Grafton, Anthony. *The Culture of Correction in Renaissance Europe.* London: British Library, 2011.

Hedges, S. Blair. "A Method for Dating Early Books and Prints Using Image Analysis." *Proceedings of the Royal Society* 462 (2006): 3555–3574.

Higgins, B. D. R. "Printing the First Folio." In: *The Cambridge Companion to Shakespeare's First Folio*, edited by Emma Smith, pp. 30–47.. Cambridge: Cambridge University Press, 2016.

Horowitz, Michael, and Jack Collins. "A Census of Copies of the First Edition of Andreas Vesalius' *De humani corporis fabrica* (1543), with a Note on the Recently Discovered Variant Issue." *Journal of the History of Medicine and Allied Sciences* 39 (1984): 198–221.

Joffe, Stephen N., and Veronica Buchanan. "The Andreas Vesalius Woodblocks: A Four Hundred Year Journey from Creation to Destruction." *Acta Medico–Historica Adriatica* 14 (2016): 347–372.

Johns, Adrian. *The Nature of the Book: Print and Knowledge in the Making.* Chicago, IL: University of Chicago Press, 1998.

Kornell, Monique. "Jan Steven van Calcar, c. 1515–c. 1546, Vesalius' Illustrator." In *Andreas Vesalius and the* Fabrica *in the Age of Printing; Art, Anatomy, and Printing in the Italian Renaissance*, edited by Rinaldo Fernando Canalis and Massimio Ciavolella, pp. 99–130. Turnhout: Brepols, 2018.

McKenzie, D. F. "Printers of the Mind: Some Notes on Bibliographical Theories and Printing-House Practices." *Studies in Bibliography* 22 (1969): 1–75.

McKitterick, David. *Print, Manuscript and the Search for Order.* Cambridge: Cambridge University Press, 2003.

Margócsy, Dániel. "From Vesalius through Ivins to Latour: Imitation, Emulation and Pictorially Repeatable Statements in the Fabrica." *Word and Image* 35 (2019): 315–333.

Margócsy, Dániel, Mark Somos, and Stephen N. Joffe. *The* Fabrica *of Andreas Vesalius: A Worldwide Descriptive Census, Ownership, and Annotations of the 1543 and 1555 Editions.* Leiden: Brill, 2018.

Needham, Paul. "Copy-Specifics in the Printing Shop." In *Early Printed Books as Material Objects,* edited by Bettina Wagner and Marcia Reed, pp. 9–20. Berlin: De Gruyter, 2010.

Nutton, Vivian. "Vesalius Revised: His Annotations to the 1555 *Fabrica.*" *Medical History* 56 (2012): 415–443.

Nutton, Vivian. "Historical Introduction." In *On the Fabric of the Human Body: An Annotated Translation of the 1543 and 1555 Editions of Andreas Vesalius' De Humani Corporis Fabrica,* edited by Daniel Garrison and Malcolm Hast, pp. lxxv–ciii. Basel: Karger, 2014.

Nutton, Vivian. "More Vesalian Second Thoughts: The Annotations to the *Institutiones anatomicae secundum Galeni sententiam.*" *Gesnerus* 11 (2015): 94–116.

O'Malley, Charles Donald. *Andreas Vesalius of Brussels, 1514–1564.* Berkeley, CA: University of California Press, 1964.

Rangel, Luz M., and Aureli Alabert. "Two Corrections to the Recent Census of the Gutenberg Bible." Unpublished Manuscript, 2019. DOI: 10.31219/osf.io/mg59f.

Rasmussen, Eric. "The Date of Q4 'Hamlet'." *Papers of the Bibliographical Society of America* 95 (2001): 21–29.

Schwenke, Paul, and Karl Preisendanz. "Zwei neue Exemplare der Ablassbriefe von 1455." *Zentralblatt für Bibliothekswesen* 36 (1919): 175–177.

Smith, Steven Escar. "'Armadillos of Invention': A Census of Mechanical Collators." *Studies in Bibliography* 55 (2002): 133–170.

Walkling, Andrew R. "Hinman, Redux." *The Collation Blog.* May 2018. https://collation.folger.edu/2018/05/hinman–redux/ (accessed 3 July 2022).

About the Authors

Dániel Margócsy is professor in the history of science, technology and medicine at the University of Cambridge.

Mark Somos holds the Deutsche Forschungsgemeinschaft's Heisenberg position at the Max Planck Institute for Comparative Public Law and

International Law, where he directs the Grotius Census Project. He published five books and over 50 scholarly articles, and co-edits *Grotiana* and the History of European Political and Constitutional Thought book series.

Stephen N. Joffe is visiting professor in history of medicine at Cedars-Sinai in Los Angeles and esteemed professor of surgery and medicine at University of Cincinnati Medical Center.

4. Meticulous Matrices: Building a Chronology of Albrecht Dürer's *Meisterstiche* Impressions through the Analysis and Documentation of Microscopic Scratches in His Engraved Plates

Angela Campbell[1]

Abstract
This chapter reviews the methods and results of an in-depth comparative study of Albrecht Dürer's three *meisterstiche* – *Melencolia I*, *Knight, Death, and the Devil*, and *St. Jerome in His Study*. As a plate is prepared for printing, hard and sharp fragments of burr residue and inclusions in the printing ink are repeatedly rubbed across the matrix: fine scratches appear on the surface of the plate almost immediately and the crisp edges of the engraved areas are gradually worn down, making the cut lines softer and less well-defined. By digitally recording and comparing the marks of deterioration on each of the *meisterstiche* impressions, this study reconstructs a chronological sequencing method for individual impressions to offer new, diverse information about Dürer's work.

Keywords: Dürer, meisterstiche, chronology, plate wear, digital documentation

1 The work undertaken toward this study would not have been possible without the indefatigable guidance and encouragement of various mentors and colleagues, including Judith C. Walsh, Dan Kushel, Nadine Orenstein, Andrew Raftery, C. Richard Johnson, Jr., Louisa Wood Ruby, and Armin Kunz. I am indebted to them all as well as to the generosity of the following institutions: the Art Conservation Department at Buffalo State College; the Andrew W. Mellon Foundation; The Metropolitan Museum of Art; the Frick Collection; and the National Park Service among many others. Thank you, too, to my family for your unwavering support, prior to and through a global pandemic.

Noyes, R.S. (ed.), *Reassessing Epistemic Images in the Early Modern World.*
Taylor & Francis Group, 2023
DOI 10.5117/9789463723350_CH04

I. Introduction

Despite Albrecht Dürer's enormous artistic output, immediate success in various media, and more than five centuries of scholarship devoted to his work, current research continues to reveal information related to his choice of materials and techniques. Given their wide distribution as prints, Dürer's three *meisterstiche* impressions – *Knight, Death, and the Devil* (1513) (Figures 4.1 and 4.2) *St. Jerome in His Study* (1514) (Figure 4.3); and *Melencolia I* (1514) (Figures 4.4 and 4.5) – have been admired and closely studied, perhaps even more than his paintings and drawings. Of the hundreds of known extant impressions of these works, many were cataloged by Bartsch (1803–1821),[2] Meder (1932),[3] and Hollstein (1954).[4] These and other studies variously describe Dürer's impressions as rich, clean, with burr, excellent, poor, lighter, darker, silvery, brownish, and/or tonal. In light of the inherent subjectivity in judgments of impression quality, this essay outlines the author's in-depth comparative study across and between impressions carried out using a modified digital camera and modern imaging technology.

As a copperplate such as those produced for the *meisterstiche* is inked for printing, the hard and sharp fragments of microscopic burr residue left by the burin and inclusions in the carbon-based ink pigment are repeatedly rubbed into its lines for hundreds of impressions, and the condition of the metal matrix gradually declines. Fine scratches appear on the surface of the plate almost immediately and the crisp edges of the engraved areas are worn down, making the cut line softer and less well-defined. The continuous accrual of these subtle signs of deterioration can be documented in successful impressions. While differences in inking, wiping, and printing will always affect the appearance of an impression, these fine aggregate lines ultimately caused the exhaustion of Dürer's exquisitely engraved plates. From this, it follows that by recording and comparing the marks of deterioration on each of the *meisterstiche* impressions, one can reconstruct an accurate chronology according to which individual impressions were printed. This chronology, while not necessarily an indicator of quality or of early modern preferences, can inform current-day scholars and admirers of Dürer's work. The results of this study are currently being developed into a website for future dissemination.[5]

2 Bartsch, *The Illustrated Bartsch.*

3 Meder, *Dürer-Katalog.*

4 Hollstein, *German Engravings, Etchings, and Woodcuts.*

5 See the forthcoming Durersdetails.com

Following its advent in the 1430s, the art of copperplate engraving experienced what was to become its defining moment in late fifteenth- and early sixteenth-century northern Europe.[6] Albrecht Dürer, followed by members of his atelier, transformed a once exclusively goldsmithing tradition into an aesthetically and financially rewarding art form.[7] In the five centuries since, artists, art historians, connoisseurs, collectors, and casual admirers, inspired by his prints' bravura artistry and intimate beauty, have tried to understand and replicate Dürer's remarkably meticulous methodologies and productions, which have inspired artists, fueled art historical research, elicited connoisseurial adoration, and stimulated collectors. Despite historical fascination with the artist and particularly his prints, the exact number of extant impressions of Dürer's three iconic *meisterstiche* engravings remains elusive.[8] Clearly hundreds were printed,[9] and Karen L. Bowen and Dirk Imhof argue that 1,000–2,000 impressions could be pulled from a well-engraved plate from the period.[10] Precious from the moment they were pulled from the press, impressions were treasured, usually in private collections.[11] However, over the course of centuries, and with the founding of the first public art collections in the eighteenth and nineteenth centuries, many were transferred from private collections to museums.[12] The present study builds on this long history, undertaking a global survey of 140 surviving impressions from private and public collections to better understand the artist and his materials generally, and document and interpret visible effects of plate degradation in the impressions specifically.

II. Materials

A print, simply put, is comprised of three parts: a printing matrix; ink; and a (typically paper) support. The engraver's burin and the printing press distinguish an engraving from other types of prints. An early practitioner of artisanal epistemology, Albrecht Dürer successfully mastered print's three basic components as well as the tools required for engraving, and a thorough understanding of these materials and methods is imperative to

6 Gascoigne, *How to Identify Prints*, section 90.
7 Harrison, "The Printed Picture in the Renaissance," p. 216.
8 McDonald, *The Print Collection of Ferdinand Columbus*.
9 Conway, *Literary Remains of Albrecht Dürer*, p. 128.
10 Bowen and Imhof, "18,257 Impressions from a Plate," p. 265.
11 Harrison, "The Printed Picture in the Renaissance," p. 239.
12 Emison, "The Simple Art and Certain Complexities in Trying to Understand It," p. 9.

building a chronology of impressions. Given Dürer's uncontested skills as an engraver, we may assume that the artist himself burined the inked lines seen in each surviving impression of his *meisterstiche*. Dürer's obsession with systematized perfection is evident from his *Four Books on Human Proportion*.[13] All this points to his effort to perfect print's constituent materials prior to and during the creation of the *meisterstiche*, seeking the smoothest and most supple copper, darkest and richest ink, and creamiest and flawless paper.

III. Matrix: Plate

Dürer's plates were made from beaten copper, unlike current-day printing plates made from rolled copper, which was not readily available until the eighteenth century, when the demand for copper sheathing on ships increased.[14] In Dürer's lifetime, copper ingots were beaten into form with a hydro-powered hammer weighing as much as 500 pounds,[15] in a process vividly described by Isabel Horovitz:

> The [preliminarily beaten] sheets were cut with water-powered shears, and often then subjected to repeated battering in stacks. A very high level of refinement was possible with [these mechanized] hammers. After a sheet was formed, it could then be smoothed by hand with a planishing hammer, which had a wide, circular, flat head, and if not too heavy, would easily flatten only the surface of the copper.[16]

Almost none of the numerous plates Dürer engraved remain in existence. The rare surviving matrix of his portrait of Philip Melanchthon occupies its own protective enclosure in Stadt Gotha Schlossmuseum (inv. no 8,24 a). The engraved image is a three-quarter profile portrayal of Melanchthon with a Latin inscription in that reads "Dürer can picture the features of the living Philip but lacks the skill to depict his mind." The plate itself is thin and somewhat curved; marks of the beating process are visible on the verso. The relevancy of the manufacturing process outlined above

13 Dürer, *Hierin sind begriffen vier bücher*. See Parshall, "Graphic Knowledge," p. 393.

14 Westermann, "Copper Production," pp. 129–130.

15 *Copper Through the Ages*, p. 32.

16 Horovitz, "The Materials and Techniques," p. 66. Though this description is furnished in the context of early modern copper supports for paintings, the method of manufacturing engraving plates was the same.

to Dürer's plates is twofold. First, beaten copper is omnidirectional and therefore easier to carve. Second, despite the fact that a beaten plate, annealed or heat-treated to reduce the internal stresses incurred during the beating process, is naturally denser and harder than a rolled plate, as will be shown below, the sixteenth-century copper used for the *meisterstiche* plates likely contained a small percentage of lead, which rendered them somewhat softer than what we would expect to find in a rolled copper plate produced today.[17]

Georg Pencz, an accomplished engraver, one who was surely aware of Dürer's materials and methods, produced several small plates preserved in The Metropolitan Museum of Art, which offer more ready access and analysis relative to the Melanchthon plate, and were consulted to shed light on the material Dürer would have accessed. X-rays captured of the plates exhibit both a distinctive pattern of concentric circles revealing the shape and impact point of the hammer(s), and a lack of grain direction within the metal. Somewhat like wood carving, engraving a rolled plate is made easier or harder depending on grain direction. By contrast, the material uniformity of The Met's period plates allows an engraver more directional flexibility. Dürer's extraordinary capacity to depict curvilinear and irregular forms like delicate curls of hair and complex folds of fabric relied heavily on the lack of grain direction in such beaten plates. Additional X-ray fluorescence spectroscopy of Pencz's engraving of *Virginia Killed by Her Father* (inv. No 2005.137) confirmed the notable presence of lead, which likely impacted both manufacturing methods and deterioration of the matrix. Engraved lines and scratches in a copperplate softened by lead would have lost their crisp edges significantly faster compared to a pure copper matrix. In further efforts to replicate the sixteenth-century plate manufacturing process, RISD professor of engraving Andrew Raftery manufactured, engraved, and printed a mock-up plate copying Dürer's *St. Paul*, selected in part because the original was created the same year as *Melencolia I*, when Dürer's hand skills had already reached their apex.[18] Over 100 impressions were pulled from Raftery's newly engraved reproduction plate, after which both plate and impressions were systematically analyzed, showing (despite numerous insurmountable inconsistencies with the historic process) that the present-day plate is both harder and more durable than a sixteenth-century plate.

17 Tylecote, *A History of Metallurgy*, p. 76.

18 A summary of the study with illustrations is available in Campbell and Raftery, "Remaking Dürer."

IV. Ink

Paper conservator Patricia Hamm aptly describes the age-old requirements for good printing ink, which had to

> be a viscous material which would neither reticulate nor spread during printing. It needed to easily transfer from the metal to the paper in a sharp manner. The ink could not: damage the paper during printing, damage the plate or metal type, damage the ink balls or the rollers or soak into the paper. Clean up had to be easy. Lastly, inks needed to dry fast enough to avoid offset, to age well and impart a fine, dense, black surface.[19]

In a fascinating essay, Kimberly Schenck differentiates Dürer's impressions, "printed with a rich, highly pigmented, black ink, producing crisp dense lines that contrast with the brightness of the paper," from impressions by highly skilled artistic contemporaries such as Lucas van Leyden.[20] The *meisterstiche* impressions specifically were consistently printed with a deep black and finely particled ink that revealed a saturated sheen when pulled from deeply cut burin lines. Despite references by Meder and many other scholars to varying ink color, any apparent variation in tone in the *meisterstiche* was likely due instead to changes that the paper substrate incurred over time, through poor storage, photo oxidation, or previous interventive treatments.[21] Further analysis by Raman spectroscopy of a heavily inked area in an impression of *Knight, Death, and the Devil* indicated the strong presence of carbon (and only carbon),[22] suggesting either so-called lamp black and/or vine black in the ink. The latter, according to Head of Paper Conservation at the Boston Museum of Fine Art Annette Manick, was derived from charcoal "made from carbonized vegetable matter commonly associated with the wine industry," during Dürer's time most frequently ranging from fine wine yeast and lees, to increasingly course vine tendrils, sprigs, and twigs."[23] Describing vine black's unique properties, Manick explains

19 Hamm, "A History of the Manufacture of Printing Ink," p. 30.

20 Schenck, "Goltzius's Use of Grey Ink," p. 186.

21 Sayre, "Introduction," p. xxiv.

22 Sylvia Centeno, Research Scientist at The Metropolitan Museum of Art, carried out the Raman spectral analysis.

23 Manick, "A Note on Printing Inks," p. xlvii.

> [v]ine black was generally used for intaglio prints because of its stiffening and thickening properties. Compared with lamp black, vine black absorbs less oil and allows for a higher pigment concentration. A plate daubed with this stiffer, thicker, less elastic ink could be wiped more easily. The ink would stay in the lines, enabling the printer to wipe the surface of the plate more cleanly.[24]

Cross-polar examination of vine black pigment at 400x magnification reveals, unsurprisingly, that the pigment is almost entirely carbon black, which is itself is a relatively soft material, but marked by small visible inclusions scattered throughout the carbon particles. Since it is unlikely that the carbon alone caused significant abrasion to the copperplate, we may assume that the additional particulate matter in the ink or on the surface of the plate resulted in microscopic scratches and general plate wear.[25]

V. Support: Paper

Dürer's preference for a materially uniform matrix, consistent use of ink, and choice of paper sheet are necessary standards to consider when examining a large body of his work. Unlike later printmakers, Dürer did not strive for artistically interesting and varied effects in his prints. Instead, his goal was to produce uniformly well-inked prints with crisp, clean lines on high-quality rag paper. Historically a rare commodity, paper became much more readily available by the fifteenth century, when plentiful linen rags allowed for the production of fine paper throughout Europe.[26] Before the introduction of wove paper moulds in 1757, all paper produced in Europe bore traces of chain and laid lines, distinctive markings of the traditional mould (so-called laid paper).[27] Dürer's *meisterstiche* sheets were no exception: in the 140 examined impressions for the present study, the papers consistently exhibited chain lines spaced about 30 mm apart, always running horizontally though the images. Five were also marked with watermarks specific to certain papermills. Dürer, always aware of his materials, accepted

24 *Ibid.*, p. xlv.

25 These inclusions are difficult to identify through Raman spectroscopy, since they are so small, and often hidden among an overwhelming presence of pure carbon particles.

26 Hunter, *Papermaking*, p. 153.

27 Baker, "The Wove Paper," p. 3.

the presence of watermarks in the sheets used for his woodcuts and some of his earlier engravings – even, notably, his *Adam and Eve* impressions (1504), often erroneously assumed to be part of his *meisterstiche*. It is clear, however, that the artist deliberately avoided watermarked sheets for printing his more intricate later engravings, especially his *meisterstiche*.[28] The vast majority of surviving *meisterstiche* sheets have margins that have been aggressively trimmed to within a quarter inch of the printed line, an unfortunate tendency of collectors needing to fit the prints into extant albums. Even in the few oversized sheets that exist, however, there is no trace of a watermark. The five impressions examined for this study found to bear watermarks were all late impressions.[29]

The papers Dürer selected for his *meisterstiche* impressions were not only vastly different from his drawing papers, but also diverged significantly from the papers used for his other engravings.[30] In the early sixteenth century, Italy and France remained the preeminent sources for fine, strong paper. The master impressions required cream-colored sheets comprised of well-beaten linen fibers, strong and supple enough to withstand the enormous pressure of the printing press against the sharp-edged copperplate. Dürer's *meisterstiche* papers were likely specially ordered from abroad and shipped in folded quires, as described by Albert J. Elen, who details the impracticalities of transporting large quantities of flat or so-called plano paper sheets: "Paper was always folded in quires and packed into reams to be transported and sold. It was seldom or never delivered as a pack of plano sheets [...] a pile of plano sheets of medium or large format would bend and therefore be difficult to handle."[31] The residual folds resulting from common shipping processes remained distinctively evident on the versos of the *meisterstiche* impressions and, where they survived, in the margins. These folds, though not exclusive to the *meisterstiche*, are reliably evident on the versos of these impressions. As trace evidence, a fold could be obliterated from the image area when the paper and plate were run through the enormous weight of the press, whereas a watermark remains undisturbed by such pressure, as the different densities in the paper caused by a watermark could never be

28 Dürer's apparent preference for un-watermarked sheets became clear in discussions with Tom Rassieur, John E. Andrus III Curator of Prints and Drawings at the Minneapolis Institute of Arts. Though Joseph Meder discusses watermarks in some of the later second-state impressions, only six of the 140 impressions examined bear a watermark and all were late, possibly posthumous impressions.

29 Campbell, "Finding Folds," p. 406.

30 Dietz, "Art and Science," p. 96.

31 Elen, "Paper Analysis," p. 199.

entirely obscured. Discernible traces of the fold, however, though erased by the press within the image area, remain forever visible on the verso of a sheet and in the margins.

A promising field of future investigation into Dürer's papers is suggested by recent work implementing software to assist in analyzing the laid and chain lines evident in early modern papers was highlighted by a 2017 study by C. Richard Johnson Jr. focused on Rembrandt's prints as part of the WIRE (Watermark Identification in Rembrandt's Etchings) project based at Cornell University.[32] The analytical methodology has since been adapted to allow for the more difficult task of evaluating papers without the use of X-rays, as described in a recent article by Johnson with William A. Sethares and Margaret Holben Ellis.[33] Without consistent evidence of watermarks, it is somewhat difficult to use this technology to analyze Dürer's *meisterstiche* papers; but it is not impossible, and experimental analysis is currently being carried out on Dürer's drawing papers as well as on a selection of *meisterstiche* papers, offering productive areas of future investigation.

VI. Technique

Rather than any sort of intentional programmatic grouping, Dürer's carefully honed engraving and printing techniques, paired with his meticulous selection of materials, make his *meisterstiche* impressions superlative. Despite vast differences in subject matter, the three prints are similarly sized and engraved with skill exceptional even within Dürer's oeuvre. *Knight, Death, and the Devil*, first printed in 1513, was the earliest; save very late impressions of the print, the work's lines are dark, clear, and full of contrast. This is particularly apparent in the outlines of the knight's horse, which unlike the more subtle shading lines remain clear throughout the life of the plate (Figures 4.1 and 4.2). In comparison, images in impressions of both *St. Jerome in His Study* and *Melencolia I* exhibit a relative softness, which may partly relate to the choice of subject.[34] Whereas the strong visual contrasts in *Knight, Death, and the Devil* are in keeping with its display of the oppositional firmness of right versus wrong, good versus evil, and dark versus light, more ambiguous themes taken up in the two later works are aestheticized

32 Johnson, *et al.*, "Chain Line Pattern Matching," 319–334.

33 Sethares, Ellis, and Johnson, "Computational Watermark Enhancement," pp. 89–96.

34 Parshall, "Graphic Knowledge," p. 404.

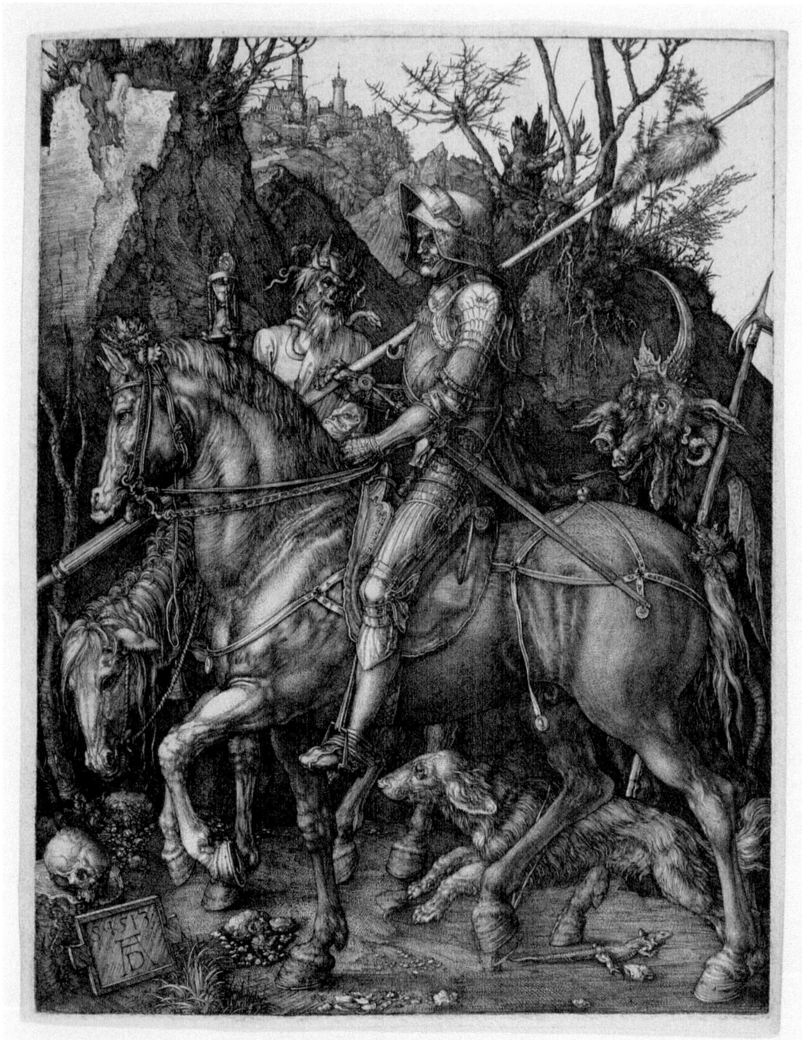

Figure 4.1 *Knight, Death, and the Devil*, early impression – Image courtesy The Metropolitan Museum of Art, Harris Brisbane Dick Fund, 1943, 43.106.2

in shades of gray. *St. Jerome in His Study*'s depiction of its protagonist deep in thought awash in peaceful, soft light, filtered through the crown-glass windows, is conducive to a less severe plate cut (Figure 4.3). The same holds true for *Melencolia*'s exploration of deep thought, double light sources, and brooding cloaked figure (Figures 4.4 and 4.5). Further still on the spectrum of fine handling, in Dürer's portrait of Philip Melanchton, the artist was able to engrave a plate with a fineness almost beyond comprehension, thanks in part to the softer metal of sixteenth-century plates, which simultaneously

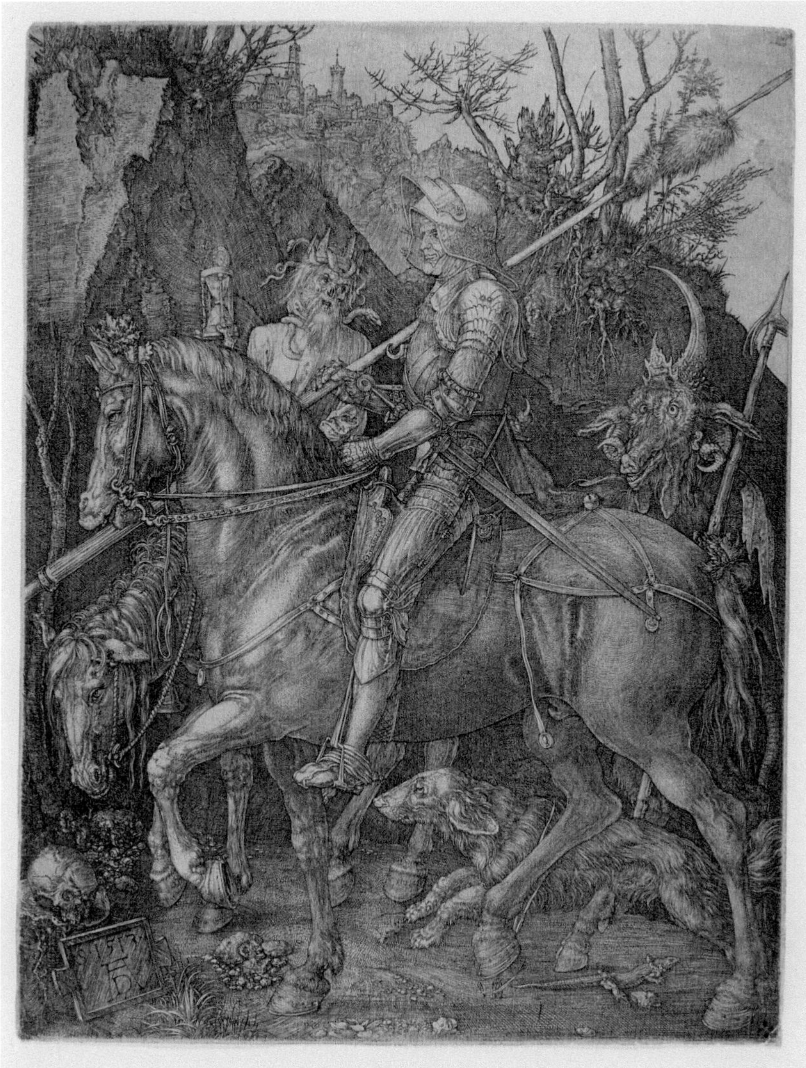

Figure 4.2 *Knight, Death, and the Devil*, late impression – Image courtesy The Metropolitan Museum of Art, Bequest of Harry G. Friedman, 1965, 66.521.95

allowed for more delicacy in an engraved line as well as more rapid plate degradation.

As the burin carved each line into the copper surface, this produced a microscopic burr of metal – an imperceptible raised edge or ridge of material – along either side of the burin's cutting stroke. Even an engraver as meticulous as Dürer would not be able to remove all of this residual burr, and as the plate was repeatedly inked, wiped, and printed, the resulting

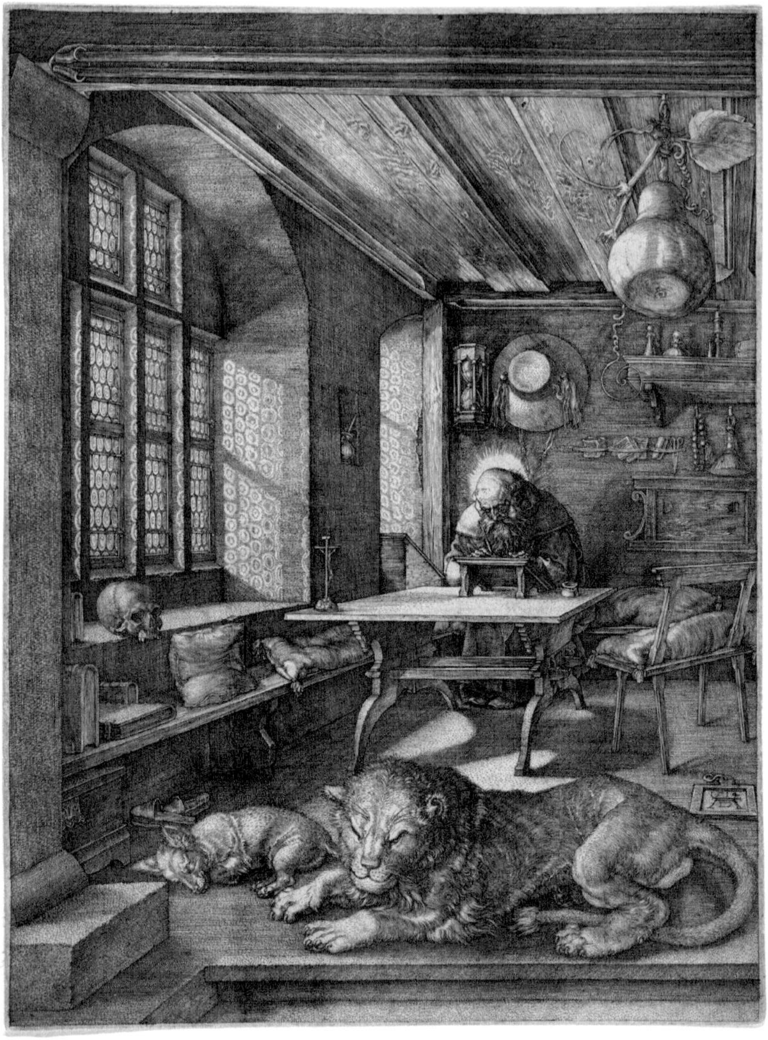

Figure 4.3 *St. Jerome in His Study* – Image Courtesy The Metropolitan Museum of Art, Fletcher Fund 1919, 19.73.68

microscopic bits of copper would scratch and wear down the plate. Ad Stijnman's recent comprehensive study of the history of intaglio printing discusses the significance of the early modern advent of the roller press, whose design concentrated the entire force of the press onto a very narrow line, thus allowing the press to force the paper deep enough into the grooves of the inked copper plate and produce a highly detailed and refined impression. Before the development of the roller press, engravings could not be produced; the platen presses used for printing books could not supply enough pressure. Conversely, once the machine was invented, woodblocks could not endure the immense pressure produced by the roller press. Copper was the preferred substrate for burin engraving; supple and durable, it was easily engraved but could also endure the weight of the press.[35]

The press, too, would contribute to plate wear, but the enormous amount of pressure issued by the roller alone did not abrade the engraved copper. However, when the press combined with the residual copper burr and ink inclusions described above, the printing process constituted a mechanism that brought about continuous and inescapable plate wear. Both presses and inks have historically borne the brunt of blame regarding plate wear; in an effort to overcome the inevitable, both have been forced into overuse to atone for their presumed sins, resulting in perceptible differences in the resulting prints compared to those pulled from fresh unworn plates. As Annette Manick explains

> [w]hen a worn plate is inked normally, the image lacks strength, intensity, and contrast because the shallow lines hold less ink [...]. In order to create a darker image using a worn plate, printers attempted to enhance, through overinking (often in combination with increased pressure), the faint lines that were most difficult to print. In the same print, however, stronger lines, especially in areas of close hatching, would lose definition owing to excess ink having been forced out of the engraved or etched grooves.[36]

VII. Parameters and Objectives of the *Meisterstiche* Survey

For the present survey of Dürer's *meisterstiche*, 140 impressions were examined and documented in print study rooms and conservation laboratories across the US and Europe: Albright Knox Art Gallery, Buffalo (2 impressions); Art Institute of Chicago (6 impressions); Ashmolean Museum of Art and Archaeology, Oxford (8 impressions); Baltimore Museum of Art (5 impressions); Bibliothèque nationale de France, Paris (4 impressions); British Museum, London (8 impressions); Brooklyn Museum (3 impressions); Connecticut College, New London (1 impression); Fitzwilliam Museum, Cambridge (5 impressions); Frick Collection, New York (2 impressions);

35 Stijnman, *Engraving and Etching*.

36 Manick, "A Note on Printing Inks," p. xlvi n4.

Germanisches Nationalmuseum, Nuremberg (10 impressions); Harvard Art Museums in Cambridge (3 impressions); Istituto Nazionale per la Grafica, Rome (4 impressions); Kupferstichkabinett–Staatliche Museen, Berlin (10 impressions); Metropolitan Museum of Art, New York (14 impressions); Morgan Library and Museum, New York (1 impression); Museum of Fine Arts, Boston (7 impressions); National Gallery of Art, Washington DC (7 impressions); New York Public Library (1 impression); Petit Palais, Paris (3 impressions); Princeton University Art Museum (2 impressions); private collection, Chicago (3 impressions); private collections, New York (6 impressions); RISD Museum of Art, Providence (1 impression); Staatliche Graphische Sammlung, Munich (3 impressions); Staatsgalerie, Stuttgart (5 impressions); Städel Museum, Frankfurt (3 impressions); Uffizi, Florence (10 impressions); Victoria and Albert Museum, London (3 impressions).[37]

The inked lines in each surviving impression of the *meisterstiche* vary dramatically in print quality and condition and have been compared verbally and visually for centuries. Neither the delicate underdrawings commonly found in engravings by artists like Hendrick Goltzius, nor faint return strokes often left by engravers at the end of a line are evident in any of the examined Dürer impressions. However, close examination of 100+ impressions disclosed the presence of other kinds of subtle marks – which would have been barely perceptible on the engraved plate and faded easily with wear – that could prove useful in identifying the relative chronology of impressions. Such marks and their implications were studied most notably by Joseph Meder in his foundational 1932 study, which categorized all of Dürer's known impressions into chronological groups classified systematically by state and by minor variations within each state according to intentional and unintentional changes to the engraved copperplates – without, however, adequate means to capture images of all the referenced changes to the print matrix. This leaves the reader entirely reliant on Meder's earnestly descriptive but highly subjective terminology.

In Meder's *Katalog*, for example, *Melencolia I* has two states, the second of which has six sub-states identified with the letters a–f. The distinguishing characteristic of a first-state impression of *Melencolia I* is a backwards-printed number nine in the number grid (Figure 4.4). In the second state, the nine has been corrected (Figure 4.5). Meder's classification system was

37 Sixteen of the impressions were examined as part of a Master's thesis completed in 2008 for the Art Conservation Department at Buffalo State College and the remaining 124 impressions were studied during an Andrew W. Mellon Fellowship in paper conservation at The Metropolitan Museum of Art from 2009–2011.

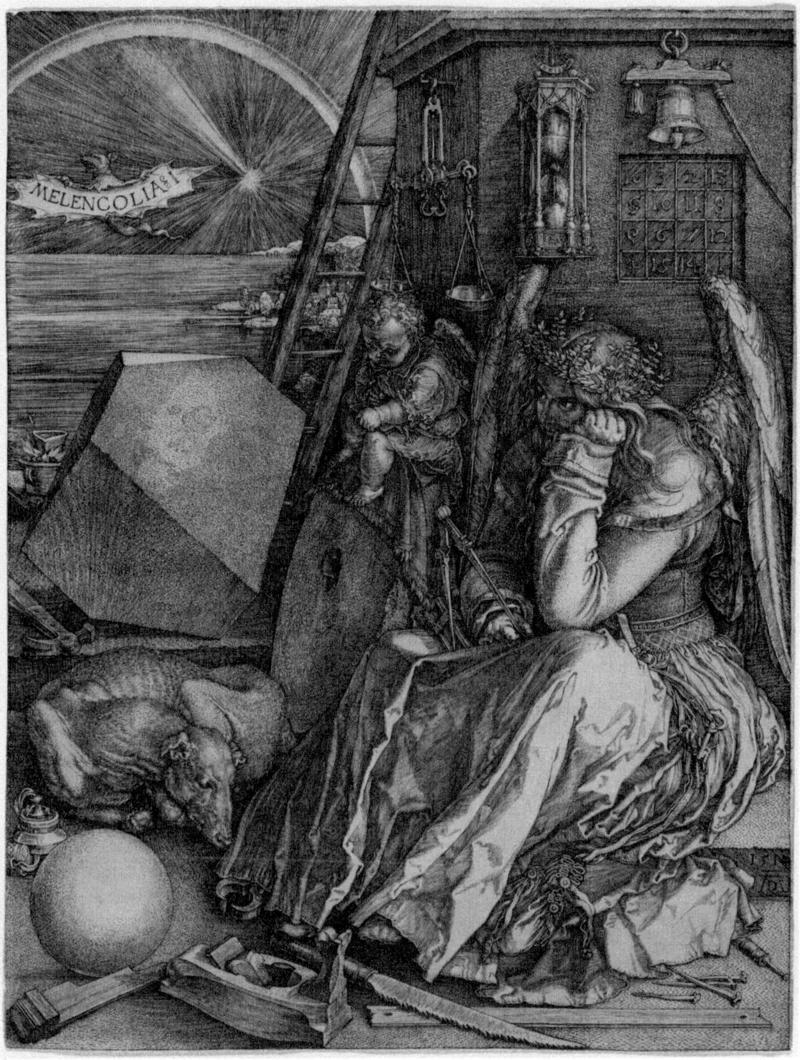

Figure 4.4 *Melencolia I*, first state – Image courtesy The National Gallery of Art, Washington DC, the Rosenwald Collection, 1943.3.3522

meant to serve as a reference for quality and/or merit, such that a first-state (Meder I) impression of *Melencolia I* is generally more desirable than a second-state (Meder II) *Melencolia I* impression.[38] Likewise, a Meder II.b impression is assumed to be earlier and more desirable than a Meder II.f impression of the same engraving, and so on. Meder described both general

38 Harrison, "The Printed Picture in the Renaissance," p. 215.

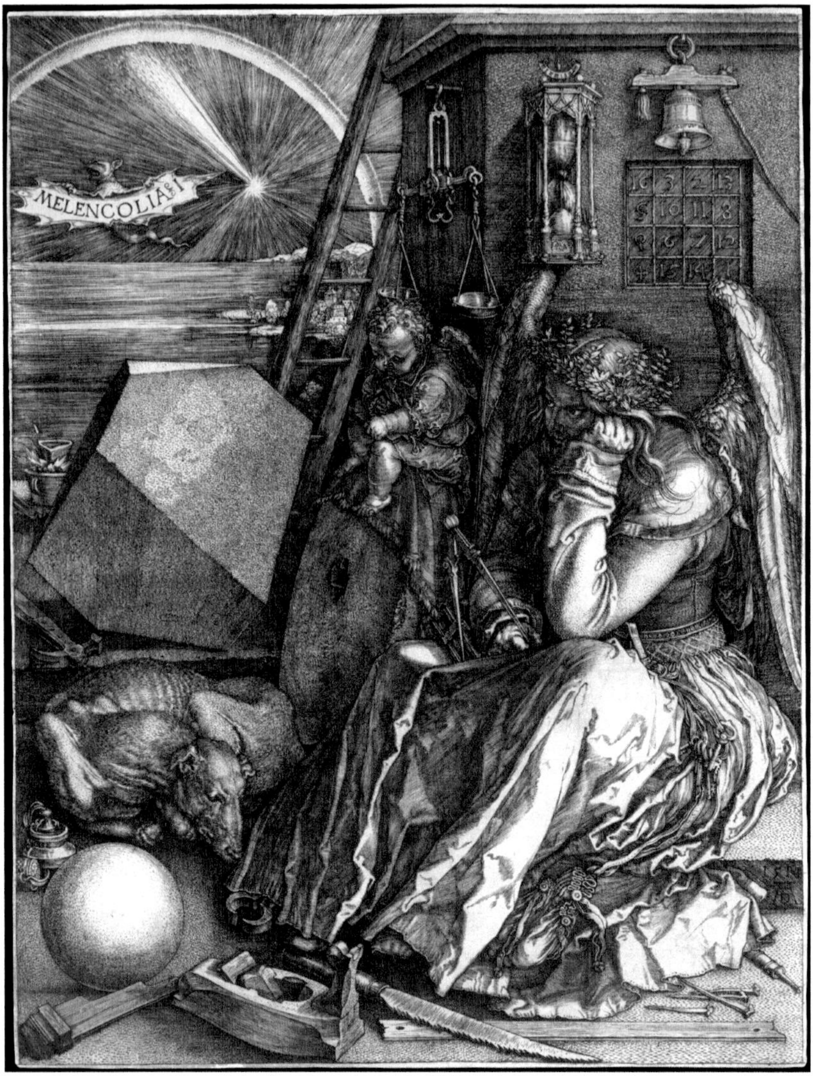

Figure 4.5 *Melencolia I*, second state – Image courtesy The Metropolitan Museum of Art, Harris Brisbane Dick Fund, 1943, 43.106.1

appearance (i.e. the impression is "silvery" vs. "brownish," or the angel has a "darker" vs. "lighter" face) and technical differences among impressions ("sharp" scratches are visible in the "ball" and/or in the angel's thigh).[39] Unfortunately, even the technical elements of Meder's classification system are notoriously imprecise: in the latter example, for instance, multiple

39 Meder, *Dürer-Katalog*, pp. 100–101.

scratches exist in both the ball and in the angel's thigh, and they appear and are worn away at different points in the plate's life. It is thus unclear to which scratch or scratches Meder was referring.

This study's goal was to re-define and refine the designation of states, and, where possible, place each impression in an item-by-item relative chronology of printing. It thus builds on curatorial work undertaken to illustrate more effectively the qualitative differences that were Meder's problematic legacy, including exhibitions featuring multiple impressions of the same engraving. Already in 1897, Sylvester Rosa Kohler's Grolier Club exhibition of Dürer prints was intended to enhance understandings of varying aesthetic quality in print impressions; later, Richard Field's 1971 exhibition "Albrecht Dürer, 1471–1528; A Study Exhibition of Print Connoisseurship" (Philadelphia Museum of Art) followed the same vein.[40] More recently, Clifford Ackley's 2009 exhibition "Albrecht Dürer: Virtuoso Printmaker" (Museum of Fine Arts, Boston) juxtaposed two impressions of Dürer's c. 1496 engraving *The Prodigal Son Amid the Swine* – one printed soon after the plate was finished around 1496, and one later c. 1525 – inviting appraisals of the subtle differences between them.[41]

In the initial stages of examining and documenting the *meisterstiche* impressions, distinguishable signs of plate wear of the sorts described in the previous section were evident from even the earliest pulls printed first in the plate's lifespan. The National Gallery of Art in Washington DC is the only collection with two first-state *Melencolia I* impressions. In both impressions, two lines substantially fainter than the burined lines are scratched across the otherwise pristine ball in the lower left quadrant of the print (Figure 4.6). This pair of lines is evidence of the process whereby as a plate is repeatedly inked, wiped, and printed, and the hard and sharp fragments of residual copper burr and inclusions in the carbon-based pigment continuously scratch and abrade the copper surface, such that the quality of the impression gradually declines, as fine scratches appear on the smooth surface. Similarly, the crisp edges of engraved areas are worn down by abrasion and printing pressure, making the cut line wider and less well-defined. While easily visible under a microscope, these marks of wear are not immediately apparent to the naked eye; as they augment and accrue, however, the macroscopic differences between an early and a late impression become incrementally apparent.

The interpretive potential of investigating incremental signs of wear was treated in S. Blair Hedges's innovative 2006 study presenting a method of

40 *Albrecht Dürer: Virtuoso Printmaker*, p. xxii.

41 *Ibid.*

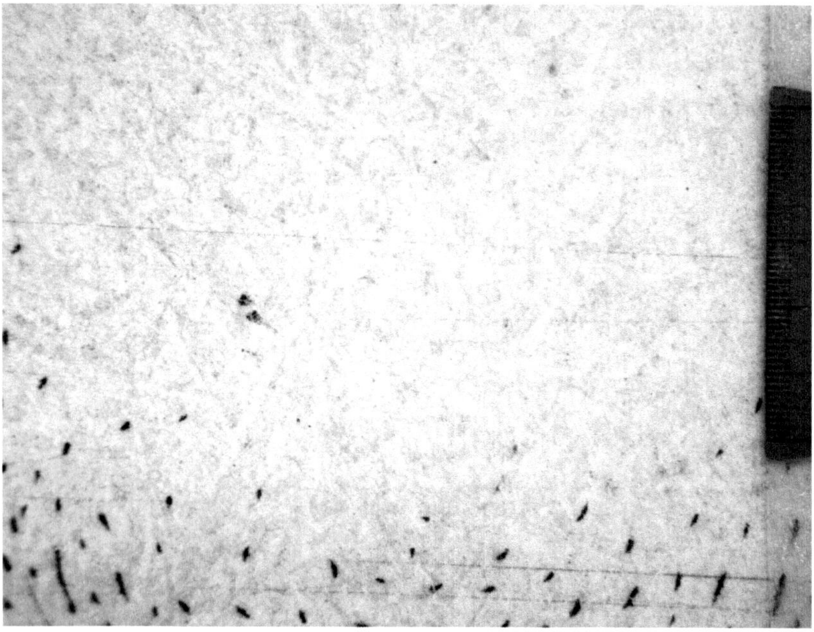

Figure 4.6 *Melencolia I*, detail, scratches in ball – Image courtesy The National Gallery of Art, Washington DC, the Rosenwald Collection, 1943.3.3522

dating early prints by measuring the change in line width of various impressions from the same plate.[42] Their method of dating posits the corrosion and repolishing of the copperplate as the main mechanism for wear, rather than the erosion of the crisp engraved edges by continuous inking, or the compression of lines by the repeated pressure of the printing press. Though Hedges's research is of great interest to the present study, it focuses on plate wear over an extended period of time. The fact that images of *Melencolia I* were greatly valued even during Dürer's lifetime strongly suggests that most of the impressions were printed between 1514 and Dürer's death in 1528. For this reason, corrosion likely has little effect on the apparent "wear" of the valuable plates, which would have been well protected and checked regularly for any signs of damage.

Another case in point are two impressions of *Knight, Death, and the Devil*: one early Figure 4.1, and one late Figure 4.2. What is notable is that the deeply engraved line outlining the horse's underbelly maintains its strength even as the plate degrades. The lines forming the interior shadow of the knight's helmet, however, are more lightly engraved, and as the plate

42 Hedges, "A Method for Dating Early Books and Prints," pp. 3555–3573.

wears, they lose their depth, resulting in a considerable loss of contrast between the knight's face and the shadow. While differences in inking and printing always affect the final impression, it is impossible to print clean lines on a well-worn plate. Consequently, recording and comparing the marks of deterioration in the 140 impressions, documented using a digital adaptation of a photography system devised by Shelley Fletcher, can facilitate the chronological reconstruction of the printing sequence for individual impressions.[43] Comparatively recent imaging technologies and the accessibility of website templates render the possibility to capture high-resolution, highly magnified images of the described plate wear, and disseminate this information efficiently through desktop, tablet, and mobile device-optimized platforms. Specifications regarding the digital camera and image-capturing system may be found in the following Appendix.

VIII. Conclusions and Further Areas of Inquiry

With a clear understanding of the artist's materials and methods, a chronology of impressions could be reliably laid out. Digital images of the 140 impressions were captured, processed, and printed, then sorted into general groups by their assigned Meder classification. In the example of *Melencolia I*, first-state impressions are identified without question by a backwards-facing number 9 in the number grid. By identifying these intentional changes to the plate, and using scratches successfully described in the *Dürer – Katalog*, such as those in the ball at the feet of *Melencolia*, a loose chronological order could be established.[44] Scratches in the ball are already evident in the first-state impressions and are deep enough to yield a printed line (Figure 4.6). In early second-state impressions, printed after the artist corrected the mistake in the number grid, the earliest scratches have faded somewhat, and a new scratch is introduced (Figure 4.7). Finally, in late impressions, all the fine scratches in the ball have been worn down and are no longer able to retain ink sufficient for representative printing (Figure 4.8). Once all the Meder-classified impressions had been organized, the remaining impressions were added to the chronology by assessing and comparing evidence of plate wear throughout the impressions, in both the lightly inked areas and in the more heavily inked shadows and monogram

43 Fletcher, "A Re-Evaluation of Two Mantegna Prints," pp. 67–77.

44 The images were captured in RAW format and processed in Adobe Photoshop. The light levels were adjusted, and the images were sized to 300dpi with 1500 pixels along the longest side.

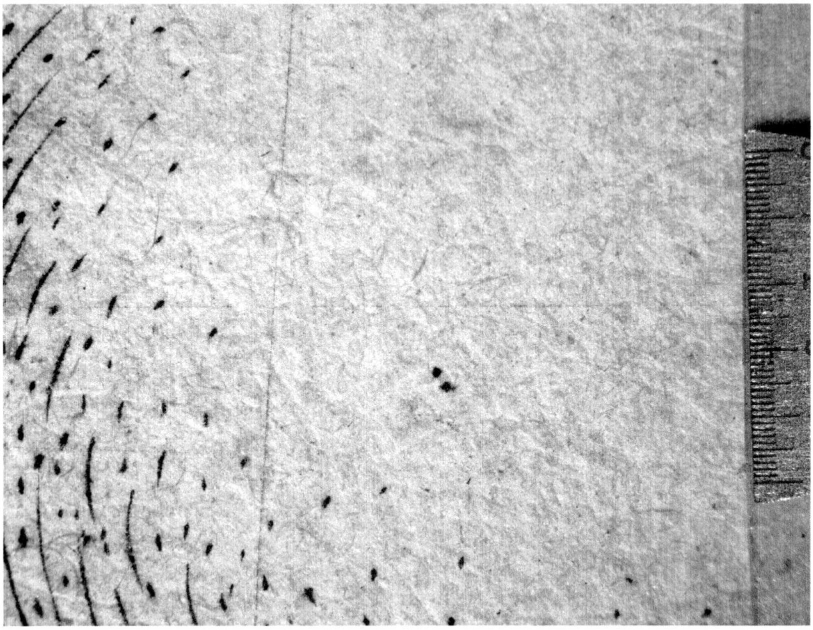

Figure 4.7 Detail of *Melencolia I*, Harvard Art Museums/Fogg Museum, Gift of William Gray from the collection of Francis Calley Gray, G1098. Image captured by author.

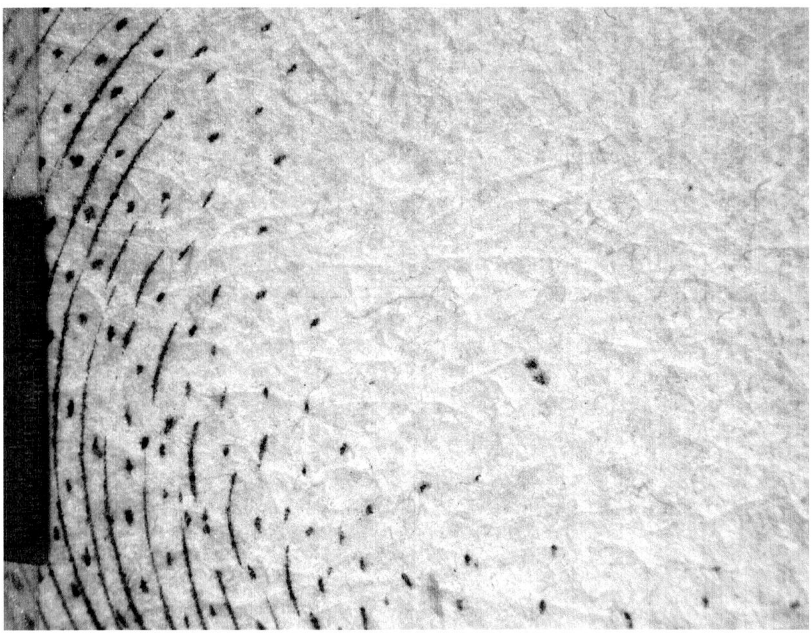

Figure 4.8 *Melencolia I*, detail, ball – museum purchase with a Centennial gift from Landon T. Clay, 68.188. Image captured by author; photograph © 2021 Museum of Fine Arts, Boston

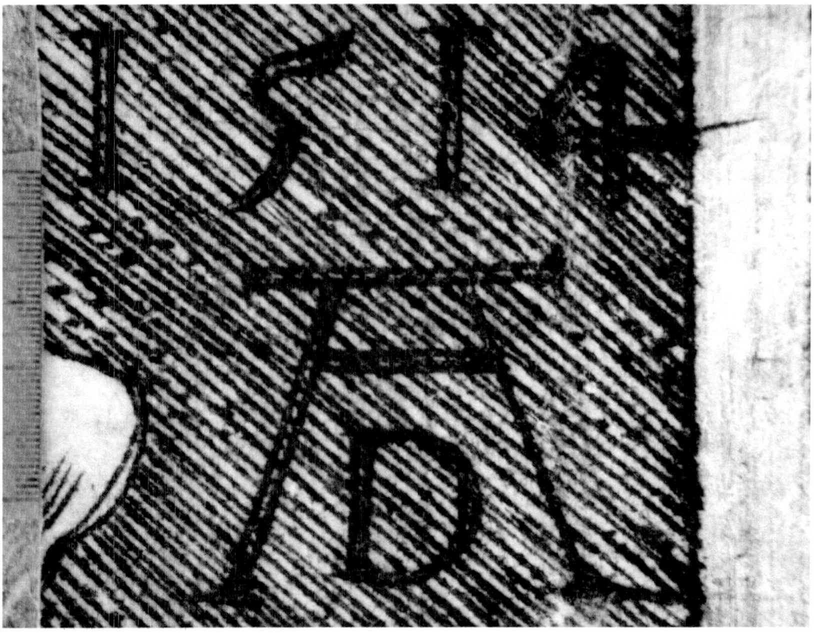

Figure 4.9 *Melencolia I*, detail, monogram – Image captured by the author with permission from The National Gallery of Art, Washington DC. Gift of R. Horace Gallatin, 1949.1.17

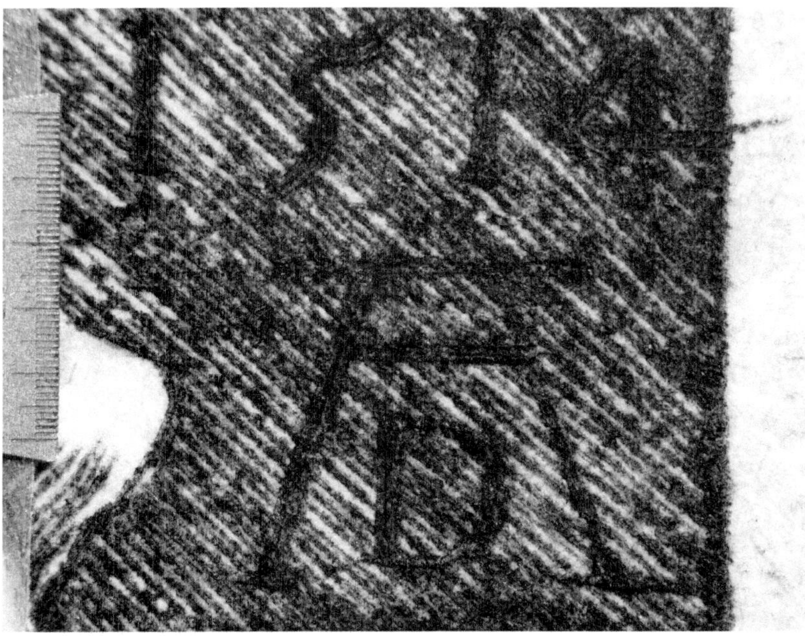

Figure 4.10 *Melencolia I*, detail, monogram – museum purchase with a Centennial gift from Landon T. Clay, 68.188. Image captured by author; photograph © 2021 Museum of Fine Arts, Boston.

(Figures 4.9 and 4.10). Unlike *Melencolia I*, neither *Knight, Death, and the Devil* nor *St. Jerome in His Study* has a clearly distinguishable first and second state. Ordering the impressions for these plates thus posed additional challenges. Similarly, in the case of all three prints, later impressions without clear scratches evident on an otherwise pristine copper surface proved difficult to arrange chronologically. They exhibit no tell-tale scratches and seem to be placed only in accordance with general aesthetic appearance. However, it is worth noting that no matter how heavily the plate is inked, the ink cannot collect in areas where the plate's incised lines have worn away.

Unfortunately, as it is impossible to include the over 1,000 detailed images of the fine scratches that determine impression chronology here, we remain reliant on additional technology. The aforementioned Durersdetails.com, a clearinghouse for the dissemination of results of this study, is a website in progress that will aim to disseminate findings and reflect additional scholarship and impression discovery. The detailed chronology, in conjunction with an understanding of Dürer's materials and engraving methods, demonstrate the effects of microscopic plate wear on his finely engraved, intricately designed *meisterstiche*. As his plates endured microscopic abrasion through repetitive inking, wiping, and printing, Dürer's clean lines and crisp details were gradually blurred and softened. These signs of plate deterioration, expressed in impressions that faithfully reflect the exact condition of the plate at each pull, allow for insight into the lives of both the artist and his masterwork. As Dürer himself notes in his dedication to Philip Melanchthon: it is possible to record what can be seen; it is much more difficult to fully understand and describe the genius behind the process.

Appendix

Calculations of the appropriate digital camera modifications were carried out largely by Dan Kushel, SUNY Distinguished Teaching Professor. A Panasonic Lumix Camera, Model DMC-LX2, was used for the project. The camera incorporates a 1/1.65" CCD sensor measuring 8.9 × 5.0mm. The camera's Leica zoom lens has a focal length range of 6.3mm to 25.2mm (equivalent to a range of 28–112 mm for 35 mm film format) with a maximum speed range of f2.8 to f6.9. All images were captured in RAW format with the camera set to a 4:3 aspect ratio and 7.5MP (maximum) resolution. Exposure metering was done using pattern mode and aperture priority at an aperture of f7.1 and an ISO of 200. Exposure times ranged from around 1/3 second to 1/500 depending on ambient light levels in the different institutions visited. While the camera is capable of capturing a 4:3 field as small as 20 mm × 15 mm, a higher level of magnification was required for this study and the camera was modified by the attachment of an auxiliary close-up lens.

An Edmund Optics, Inc. achromatic doublet lens (cat# NT32-323) was purchased and secured to the front of the camera's lens. The lens had a focal length of 50 mm (+20 diopters) and a diameter of 25 mm. It allowed the camera to be focused close enough to capture a field of view as small as 12.7 × 9.5 mm, a level of magnification deemed sufficient for this study. To mount the auxiliary lens onto the camera's lens, a 3-mil Mylar strip was first adhered with double-stick tape around the auxiliary lens' circumference. The strip's 11 mm width extended slightly beyond the rear surface of the lens allowing the auxiliary lens to rest comfortably in the ridges of the interior casing of the camera's lens. Once it was properly placed, six 5 mm × 15 mm strips of black photographic tape were used to secure the auxiliary lens to the camera's lens housing. To reduce flare, a 10 mm wide strip of black tape was applied around these smaller strips thus encasing the auxiliary lens in black. This method of mounting did not interfere with the camera lens' extension movements.

A Sharpics, Inc. model MP-16 two-section, 16-inch, tabletop monopod with 2-inch base clamp was used to support the camera over the prints. The arm was secured to the table with the base clamp and the arm was set to its maximum reach. The camera was then leveled with a spirit level. The prints were photographed using ambient room illumination.

To ensure that exactly the same areas would be photographed in each of the impressions studied, a template was constructed from 3-mil Mylar. A 20. mm × 10 mm rectangular cut-out was made for each of the nine areas to be compared, and the cut-outs were used to assist in positioning of the camera. The surface of the Mylar surrounding each cut-out was sanded slightly so that the openings in the template could be easily located. Registration marks were made on the Mylar with permanent marker to ensure that the template would be accurately placed on the prints. The nine rectangular areas were labeled "A" through "I" to assist in identification of the images.

Bibliography

Baker, Cathleen A. "The Wove Paper in John Baskerville's Virgil (1757): Made on a Cloth-Covered Laid Mould." In *Papermaker's Tears: Essays on the Art and Craft of Paper*, edited by Tatiana Ginsberg, pp. 2–44. Ann Arbor: The Legacy Press, 2019.

Bartsch, Adam von. *The Illustrated Bartsch*, 50 vols. Edited by Walter L. Strauss. New York: Abaris Books, 1978.

Bowen, Karen L., and Dirk Imhof. "18,257 Impressions from a Plate." *Print Quarterly* 22, no. 3 (2015): 265–279.

Campbell, Angela. "Finding Folds: Albrecht Dürer's *Meisterstiche* Papers." *Print Quarterly* 24, no. 4 (2012): 405–411.

Campbell, Angela, and Andrew Raftery. "Remaking Dürer: Investigating the Mater Engravings by Masterful Engraving." *Art in Print* 2, no. 4 (2012): 15–21.

Conway, Sir William Martin. *Literary Remains of Albrecht Dürer.* Cambridge: Cambridge University Press, 1889.

Copper Through the Ages. London: Copper Development Association, 1955.

Dietz, Georg Josef. "Case Studies in Art and Science: Analysing Dürer's Early Pen-and-Ink Drawings." In *The Young Durer: Drawing the Figure*, pp. 96–97. London: The Courtauld Gallery, 2013.

Dürer, Albrecht. *Hierin Sind Begriffen Vier Bücher von Menschlicher Proportion durch Albrechten Dürer von Nürnberg Erfunden und Beschriben….* Nuremberg: Hieronymus Andreae, 1528.

Elen, Albert J. "Paper Analysis in Italian Drawing-Books of the 15th and 16th Centuries." In *Le Papier au Moyen Âge. Historie et Techniques,* edited by Monique Zerdoun Bat-Yehouda, pp. 193–202. Turnhout: Brepols, 1999.

Emison, Patricia. "The Simple Art and Certain Complexities in Trying to Understand It." In *The Simple Art: Printed Images in an Age of Magnificance*, edited by Patricia Emison, pp. 1–19. Durham, NC: The Art Gallery, University of New Hampshire, 2006.

Field, Richard. *Albrecht Dürer 1471–1528: A Study Exhibition of Print Connoisseurship.* Philadelphia, PA: Philadelphia Museum of Art, 1970.

Fletcher, Shelley. "A Re-Evaluation of Two Mantegna Prints." *Print Quarterly* 14, no. 1 (1997): 67–77.

Gascoigne, Bamber. *How to Identify Prints.* New York: Thames and Hudson, second edition 2004.

Hamm, Patricia. "A History of the Manufacture of Printing Ink from 1500–1900 with Notes for the Conservator." In *Conference Papers, Manchester 1992*, edited by Sheila Fairbrass, pp. 30–34. Institute of Paper Conservation, 1992.

Harrison, Charles. "The Printed Picture in the Renaissance." In *Making Renaissance Art, Vol. 1,* edited by Kim W. Woods, pp. 211–247. New Haven, CT: Yale University Press, 2007.

Hedges, S. Blair. "A Method for Dating Early Books and Prints Using Image Analysis." *Proceedings of the Royal Society* 462 (2006): 3555–3574.

Hollstein, F. W. H. *German Engravings, Etchings, and Woodcuts ca. 1400–1700.* Amsterdam: M. Hertzberger, 1954.

Horovitz, Isabel. "The Materials and Techniques of European Paintings on Copper Supports." In *Copper as Canvas: Two Centuries of Masterpiece Paintings on Copper 1575–1775*, pp. 63–92. New York: Oxford University Press, 1998.

Hunter, Dard. *Papermaking*. New York: Dover Press, 1978.

Johnson, C. Richard Jr., *et al.* "Chain Line Pattern Matching in Rembrandt's Prints." In *Rembrandt and His Circle: Insights and Discoveries*, edited by S. Dickey, pp. 319–334. Amsterdam: Amsterdam University Press, 2017.

Koehler, Sylvester Rosa. *A Chronological Catalogue of the Engravings, Dry-Points and Etchings of Albert Dürer*. New York: The Grolier Club, 1897.

Manick, Annette. "A Note on Printing Inks." In *Italian Etchers of the Renaissance and Baroque*, edited by Sue Welsh Reed and Richard W. Wallace, pp. xliv–xlvii. Boston, MA: Museum of Fine Arts, 1989.

McDonald, Mark P. *The Print Collection of Ferdinand Columbus (1488–1539): A Renaissance Collector in Seville*, 3 vols. London: British Museum Press, 2004.

Meder, Joseph. *Durer-Katalog. Ein Handbuch uber Albrecht Durers Stiche, Radierungen, Holzschnitte, Deren Zustande, Ausgaben und Wasserzeichen*. Vienna: Verlag Gilhofer & Ranschburg, 1932.

Parshall, Peter. "Graphic Knowledge: Albrecht Dürer and the Imagination." *The Art Bulletin* 95, no. 3 (2013): 393–410.

Sayre, Eleanor A. "Introduction." In *Albrecht Dürer Master Printmaker*. Boston, MA: Museum of Fine Arts, 1971.

Sethares, William A., Margaret Holben Ellis, and C. Richard Johnson, Jr. "Computational Watermark Enhancement in Leonardo's *Codex Leicester*." *Journal of the American Institute for Conservation* 59, no. 2 (2020): 87–96.

Schenck, Kimberly. "Goltzius's Use of Grey Ink." *Print Quarterly* 15, no. 2 (1998): 186–190.

Stijnman, Ad. *Engraving and Etching, 1400-2000: A History of the Development of Manual Intaglio Printmaking Processes*. London: Archetype, 2012.

Tylecote, R.F. *A History of Metallurgy*. London: The Metals Society, 1976.

Westermann, Ekkehard. "Copper Production, Trade, and Use in Europe from the End of the Fifteenth Century to the End of the Eighteenth Century." In *Copper as Canvas: Two Centuries of Masterpiece Paintings on Copper 1575–1775*, pp. 117–130. New York: Oxford University Press, 1998.

About the Author

Angela Campbell is an independent scholar and the Book and Paper Conservator/Project Inspector for the Northeast Region of the National Park Service. She received her Master's, C.A.S. from Buffalo State College in 2009. Prior to joining NPS, Angela was an Assistant Paper Conservator at the Metropolitan Museum of Art.

5. Digital Resuscitation of the Officina Plantiniana's Woodblock Collection: Goals, Approaches, and Results

Jolien Van den Bossche

Abstract

This chapter addresses an often-overlooked aspect of the study of early modern illustrations: the woodblock matrices themselves. This object-based approach provides new insights, exposing elements that are not visible when looking at the printed image. Investigation of the recessed areas of the block can render information about the artist's hand, often cancelled out in the print. The Museum Plantin-Moretus in Antwerp owns an impressive collection of nearly 14.000 woodblocks, used in the *Officina Plantiniana*, which is now digitized. The material research of the blocks benefits from digitization through multi-image comparison, high resolution zooming on miniscule details and manipulation of size, colour and contrast to enhance particular features. This article discusses the *work-in-progress* of the digitization process and the first findings.

Keywords: woodblock, printing matrices, digital methods, Plantin-Moretus

I. Introduction: The Plantin-Moretus Museum and Its Woodblock Collection

This chapter reviews the aims, methods, and preliminary results of the recent digitization project for the *Officina Plantiniana's* woodblocks in the typographical collections of the Plantin-Moretus Museum, Antwerp. An extraordinarily rich but long overlooked primary resource, the blocks stand to benefit from recent research investigating the role of material culture and visual representation in early modern science by examining the emergence,

---

Noyes, R.S. (ed.), *Reassessing Epistemic Images in the Early Modern World.*
Taylor & Francis Group, 2023
DOI 10.5117/9789463723350_CH05

production, and role of printed images as tools in scientific texts. Making an online clearinghouse of digital images of the blocks linked to basic descriptions in an online catalog, the project facilitates open access to this exceptional and unique collection to enable emerging object-based research that investigates not merely the printed image, but also the matrices used to print them, shedding new light on the Press's history specifically and early modern printing and knowledge production more broadly.

Christoffel Plantin was born in France c. 1520 and was taught printing and book-binding before moving to Antwerp in 1548 for religious reasons, where he opened his printing office the *Officina Plantiniana* in 1555. Appointed *Prototypographe* or chief printer of the city in 1570 by the king of Spain, he was regarded as the period's foremost printer.[1] By 1575, his operations were based in buildings that now constitute the Plantin-Moretus Museum and boasted sixteen presses, 32 printers, twenty compositors, three proof-readers, and one unskilled laborer, in addition to numerous skilled artisans to whom he contracted out work.[2] According to Leon Voet's extensive bibliography, between 1555 and 1589, Christoffel Plantin printed and published approximately 2,500 works in Antwerp and Leiden, ten per cent of which were scientific publications in the fields of botany, engineering, mathematics, natural history, medicine, astronomy, and geography.[3] Within these nearly 400 editions were illustrated with images printed from woodblocks, not including works that contained only woodcut initials and head- and tailpieces. Dirk Imhof documents an additional 180 such editions with woodblock illustrations for the period 1589–1610.[4] After Plantin's death in 1589, his successors found 7,350 woodblocks; following the sale of the Plantin-Moretus house in 1876 to the city of Antwerp, a total of 13,793 woodblocks was counted.[5] By the mid-twentieth century, the entire collection had been reprinted and organized according to subject matter in a five-volume catalog.[6] According to topical categories, at least 35 per cent of the blocks were illustrations of scientific subjects.[7]

1 For an overview of the history of the Press, see Voet, *The Golden Compasses*. On subsequent Moretus generations from the late seventeenth century, see Van den Wijngaert, "De late Moretussen en de boekillustratie."

2 Selleslach, "Discipline en solidariteit in de Officina Plantiniana." On his working practices, see also Voet, *The Golden Compasses*.

3 Voet, *The Plantin Press, 1555–1589*.

4 Imhof, *Jan Moretus and the Continuation of the Plantin Press*.

5 Lemli, "De Officinae Plantinianae."

6 *Afdrukken houtblokken Museum Plantin-Moretus*.

7 Voet, *The Golden Compasses*, p. 208. For further information on the book production and illustrations, see Depauw and Imhof, *De boekillustratie*.

The *Officina Plantiniana*'s business archives indicate that the production of these blocks was a collective effort. Illustrations were first drawn by artist-designers on the block, then cut by specialized woodcutters, some of whom were also designers.[8] Plantin had different artists, both designers and woodcutters, working for him outside the premises of his printing house.[9] According to Dodoens's preface in his *Frumentorum*, Plantin insisted on having new woodcuts made "drawn on the basis of living plants."[10] Archives show that the anonymous block cutter was better paid than the designer, artist Pieter van der Borcht.[11] Given that much block cutting was done in the workshop supervised by the paid cutter, ascribing a block to a certain master solely on the grounds of style or even signature is not without risk and should always be thoroughly substantiated.[12]

II. Object-Based Research: Project Goals and Parameters

Despite increasing interest in object-based research since the "material turn" in historical studies, little research and analysis has been done with the Plantin-Moretus Museum's woodblock collection, due to potential health risks resulting from the blocks' treatment with the poisonous pesticide Xylamon in the 1960s to prevent further woodworm damage, making digital access essential. The Museum Plantin-Moretus undertook the digitization project with three primary objectives. First, to circumvent the challenge of the necessary limits on the close physical consultation of the blocks given their toxicity by providing as much visual information and metadata as possible, with consistent and qualitative digital photography of all blocks (at least front and back). Digitization allows multi-image comparison, high resolution zooming on miniscule details, and manipulation of size, color, and contrast to enhance particular features. Digitization is likewise beneficial for conservation purposes, because the images clearly record the physical

8 Voet, *The Golden Compasses*, p. 196.

9 On the division between designer and woodcutter, see Delen, "Les artistes collaborateurs de Christophe Plantin" and *Idem*, "Les illustrateurs français de Christophe Plantin."

10 Gilias, Van Tilburg, and Van Roy, *Rembert Dodoens*, p. 46.

11 The payment records can be found in Antwerp, Plantin-Moretus Museum Archives, Arch 31, fos. 29, 85, 108. A summarized account of these records can be found in Voet, *The Plantin Press*, 2: 828. They are also discussed in Wei-Hsuan Chen, "A Woodblock's Career."

12 Voet, *The Golden Compasses*, p. 196, refers to an example found in the archives: Plantin-Moretus Museum Archives, Arch. 38, fol. 96v, regarding Arnold Nicolai, woodcutter: "Le 17 mars baillé la Chyromantie de Tricasse à tailler par son garson les figures et luy mesmes la lectre a $3^{1/2}$ st. pièce." But these divisions have not always been recorded.

condition of the blocks at a given moment in time. Second, to expand access to the collection for scholars, artists, and the general public, promoting research and creative reuse of the data, cataloging all the woodblocks in an Adlib database (minimal and basic registration) and linking metadata to digital files in the Museum's Digital Asset Management System (DAMS), an image library with basic descriptive metadata.[13] Multi-image comparison of high quality images enabled by the site Metabotnik (furnishing the capacity to upload a large amount of files to create one large file) stands to prove useful for the investigation within and across a large amount printing materials.[14] Third, to provide an adaptable model for online registration and digitization, communicating about and disseminating both the work process and results for future iteration by other museums and cultural institutions with similar collections, using appropriate means. To this end, both the digitization and registration process are available through free online manuals on the website of Project CEST.[15]

A note on some material exigencies necessitating an adaptable, pragmatic approach to the studio and camera set-up: small variations in block thickness required blocks to be organized by size and thickness in advance, since both required different camera settings. The fact that most of the blocks are significantly darkened from years of use, with some close to black, made it difficult to retain contrast on different blocks with a neutral background and under similar lighting conditions. Parallel analogous issues stem from, on the one hand, some blocks having shinier surfaces than others – either because of a particular ink that was used or the way the blocks were cleaned after printing – and, on the other hand, from significantly smaller blocks. Both have the tendency to be overexposed under the camera settings employed for other blocks, thus necessitating altering the aperture setting of the camera. For example, most blocks could be photographed with an aperture of f/11, but with smaller or more reflective blocks, an aperture of f/13 or f/15

13 The digital asset management system is available through the following link: https://dams.antwerpen.be/ (accessed 7 March 2021).

14 Miller, "The Magical Number Seven, Plus or Minus Two."

15 Cultureel Erfgoed Standaarden Toolbox, translated as "Cultural Heritage Standardization Toolbox." Originally a temporary project commissioned by Flanders Heritage Agency to define standards for archiving, describing and exchanging museum objects, CEST is now an up-to-date website with software packages, manuals, guidelines and glossaries for different types of cataloging programs (including Adlib) that enables users to create a specific collection profile that can be used by other catalogers as a reference for their own collections. Every woodblock gets a description in Adlib based on our adapted registration profile for CEST: https://www.projectcest.be/wiki/Publicatie:Invulboek_objecten/Profielen/Museum_Plantin-Moretus (accessed 7 March 2021).

was chosen to stop down the lens. While this inevitably meant some loss of depth on the photos, all relevant information on the four sides was recorded for registration in Adlib.

III. Digitization

Following the example of the Rijksmuseum manual for the photography of 3D objects, an in-house studio was set up in the Plantin-Moretus Museum and a detailed workflow was established with specific photographic instructions – detailing lighting setup, camera perspective, and quality parameters – accompanied by schematic illustrations, a course of action for editing the digital images in Adobe Photoshop Lightroom Classic©, and an outline of the process of changing file names in bulk in IrfanView and the uploading process in DAMS.[16] Streamlining in a timely manner the time-consuming online registration and digitization of nearly 14,000 blocks without a loss of quality required implementation of specific quality standards.[17] Woodblocks were shot with an overhead camera angle, resting on a table with a height of 86 cm. The camera was set up perpendicular from above, at a height of 29.75 cm to the top of the block. To ease the manipulation and substitution of the blocks, the repro stand's column with arm is directed away from the photographer. Because of this placement, the camera is now in a downwards position relative to the photographer. Therefore, the woodblocks and QP reference card are turned in the same direction as the camera. Raking light from a height of 110 cm (from center of the lamp to the floor) and a distance of 70 cm from the center of the woodblock, was selected to accentuate surface texture and exaggerate shadows. Depending on the depth of the recessed areas and the reflexion of the blocks, the height of the lamp could be altered to hit the blocks from a different angle. Raking light

16 The Rijksmuseum Manual for the photography of 3D objects gives an overview of the way in which the Rijksmuseum studios photograph object groups. The manual provides instructions about the different lighting setups for photographing objects which fall into one of two categories: standing objects or lying objects. It is available on their website through this link: https://www.rijksmuseum.nl/en/2d3d-2017/rijksmuseum-manual-for-the-photography-of-3d-objects (accessed 7 March 2021).

17 1) objects are photographed with a neutral grey background; 2) objects' specific characteristics are clearly visible and legible; 3) objects are photographed from the same perspective; 4) objects are photographed from the front and back; 5) images have a resolution of 600 dpi; 6) ColorChecker Passport Camera Calibration© is used to obtain neutral, consistent color; 7) objects are photographed with a QP-neutral greyscale card with millimetre scale at the same height of a given block, as reference for white balance and dimensions.

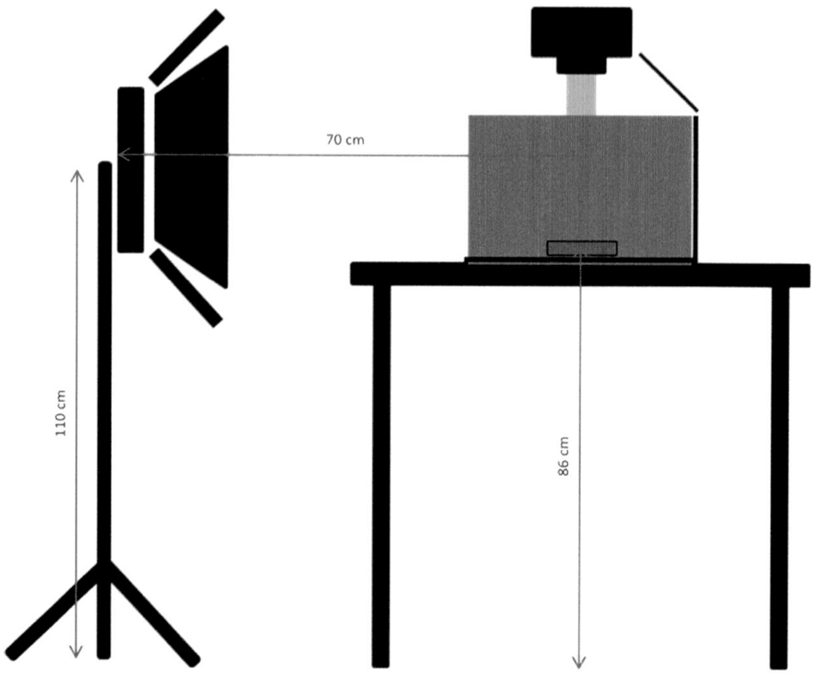

Figure 5.1 Schematic representation of in-house studio setup (Collectie Stad Antwerpen, Museum Plantin-Moretus)

caused some textured areas to appear completely dark. To fill these shaded areas with light and to reveal more characteristics, a reflective box was made with four white reflective panels. Three white sides ensured gradual tapering of the light. The fourth reflective panel rested on top of the box to limit large shadows of the object being cast onto the neutral background. The panel closest to the photographer could be opened when the blocks were changed (Figure 5.1).

To obtain a quality of 600 dpi (dots per inch), a woodblock with an average thickness of 2.2 cm was placed underneath the camera next to a ruler. The camera was set to a certain height, in this case 29.75 cm, and a picture was taken (lens in AF, does not have to be sharp). This picture was then uploaded in Photoshop CC© and a square of 10 cm × 1 cm was drawn, using the ruler on the image as a reference. The number of pixels per 10 cm was counted in Photoshop© and converted to inches by multiplying the result per centimeter with 2.54. For example, it read 2,359 pixels per 10 cm, thus 235.9 pixels per 1 cm. To define the pixels per inch, this number was multiplied by 2.54 cm (1 inch), resulting in 599.18 dpi. If this result was above 600 dpi, the camera was lowered; if below 600, the camera was raised. After sharpening the image manually, these studio settings could be used for all blocks with the same average thickness and size. They were mostly photographed with aperture f/11, exposure time 1/13s, ISO 100, lens 60 mm, image format NEF and image size 6016 × 4016.

The fact that each camera reads color differently and different lenses and changing lighting conditions can also impact color and give rise to variations made color management and related parameters an important part of the workflow. The studio was closed from natural daylight and other light sources to ensure consistent lighting to the highest degree possible. Using Adobe Photoshop Lightroom Classic©, a lens correction was applied according to the type of camera and lens used.[18] Secondly, a color correction was applied according to a profile generated with ColorChecker Passport Camera Calibration© software, an industry standard reference with 24 solid tone colored squares that was photographed in the same position and under the same conditions as the woodblocks. The resulting image was then uploaded to Lightroom CC©, where it was used to create a custom DNG profile with the ColorChecker Camera Calibration Software©, which could thereafter be selected as a custom profile per Camera Calibration to apply it to all images of the blocks under similar lighting conditions. Once photographed, individual digital images of the blocks were cropped and thereafter renamed in IrfanView according to their object number, then uploaded in bulk in DAMS using a CSV-file that interlinks the digital image to the asset in Adlib. To optimize the collection's searchability, the reprinted images produced for the aforementioned mid-twentieth-century multi-volume catalog of the woodblock collection were used to represent the primary search result when searching for a specific woodblock by object number on the museum's search site or on DAMS,[19] followed by the recto view of the actual block and then the block's verso.

IV. Registration and Adaptations

For registration in Adlib, rather than working from the woodblock itself, the twentieth-century reprinted images were used for reference, given their greater legibility and retrievability relative to the fronts of the blocks. Untitled blocks were furnished with unique names according to the publications

18 In this case, we used a Nikon D750 camera and a AF Micro-Nikkor 60mm f/2.8D lens.

19 See *Afdrukken houtblokken Museum Plantin-Moretus*.

for which they were used.[20] That the same block was often used repeatedly throughout several publications automatically generated a time sequence for certain blocks within Adlib.[21] For the fundamental task of screening all Plantin publications with woodblock illustrations to link each print with its respective woodblock, we turned to a number of foundational studies on the collections, firstly that of former curator of the Museum Plantin-Moretus and Print Cabinet Leon Voet, who prepared an important bibliography of the works printed and published by Christopher Plantin at Antwerp and Leiden from 1555 through mid-1589 (the *terminus ante quem* furnished by the date of Plantin's death in July). Voet's survey supplied the following information for each publication for which a copy was preserved: short title; title-page transcription; collation; content; illustrations (when present), locations of known surviving copies; and references and notes (i.e. archival references).[22] Voet's extensive bibliography was consulted together with that prepared for the oeuvre of Jan Moretus I by Dirk Imhof, resulting in a list comprising all publications issued by the Plantin-Moretus presses containing woodblock illustrations, excluding those that contained only woodblock initials or tailpieces.[23] This list was then consulted in the time-consuming page-by-page screening of all publications with woodblock illustrations for the period 1555–1610. The scope of this screening entailed the recruitment of the use of Artificial Intelligence Software (AI) to first detect and extract automatically illustrations from surrounding printed text, then match the illustrations to a dataset comprised of the modern reprints in a search engine.[24] Such an automated process seems especially

20 Per the recommendation of the project's international advisory committee.

21 A good example of this is the woodblock with inventory number MPM.HB.04094, which was given the vernacular name "Bolderik [Agrostemma githago]." It appears first in Dodoens, *Florum*, under the name "Nigellastrum." In de l'Obel, *Plantarum, seu Stirpium historia*, it is depicted as "Pseudomelanthium." In the same author's *Kruydtboeck* it appears under the name "Cooren-roosen." This same plant appears in fourteen publications between 1568 and 1647 under different names. A complete overview of these publications can be found on https://search.museumplantinmoretus.be/Details/collect/391586 (accessed 7 March 2021), or by searching on the inventory number in the Museum's online catalog.

22 Voet, *The Plantin Press*. For the continuation of the press under Jan Moretus, see Imhof, *Jan Moretus and the Continuation of the Plantin Press*.

23 Due to his background in book-decorating, Plantin was keen on using decorative initials and finery in nearly all his publications. Given that Plantin published approximately 2,500 works in Antwerp and Leiden alone between 1555 and 1589, it would be nearly impossible to match all initials and tailpieces with their woodblocks manually. The initials take up 30 per cent of the entire woodblock collection of 13,794 blocks.

24 Tests are being performed using the artificial intelligence software provided by the Visual Geometry Group at the Department of Engineering Science at Oxford University.

useful for the woodblock initials widely used in manifold variations across numerous editions.

In addition to an assigned title for each woodblock, every block was also assigned a basic registration in Adlib (object number, title, object name, institution, creator, production date, description, material, technique, acquisition date, acquisition method, and literature), and the entire registration workflow was published online on CEST. In the early modern context of Plantin's publishing block collection, neither the woodblocks nor their resulting prints were considered art objects *per se*, but rather were overwhelmingly produced to illustrate or ornament an accompanying text to which they remained subsidiary. In these circumstances, the blocks were mostly untitled; if they were assigned titles, those titles often changed across different publications; this was frequently the case for emblematic and scientific illustrations. A given plant could be assigned as many as ten different Latin names in the botanical works published by Plantin.[25] Since such cases foreclose on the choice of just one of these botanical names as the object title in Adlib, we decided to assign the modern vernacular (Dutch/English) together with the scientific name in Latin to the block for each plant as the object title, drawing on previous work identifying plants in some editions, and ensuring that different blocks of the same plant received the same title.[26] However, given that plants' modern names often are not relevant for historical research, we also indicated both specimens' different historical names in various publications and the respective pages where they were printed, within a new field in Adlib labeled "printed in." These data – plants' historical names and the publications wherein their representative woodblocks were used – are also searchable online.

For the registration of woodblock initials in the collection, we recovered some of the work by Kristof Selleslach for TORAD (Typografische Ornamenten Repertorium van Antwerpse Drukkers 1541–1600), a proprietary database of typographical ornaments from Antwerp printers.[27] Created by systematically screening hundreds of Plantinian publications page per page for ornaments, TORAD was divided into fleurons (flower-shaped motifs), devices, initials, and vignettes, and each illustration was described using iconographical and stylistic drop-down lists. Drawing on both this database

25 Wei-Hsuan Chen, "The Blocks, the Tools, and the Printed Plants." See also van Zanen, *Planten op papier*.

26 See e.g. identifications for plants in historical publications: https://leesmaar.nl/ (accessed 7 March 2021).

27 TORAD is a local database and is not available online. As of 2021, limited access can be granted upon request to researchers on site at the Plantin-Moretus Museum.

and Stephen Harvard's published catalog of woodcut initials used by Plantin, we developed a descriptive registration system for ornamental initials in the collection, employing the blocks' dimensions as a distinguishing factor and adding a series title to link individual initials to their alphabets.[28] In the case of different blocks not necessarily made for the same publication, but exhibiting similar characteristics or themes, rather than creating a series title we associated them by using a broader term in the title – i.e. "medical illustration," "cosmographical illustration," "scientific illustration," etc. – followed by a colon and more specific description. In other cases, to make certain sets of woodblocks retrievable required a series title or more elaborate description. For example, all the woodblocks in the *In C. Suetonii Tranquilli XII caesares commentarii* by Laevinus Torrentius (Lieven van der Beke, 1525–1595) published by Plantin depict ancient Roman coins.[29] A description of these blocks according to our registration profile would have resulted in similar titles (i.e. "Silver Roman coin: denarius of Augustus"), making it challenging to search for a specific block. Therefore, for these blocks the description field is used as a distinguishing factor, wherein inscriptions on and a basic iconography for each coin were added. These cases illustrate the adaptive nature of the registration process, which remains ongoing beyond the lifespan of the project, as further information is subsequently added in Adlib.

V. Project Results and Areas of Further Inquiry

While the project is chiefly intended to furnish a platform and information clearinghouse enabling subsequent further research, the digitizing process has yielded some initial noteworthy case studies, reviewed here by way of conclusion. Firstly, there is the case of the adaptive transformations wrought upon irregularly and unusually shaped woodblocks that feature diverse cut-outs excised from the block (exceeding the carving of the block's printed design), which range from L-shaped and cross-shaped to frames and completely asymmetrical examples (see specific illustrated examples below). Comparing these blocks against their resulting printed illustrations in Plantinian publications reveals that the various cut-outs to the wood were made in order to enable the closer printing the text relative to the image. For example, the irregular block for printing a representation of

28 Harvard, *Ornamental Initials*.
29 Torrentius, *Laeuini Torrentii*.

the flowering plant known as White Dog's Tooth shows three separate rectilinear areas completely cut out of the block, with its lower right corner (recto) removed and two substantial notches cut from its left edge (also recto).[30] By comparing the block to printed illustrations from Plantin's press, a progressive adaptation of the block can be determined. First, for page 232 in Plantin's 1581 printing of the Dutch botanical treatise by Flemish physician and botanist Matthias Lobelius (Matthias de l'Obel, 1538–1616), a cut-out was made to the block's lower corner so that text could be added next to the plant's detached leaf (on the right in the block; on the left in the printed image).[31] Later, for page 196 in Lobelius's subsequent work printed in 1591, two more cut-outs were made to the block's edge, for the addition of further text next to details of the plant's flowers.[32]

In this way, the material state of the woodblock became an index of the progressive transformations undergone by Lobelius's botanical oeuvre. Similar approaches were also identified in other works, such as the case of blocks printed in Plantin's various editions of the popular emblematic compendium by Claude Paradin (c. 1510–1573).[33] Today, the woodblock cut with a design for the emblem *Immensi tremor Oceani* ("Trembling fear of the immense Ocean"), featuring a circular interlinked shell chain suspending a medallion figuring St. Michael the Archangel, shows a central square cut-out in such a way that the type could not only be set very close to the chain, but also within it (Figure 5.2).[34] However, that such a type setting only appears in Plantin's later emblematic editions published after 1567 suggests that this drastic excavation of the block was done about a decade (and perhaps more) after it was initially cut, to meet changing conditions.[35]

In addition to cut-outs that completely removed chunks of the woodblock from back to front, further kinds of adaptations of the blocks were often identified, including gouges in the wood surface or other parts that were cut away, which raise interpretive issues taking up questions of whether such

30 Object number MPM.HB.05583. https://search.museumplantinmoretus.be/details/collect/393070 (accessed 7 March 2021).

31 Lobelius, *Kruydtboeck*.

32 *Idem*, *Icones stirpium*.

33 See Voet, *The Plantin Press*, 4: 1812–1819.

34 Object number MPM.HB.01197, https://search.museumplantinmoretus.be/details/collect/390364 (accessed 7 March 2021).

35 See e.g. Paradin, *Symbola heroica*, p. 24. More information about the emblems in Claude Paradin's work can be found through the search site of the Museum Plantin-Moretus, on https://search.museumplantinmoretus.be/home (accessed 7 March 2021).

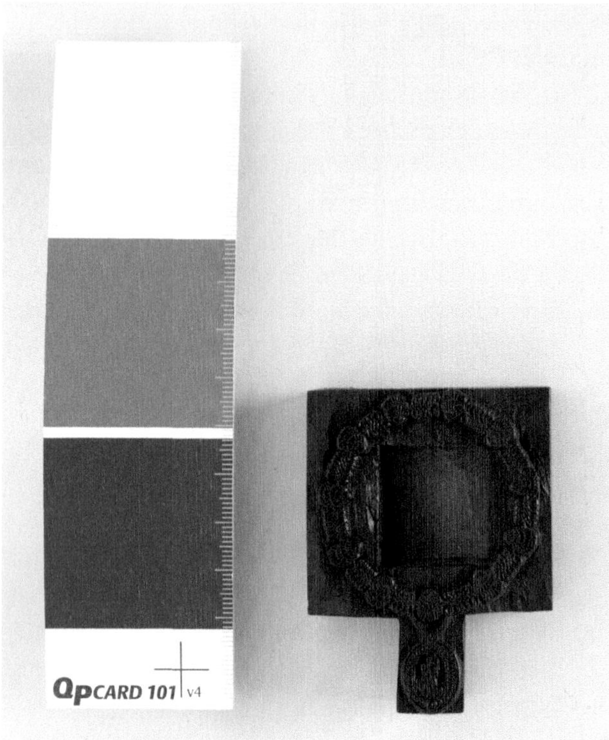

Figure 5.2 Example of an irregularly shaped block used in Claude Paradin's *Les devises héroiques*, object number MPM.HB.01197 (Collectie Stad Antwerpen, Museum Plantin-Moretus)

interventions represent adjustments made after proof printing to prevent paper from catching ink that coalesced in larger recessed areas, or "editorial" corrections or deletions of certain details based on new knowledge or a patron's or author's behest. Recent studies, such as that by Iris Kockelbergh investigating the reuse and adaptation of a Plantinian botanical woodblock cut with a depiction of a clove tree demonstrate that such questions can be sometimes worked out by comparative analytical approaches mapping blocks and their material transformations against printed illustrations across different publications.[36]

Similarly, such comparative analyzes allows detection and exploration of a range of diverse period technical approaches to carving blocks, as Jessie Wei-Hsuan Chen's recent article highlights a minimum of four different cutting techniques (referred to as "Styles") within a group of botanical blocks:

36 Kockelbergh, "The Rich Collection of Woodblocks."

two in the original blocks (Styles 1 and 2), and two in a set of duplicate blocks cut to reproduce the originals (Styles 3 and 4).[37] Our digitization supported and expanded these findings, which is evident most clearly from studying the blocks rather than the printed images. Following the classification of styles as outlined by Chen, in Style 1, the block's corners and edges were cut lower than the rest of the block, resulting in a ramp-like effect whereby the block's higher, thicker center gradually tapers to lower, thinner edges.[38] In Style 2, the recessed areas of the block are almost perfectly smooth, evidencing a high degree of polish more commonly associated with finished artworks.[39] Style 3 exhibits a distinct carving technique with small, regular cutting marks; blocks from this type are often lighter in color and have different worm holes in the back, suggesting a different wood type.[40] Style 4 shows a strikingly different more aggressive and less finished cutting technique marked by wide gouges and large pieces of wood chiseled off in chunks, with reddish-hued blocks rife with wormholes.[41] That on these last blocks the cuts often run horizontally to the right of the raised image, design, or motif, and vertically to the left, could be indicative of the handedness of the cutter and the way he turned and re-positioned the block to excavate the wood without damaging the design.

Hardly unique to the botanical blocks, this relative diversity of cutting techniques can be found across other parts of the collection, and constitutes a fertile field for further art historical and archival research that could for example attribute certain blocks to a particular hand or workshop on the basis of technical correlations traceable in the blocks' cut surfaces. Along the lines of attribution, a small initial "C" was incised in the recessed areas of some of the woodblocks produced to illustrate the description in verse and prose of the Archduke Mathias of Austria's politically important entry into Brussels on 18 January 1578, composed by Brabantine humanist Jan (or Johan) Baptist Houwaert (1533–1599) and published by Plantin in 1579.[42] That this initial was not cut in reverse signifies the letter was not intended

37 See Wei-Hsuan Chen, "A Woodblock's Career."

38 Wei-Hsuan Chen, "A Woodblock's Career," fig. 1.

39 *Ibid.*, fig. 2.

40 *Ibid.*, fig. 3.

41 *Ibid.*, fig. 4.

42 Houwaert, *Sommare beschrijvinghe*. See e.g. woodblock nrs. MPM.HB.02836 https://dams.antwerpen.be/asset/E1iUFqlg9eWcRdCOMiQQRKw5 (accessed 7 March 2021), MPM.HB.02843 https://dams.antwerpen.be/asset/Q2DRFOLShTPcTgKuPHXhORld (accessed 7 March 2021), MPM.HB.02841 https://dams.antwerpen.be/asset/LRfKbgJHJJQXVaZTC2Dk7exz (accessed 7 March 2021), MPM.HB.02842 https://dams.antwerpen.be/asset/q1cICXmJZfdNaVDakgmTYJUe (accessed 7 March 2021).

to be seen in the printed image, but remained an attribute of the block. It is possible these letters may have served as a guide for the printers in preparing the heavily illustrated work, or constituted a practicality for the owner of the blocks for safekeeping, transport, or payment of the cutter. Given the significance of both the publication and event it commemorated, it is also possible that the block cutter took pride in their work and assigned his name to them: while Voet suggested the woodblocks for this publication were perhaps made by Antoon van Leest based on his previous signed works for Houwaert, the letter C could well be a reference to Cornelis Muller, who often signed his blocks in a similar way on the raised, printed surface.[43]

Similarly intriguing and potentially promising for future investigation are the many composite or assembled blocks in the collection, often cobbled together with nails, such as that figuring the Michoacán or jalap root (*Ipomoea purga*), traded from the Americas and valued in early modern medicine for its purgative properties (Figure 5.3).[44] This block presents an especially interesting physical aspect, since it actually consists of two blocks that were attached not butted directly against each other edge to edge, but rather have been assembled according to wood-joining techniques resulting in a stronger composite block more resistant to wear. An unfinished lower portion of the larger top block, carved with a tangle of the plant's leaves and tendrils, extends downwards; the smaller bottom block, cut with a profile and cross-section representation of the bifurcated tuberous root, was overlaid onto this extension and fixed with nails. Between the two coupled blocks a clear difference can be seen in cutting technique between the respective recessed areas of the bottom and top portions: while the cutaway areas on upper block are quite smooth, the recessed parts on the bottom block are much rougher, as if large chunks of wood were hastily chiseled away.

Cross-referencing this composite block against Plantinian publications demonstrates that when this plant illustration first appeared in Plantin's shop in the context of Matthias Lobelius's 1576 herbarium, it apparently consisted only of the bottom block depicting the root, which, judging from the print, initially included an illusionistic detail of a bit of cut stem emerging from the root.[45] When the plant re-appeared in the illustrations for the 1581

43 Voet, *The Plantin Press*, nr. 1411. Further archival research should be conducted before assigning the blocks to a cutter.

44 Object number MPM.HB.04557, https://search.museumplantinmoretus.be/details/collect/392050 (accessed 7 March 2021). For the plant's period medicinal applications see Gänger, "World Trade in Medicinal Plants."

45 Lobelius, *Plantarum seu Stirpium*, p. 345.

Figure 5.3 Example of an assembled block, object number MPM.HB.04557 (Collectie Stad Antwerpen, Museum Plantin-Moretus)

Dutch-language *Kruydtboeck*, the first block had been not only re-cut to remove the piece of stem, but also affixed to the second top block carved with the tendrils and leaves, their joining betrayed by a thick black zone on the page just at the area where root meets stems, indicating that an amount of excess printer's ink had seeped into the crack at the joint line between the two blocks and then transferred to the paper.[46]

At this point, Lobelius presumably found it important to illustrate the leaves as well as the root, perhaps to signal his own increased understanding of the immensely important medicinal plant (many tons of which were shipped to Europe from the Americas annually) through ostensible first-hand knowledge gathered either directly by observing imported specimens or indirectly by copying drawings or reproducing the semblance of vine-like foliage from written descriptions; or perhaps to convey additional knowledge to enterprising and medically-minded travelers to the Americas about the above-ground appearance of the plant (in fact a type of flowering crawling vine) to aid in its identification and extraction (or some combination

46 *Idem*, *Kruydtboeck*, p. 747.

Figure 5.4 Example of an Aguilonius block with cavities for moveable type, object number MPM.HB.03328 (Collectie Stad Antwerpen, Museum Plantin-Moretus)

thereof).[47] By the time the two-block assemblage was reprinted in the 1591 Latin edition published by Plantin, the printed image disclosed the wear-and-tear of time and repeated pressure of the presses on the composite block: what had been a tight joint marked by an excess of ink a decade earlier had since spread to become a gap too wide to collect ink that registered as a thin blank line of paper between the top of the root and the stems, a line equally visible in the twentieth-century reprint registered in DAMS.[48]

Like composite blocks, blocks with cavities for the insertion of movable metal type, or those bearing particularly informative physical traces attesting to the creative and technical processes inhering in designing and cutting prove especially promising objects for further exploration in the cultural contexts of their production and use. In the first case, many such blocks adapted for type belong to the sub-collection of over 500 geometric diagrams cut by carvers Gillis, Willem, and Jan Claessens to illustrate the massive 1613 volume on optics authored by Ibero-Brabantine Jesuit

47 Gänger, "World Trade in Medicinal Plants," p. 47.
48 Lobelius, *Icones stirpium*, p. 625.

mathematician Franciscus Aguilonius (Francisco de Aguilón or François de Aguillon, 1567–1617).[49] Printed on 684 folio pages and more famously furnished with copperplate engravings by Theodore Galle after designs by Peter Paul Rubens, this costly work demanded a production process that economized wherever feasible; including woodblocks that could accommodate type reduced the costs of woodblock cutting, since one block could be reused in different contexts.

To cite but one example, a single Aguilonius block was printed on page 377 and page 398, in each instance with different cast metal type pieces, which were inserted into rectangular holes and a V-shaped notch cut clear through the block and visible from both recto and verso (Figure 5.4).[50] In the second case, the backs or sides of blocks can reveal new information, including non-representational cuts on the verso suggestive of the carver testing tools and/or wood (such marks are found on the composite botanical block discussed above), as well as cut-out asterisks, written and incised numbers and annotations, and even uncut images, like the drawing on the backside of a block cut with the portrait of biblical scholar Guy Lefèvre de la Boderie (1541–1598), who assisted in the Syriac translation of the New Testament of the famous Antwerp Polyglot Bible printed by Plantin between 1568 and 1572 (Figure 5.5).[51] The block's verso still bears traces of a second sensitive, highly-finished portrait, also of Boderie, encircled by the same text writ in reverse and thus clearly prepared for cutting but never cut, its rendering of the sitter evincing some resemblance to an engraved portrait of Boderie in the British Museum.[52]

Taken together, all of these cases are highly suggestive of a fundamental reassessment of the blocks and the printed images they engendered not as fixed and stable entities, but rather enduring and continuously evolving organisms, ever adapting in punctuated equilibrium within the larger dynamic ecosystem of the Plantin house.

49 Aguilonius, *Opticorum libri sex*. For this massive project the Officina Plantiniana paid Jan (94 blocks), Gillis (275 blocks) and Willem or Guilliam Claessens (135 blocks).

50 Object number MPM.HB.03328, https://search.museumplantinmoretus.be/details/collect/382589 (accessed 7 March 2021).

51 Object number MPM.HB.02085, https://dams.antwerpen.be/asset/cYdFdhCah8eXKeVnLEDseypV (accessed 7 March 2021).

52 For the cut portrait on the woodblock, see: https://dams.antwerpen.be/asset/toWTfUZoTkQfoRwKXdNbJgVT, (accessed 7 March 2021). For the portrait in the British Museum, object number 1928,0313.189, see: https://research.britishmuseum.org/research/collection_online/collection_object_details/collection_image_gallery.aspx?assetId=489782001&objectId=1456130&partId=1 (accessed 7 March 2021).

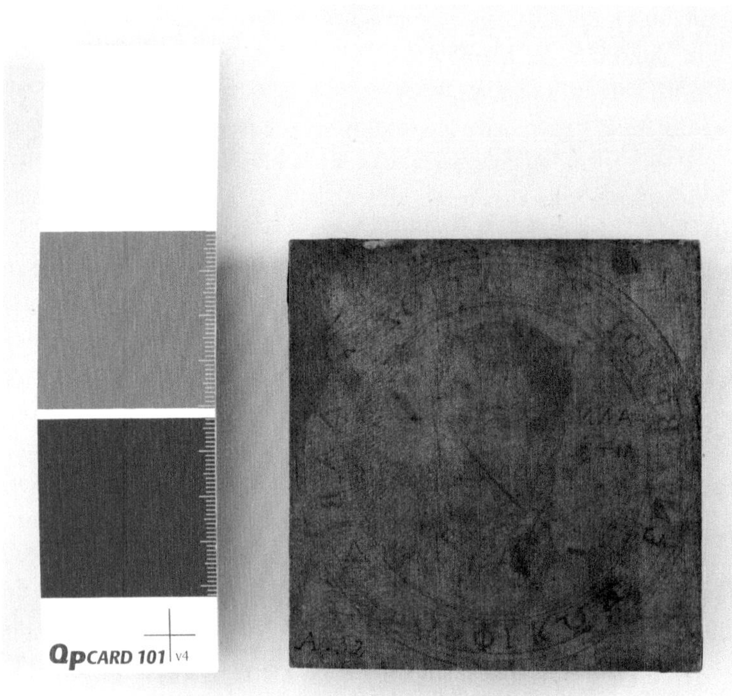

Figure 5.5 Uncut portrait of Guy Lefèvre de la Boderie on the back of woodblock, object number MPM.HB.02085 (Collectie Stad Antwerpen, Museum Plantin-Moretus)

Bibliography

Archival sources

Antwerp, Plantin-Moretus Museum Archives, Arch. 31.

Plantin-Moretus Museum Archives, Arch. 38.

Early modern printed sources

Aguilonius, Franciscus. *Opticorum Libri Sex Philosophis Iuxtà ac Mathematicis Utiles*. Antwerp: ex officina Plantiniana, apud viduam et filios Jo. Moreti, 1613.

Houwaert, Johan Baptist. *Sommare Beschrijvinghe vande Triumphelijcke Incomst vanden … Aerts-Hertoge Matthias….* Antwerpen: Christoffel Plantijn, 1579.

Lobelius, Matthias. *Kruydtboeck oft Beschrijvinghe van allerleye Ghewassen, Kruyderen, Hesteren, ende Gheboomten*. Antwerp: Christoffel Plantyn, 1581.

Lobelius, Matthias. *Icones Stirpium, seu Plantarum tam Exoticarum, quam Indigenarum….* Antwerp: ex Officina Plantiniana, 1591.

Paradin, Claude, and Gabriele Simeoni. *Les Devises Heroiques*. Antwerp: De l'imprimerie de Christophle Plantin, 1561.

Paradin, Claude. *Symbola Heroica M. Claudii Paradini Belliiocensi Canonici, et D. Gabrielis Symeonis Multo quam Antea Fidelius de Gallica Lingua in Latinam Conuersa*. Antwerp: ex officina Christophori Plantini, 1583.

Torrentius, Laevinus. *Laeuini Torrentii in C. Suetonii Tranquilli 12. Caesares commentarii*. Antwerp: ex officina Christophori Plantini, 1578.

Secondary sources

Afdrukken Houtblokken Museum Plantin-Moretus. 6 Volumes. Antwerp: Museum Plantin-Moretus, 1960.

Delen, A.J.J. "Les Artistes Collaborateurs de Christophe Plantin." In *Sept Études Publiées à l'Occasion du Quatrième Centenaire du Célèbre Imprimeur Anversois Christophe Plantin, 1520–1920*, edited by Musée Du Livre Bruxelles, pp. 85–123. Brussels: J.E. Buschmann, 1920.

Delen, A.J.J. "Les Illustrateurs Français de Christophe Plantin." *Le Bibliophile* 2 (1932): 63–74 and 175–183.

Depauw, Carl, and Dirk Imhof. *De Boekillustratie ten Tijde van de Moretussen*. Antwerp: Stad Antwerpen, 1996.

Gänger, Stefanie. "World Trade in Medicinal Plants from Spanish America, 1717–1815." *Medical history* 59, no. 1 (2015): 44–62.

Gilias, Guy, C. R. Van Tilburg, and Vincent Van Roy. *Rembert Dodoens. Een Zestiende-eeuwse Kruidenwetenschapper, Zijn Tijd- en Vakgenoten en Zijn Betekenis*. Antwerp–Apeldoorn: Garant, 2016.

Harvard, Stephen. *Ornamental Initials: The Woodcut Initials of Christopher Plantin: A Complete Catalogue*. New York: American Friends of the Plantin-Moretus Museum, 1974.

Imhof, Dirk. *Jan Moretus and the Continuation of the Plantin Press: A Bibliography of the Works Published and Printed by Jan Moretus I in Antwerp (1589–1610)*. Leiden: Brill, 2014.

Kockelbergh, Iris. "The Rich Collection of Woodblocks of the Museum Plantin-Moretus and Its Use in the Officina Plantiniana." In *Ulisse Aldrovandi. Libri e Imagini di Storia Naturale nella prima Età Moderna*, edited by Guiseppe Olmi and Fulvio Simoni, pp. 109–118. Bologna: Bononia University Press, 2018.

Lemli, Jozef. "De Officinae Plantinianae te Antwerpen en te Leiden en hun Botanische Edities van 1589 tot 1647." In *De Botanica in de Zuidelijke Nederlanden (Einde 15de Eeuw – ca. 1650)*, edited by Dirk Imhof and Francine de Nave, pp. 57–59. Antwerp: Museum Plantin-Moretus, 1993.

Miller, George. "The Magical Number Seven, Plus or Minus Two: Some Limits on Our Capacity for Processing Information." *The Psychological Review* 63 (1956): 81–97.

Selleslach, Kristof. "Discipline en Solidariteit in de Officina Plantiniana." Lecture, Friends of the Plantin-Moretus Museum, Antwerp, 11 September 2019.

van den Wijngaert, Frank. “De late Moretussen en de Boekillustratie.” *De Gulden Passer* 25 (1947): 186–240.

van Zanen, Sylvia. *Planten op papier: het pionierswerk van Carolus Clusius (1526–1609)*. Zutphen: WalburgPers Algemeen, 2020.

Voet, Leon. *The Golden Compasses. The History of the House of Plantin-Moretus.* Amsterdam: Vangendt & Co, 1969.

Voet, Leon. *The Plantin Press, 1555–1589: A Bibliography of the Works Printed and Published by Christopher Plantin at Antwerp and Leiden.* Amsterdam: Van Hoeve, 1982.

Wei-Hsuan Chen, Jessie. “The Blocks, the Tools, and the Printed Plants: The Making and Printing of Botanical Woodblocks at the Early Modern *Officina Plantiniana.*” Master’s thesis, Utrecht University, 2017–2018.

Wei-Hsuan Chen, Jessie. “A Woodblock’s Career: Transferring Visual Botanical Knowledge in the Early Modern Low Countries.” *Nuncius: Journal of the Material and Visual History of Science* 35, no. 1 (2020): 20–63.

About the Author

Jolien Van den Bossche (MA, Painting Restoration) was project assistant within the typographical collections of the Museum Plantin-Moretus in Antwerp, Belgium. She is currently advising curator at the Mercatormuseum in Sint-Niklaas.

Part 2

Imprints as Instruments

6. Academic Print Practices in the Southern Netherlands: Allegory and Emblematics as Epistemic Tools

Gwendoline de Mûelenaere

Abstract
This chapter demonstrates that student life constituted an important source for the creation and circulation of printed images by considering a visual domain that flourished in the Southern Netherlands in the seventeenth century: illustrated college notebooks. The well-preserved corpus presents an ambivalent relationship between knowledge and imagination. This essay focuses on the symbolic language applied to lecture notebooks, the usage of which was encouraged in similar academic practices such as Jesuit *affixiones* and thesis prints. It also addresses the materiality of prints and their manipulation by students. My hypothesis is that the sphere of learning was, via the printing world, bathed with Jesuit culture and visual language that perpetuated in higher education.

Keywords: lecture notebooks, allegory, emblematics, materiality of prints

I. Introduction: The Visual Languages in Illustrated Lecture Notebooks

The curriculum at the Old University of Louvain (founded 1425) remained that of a typically medieval institution in the early modern period. Students started at the *Studium Generale* of the Faculty of Arts of Louvain before being (possibly) admitted to one of the higher faculties of medicine, law, or theology. The *magister artium* title required nine months' training in logic or dialectic, nine months of physics, four months of metaphysics and ethics, and, finally, three months of rehearsal exercises (*repetitiones*).

Noyes, R.S. (ed.), *Reassessing Epistemic Images in the Early Modern World.*
Taylor & Francis Group, 2023
DOI 10.5117/9789463723350_CH06

In the *Artes* faculty, students took *verbatim* notes of the text dictated by teachers during class.[1] Taking and saving notes was a widespread habit among humanist scholars, scientific observers, and travelers.[2] In many early modern universities, the professor spoke at dictation speed, which resulted in fairly reliable copies of the course.[3] It also happened that the text was copied or even bought in advance, on the basis of course notes from previous years.[4]

Louvain's well-preserved volumes of lecture notes or *dictata* were often embellished by students with drawings and then engravings.[5] This tradition of illustrating lecture notes flourished in the first half of the seventeenth century, continuing until the mid-eighteenth century.[6] Visual material includes logical diagrams, arborial diagrams, tables, "fill-in-the blank" title pages, humoristic doodles, ornamental initials, landscapes and views of cities, portraits of rulers, geometric patterns, scientific instruments, astronomical representations, anatomical plates, attributes and personifications, religious and mythological narratives, satirical or grotesque figures, frames and cartouches surrounding handwritten titles, drawn head pieces and scrolls, object-shaped text, small devotional vignettes, illustrated colophons or chronograms, and symbols of the four pedagogies.[7] Images inserted in these manuscripts could be divided into two mains categories: scientific imagery directly aiming at fostering understanding of the subject, and figurative iconography pertaining to allegory, emblematics, and symbolism.[8] The two distinct but sometimes complementary modalities of the image were marshaled in service of knowledge acquisition and transmission.

1 The content of the Faculty's courses was established by the statutes of 1567–1568, which stipulated a strict adherence to the traditional corpus of Aristotelian treatises. The statutes were renewed in 1639 without major changes (Vanpaemel, "Louvain printers," p. 243).

2 Blair, "The Rise of Note-Taking," pp. 303–316.

3 *Ibid.*, pp. 308–309.

4 De Ridder-Symoens, "De praktijk van kennisoverdracht," p. 11.

5 Most of the college notebooks cited or analyzed in this essay can be found in the online database Magister Dixit, a project led by the Katholieke Universiteit Leuven (hereafter KUL): http://heron-net.be/lectio/magister-dixit-collection (accessed 3 July 2022). It contains 570 digitized manuscripts kept at the Royal Library of Belgium (KBr), KUL, M Museum, Université catholique de Louvain (UCL), and Abdij van Berne Heeswijk (Berne). 43 additional illustrated volumes are held at the University Library of the ULg (Liège).

6 However, this tradition was already established in the fifteenth century. See Smeyers, "Een collegeschrift," pp. 243–303.

7 The Lily, the Pig, the Castle, and the Falcon were the four pedagogies, or colleges, in which students of the *Artes* faculty were divided.

8 About the scientific images inserted in lecture notebooks, see Vanpaemel, "Louvain printers," pp. 241–254.

This corpus furnishes the possibility to study engraved images from the perspective of reception. Students were responsible for the material preparation of their manuscripts, and they could choose images, where and how to arrange them, and with which text or other images to combine them. *Bricolage*, reuse, recycling, appropriation of images, and adaptation according to the content taught represent some of the many modes of creation of these visual materials, which fulfilled aesthetic and epistemological, mnemonic, and socio-political intentions.[9] This essay focuses on the symbolic language applied to lecture notebooks, whose usage was encouraged by the Jesuits in similar academic practices such as *affixiones* and thesis prints. It also addresses the materiality of prints and their manipulation by students, who were at the same time producers and readers of their own manuscripts.

II. Engraved Series, Allegories, and Personifications

Louvain printers and booksellers played an important role in the selection of engravings inserted in college notebooks.[10] Michael Hayé (?–1676), Lambert Blendeff (c. 1650–1721), and Petrus Denique (1683–1746) produced and sold printed images.[11] These were recuperated from earlier publications not specifically created for the course matter taught in the faculty; most were reprints from existing (and sometimes well-known) plates or copies after older compositions, and marketed to student customers. Series of engravings were particularly favored by students as they induced a sense of continuity

9 About this aspect of *bricolage* in emblematics, see Russel, *Emblem and Authority*; *Idem*, *Emblematic Structures*, p. 7.

10 See Vanpaemel, "Louvain printers," pp. 245–248.

11 Hayé, Blendeff, and Denique had an artistic background and were not attached to the University as official book printers. Michael Hayé, a Flemish printmaker, draftsman and publisher, was admitted in 1661 to the Guild of Saint Luke in Antwerp, but moved in the following years to Louvain. In 1665, he made a request to Louvain University to be admitted as a member, in order to benefit from special academic privileges, but his request was denied (Vanpaemel, "Louvain printers," p. 245). In 1666, Hayé is mentioned in the university registers (*matricules*), which means that he paid his tuition fees and was therefore regularly enrolled at the university. In the following years, Hayé specialized in engravings for students, as will be discussed below. Lambert Blendeff, from Liège, was mostly known as a painter of religious scenes and portraits. He settled in Louvain around 1676. A year later he was appointed city painter, and in 1684 he was appointed painter or "iconograph" of the University (Van Even, "Blendeff," col. 470–471). Petrus Denique, the son of a well-known family of printers in Louvain, presumably bought the original engravings from his predecessors. He graduated from the Faculty of Arts in 1705, and held an official administrative position in the Faculty (Vanpaemel, "Louvain printers," p. 245).

through the chapters and subchapters of a single volume. Placed under or facing titles, they lent rhythm to reading, marking out the volume's different components and visually structuring the manuscript, or introducing popular topics, humor, or academic content.[12] In a treatise on the elements from a notebook of Henricus Joannes Van Cantelbecke (Ms. 207, *Physica*, 1669–1670), two humoristic series of prints are brought into dialog to illustrate each of the four elements (*Ignis, Aer, Aqua, Terra*). Under the heading *Aqua* the student placed an engraving by Crispijn II de Passe of a bearded doctor examining a urinal by candlelight opposite an etching by Jacques Callot depicting a fireworks display on the Arno River from the series *Capricci di varie figure* (Figure 6.1). The other elements are accompanied by similar juxtapositions. In a physics manuscript of Van Cantelbecke (Ms. 208), two series of prints by the same artists are paralleled to illustrate *Visus* and *Auditus*.[13] Likewise, in the notebook of Martinus Verdonck (Ms. K6), engravings of the senses with Dutch captions face corresponding handwritten Latin titles, accompanied by cut ornamental engravings: angels with cartouches; putti surrounding a pedestal; an eagle holding a banner. On fol. 228, the angel's outline is surrounded by fine pen strokes, and the drawing of a shield containing the title of the section, *De Auditu*, has been added under his hand.[14]

Collectable and marketable, printed suites of allegorical images depicting allegories were produced in large numbers in this period, broadcasting and fostering interest in new ideas and discoveries based on observation and hands-on experience.[15] In the domain of allegory, personifications were a recurring device in serial prints, easily recognizable, but often accompanied by captions and titles.[16] Taken up by the students and adapted to course subject matter, the addition of self-evident allegories to the philosophical content and its juxtaposition with other prints presenting a different mode of pictorial depiction (descriptive, satirical) fulfilled a mnemonic and didactic role.

12 Berger, *The Art of Philosophy*, p. 125.

13 In total, 68 etchings taken from three series by Jacques Callot (*Capricci di varie figure*, *Balli di Sfessania* and *Varie figure Gobbi*) were used by Van Cantelbecke, and are distributed into his six college notebooks (KUL, Ms. 204 to 209). Van Vaeck and Verberkmoes discuss the link between Callot's etchings and the subject matters ("Humor in collegedictaten," pp. 196–204).

14 Martinus Verdonck, *Physica*, 1669: http://lectio.ghum.kuleuven.be/lectio/items/show/1771 (accessed 3 July 2022).

15 Park, "Allegories of Knowledge," pp. 357–358.

16 See i.a. Baskins and Rosenthal, *Early Modern Visual Allegory*; Melion and Ramakers, *Personification*; Remmert, *Picturing the Scientific Revolution*.

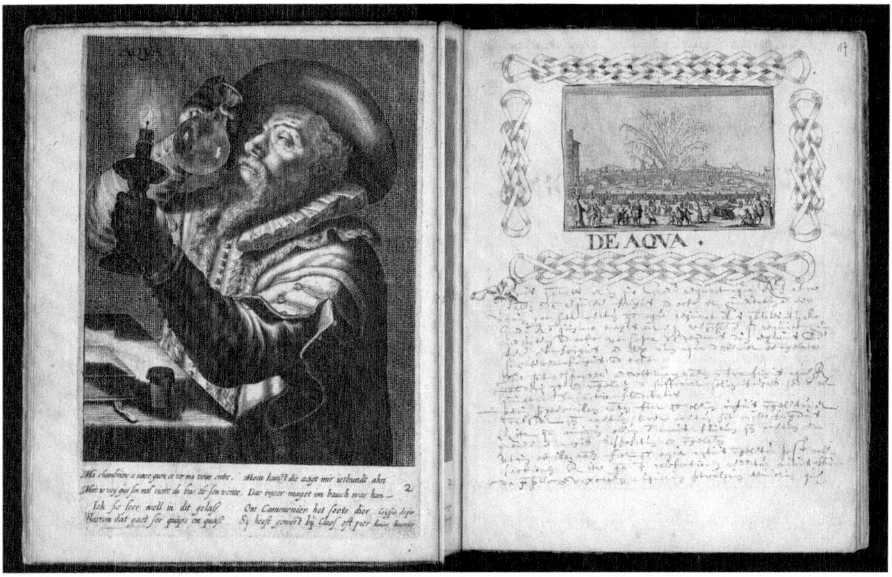

Figure 6.1 Unknown artist after Crispijn II de Passe, *Aqua*, from the series "De Vier elementen," c. 1620–1669; and Jacques Callot, *Crowd Gathered on the Bank of the Arno River, and Watching Fireworks Display*, from the series "Capricci di varie figure," c. 1617. Engraving inserted and etching pasted on in KUL, Ms. 207, *Physica*, Henricus Van Cantelbeke, 1669–1670, fols. 16v–17r.

III. Emblematic Language in Academic Productions: The Jesuit Influence

Printed series were also borrowed from emblematic literature. The most widespread sets of emblematic devices reused in Louvain *dictata* are the plates originally conceived for *Lux Evangelica* (1648 and later editions) and Carmelite father *Firmamentum symbolicum* (1652), by Jesuit father Hendrik Engelgrave and Sebastianus a Matre Dei, respectively. Each of the Latin sermons collected in *Lux Evangelica* is preceded by an emblem composed of one or two *motti*, an image and a short prose text summarizing the sermon structure; the copperplates were sold to Michael Hayé, who sold offprints on loose leaflets to university students.[17] *Firmamentum symbolicum* presents 50 laudative essays on the Virgin Mary combined with emblems. In total, 87 compositions from *Lux Evangelica* and 38 images from *Firmamentum*

17 For an overview of the recuperation process of the copperplates by Michael Hayé, see Van Vaeck, "Printjes," pp. 154–160. According to Van Vaeck, these series were used in Louvain lecture notebooks until the reform of the curriculum of the Faculty of Arts in 1764. There are some rare occurrences after this date and until the suppression of the university in 1797.

symbolicum were inserted in Louvain notebooks. From at least 1677, the engravings were published in a second state including the title of a course chapter or section.

An instructive case study in the role of emblematics in the academic milieu involves mapping the trajectory of a single motif, such as that of moths flying into the flame of a candle, traditionally used to figure the dangerous attractions of the court. Gilles Corrozet's *Hecatomgraphie* (Paris: Denis Janot, 1540) combined the moth composition with the title "War is Sweet to Those Who Have Never Experienced It,"[18] inflecting Erasmus's essay on the adage *Dulce bellum inexpertis* (IV, I, 1) meditating on the subject of war.[19] This theme appears in a collection of *affixiones* from the Jesuit college of Brussels, dated 1633 (Figure 6.2).[20] *Affixiones* were annual emblematic exhibitions produced by Jesuit pupils, preserved in the form of commemorative manuscripts made by professional artists. The practice, influenced by printed emblematic literature, was in use at the Brussels Jesuit College from 1630 to 1685. In the 1633 Brussels *Affixiones*, folio 105r displays the same visual device and the title *Qui amat periculum peribit in illo* ("Whoever dallies with danger will perish in it") from *Ecclesiasticus* 3:27.[21] *Affixiones* subjects could be selected according to topical issues, particularly political circumstances, as is the case here: in the 1630s, the Spanish Netherlands particularly suffered from the conflict with the Dutch Republic and the 1633 compendium offers numerous irenic arguments. The moth motif later illustrated amatory suffering.[22] In Otto Van Veen's *Emblemata aliquot selectiora amatoria* (Amsterdam: Willem Blaeu, 1618), emblem 33 features the moth, the title *Brevis & damnosa voluptas* ("How short this pleasure is, and how dangerous it is"), and a trilingual text warning

18 The text reads: "Butterflies will burn / To the candle that glows. / Like one wants to the battle to go / Who does not know how much war damages" (my translation). About this motif, see Russel, *Emblem and Authority*, p. 83.

19 In the first version of the *Adagiorum princeps* (1508), the brief commentary on this adage emphasizes inexperience, especially that of youth. From the following editions on, the text undergoes considerable development, and deals with the problem of war and peace. See Erasmus, *Dulce bellum inexpertis*.

20 *Emblemata de bello, peste, fame affixa a rhetoribus et poetis collegii Societatis Jesu Bruxellis anno 1633* ("Emblems on war, disease and hunger exhibited by the rhetoricians and poets of the College of the Society of Jesus in Brussels in the year 1633", KBr, Ms. 15.683). Another *affixio* of the volume combines the motto *Dulce inexpertis* with another *pictura*, a soldier fishing (fol. 23r). About that manuscript and its historical context, see Ems, "Mises en scènes," pp. 707–708; Porteman, *Emblematic Exhibitions*, pp. 34 and 86–87.

21 The composition is based on that of Otto Van Veen (see below).

22 Coleman, *An Illustrated Love "Canzoniere"*, pp. 57–58.

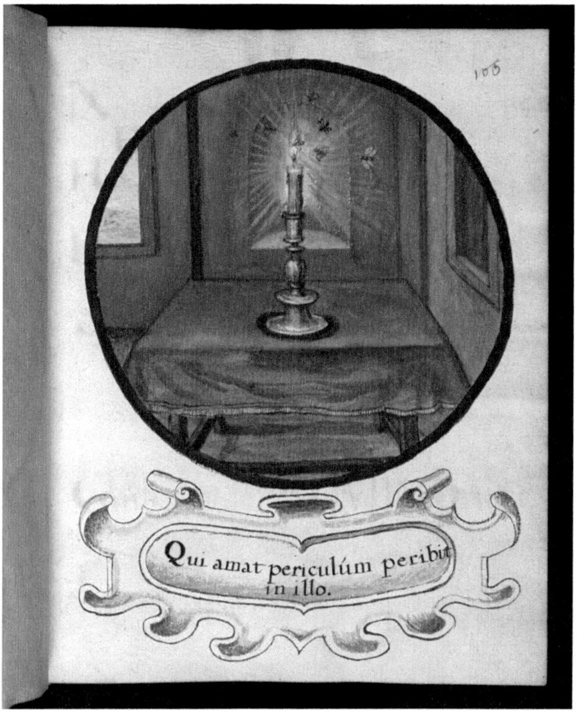

Figure 6.2 "Qui amat periculum peribit in illo," in *Emblemata de bello, peste, fame*, 1633. KBr, Ms. 15.683, fol. 105r.

against the risks of love. In Engelgrave's *Lux Evangelica* (emblem XLI, p. 367), the motif is accompanied by a quote from Luke 17:12, *A longe* ("At a distance"), and the motto *Tetigisse perisse est* ("Having touched is to lose oneself").[23] The text further develops these ideas: "The occasion to sin has to be greatly avoided, in order that we do not run towards immediate danger like it is customary for midges that are flying around a lamp."[24] When adapted from *Lux Evangelica* and reused by students for illustrating their manuscripts in physics, however, this motif was divested of these moralizing overtones. The engraving inserted in the physics notebook of Norbertus Josephus Ligiers is removed from its emblematic apparatus and accompanied by the caption *De Sphera activitatis*, that is, the extent to

23 "As he was going into a village, ten men who had leprosy met him. They stood at a distance and called out in a loud voice, 'Jesus, Master, have pity on us!'"

24 Longè peccandi occasio fugienda est: ne culicum in morem lucernam circumvolitantium, praesens periculum incurramus. § I. Qui amat periculum, in illo peribit. Eccle. 3. § II. Longè à periculis recedendum ("Leave perils far behind you"). § III. Peccandi occasiones rescindenda ("Occasions to sin must be cut off").

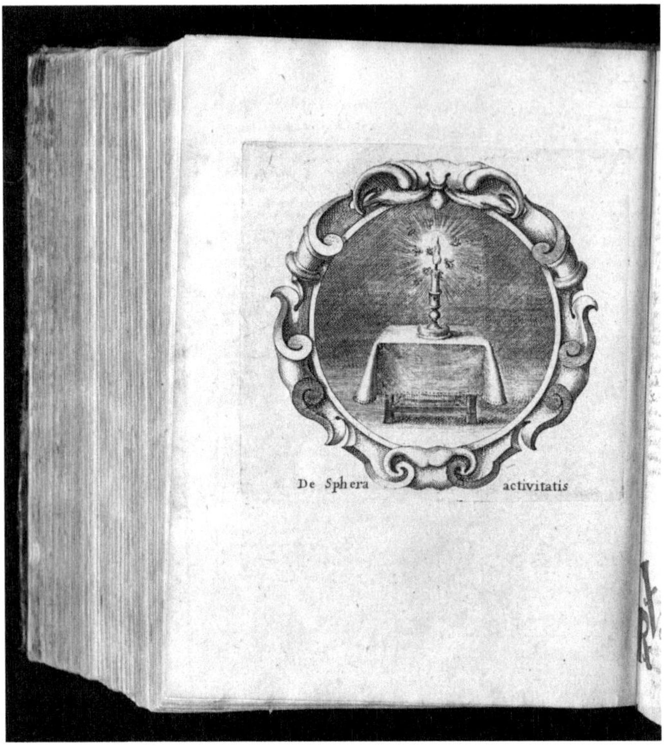

Figure 6.3 *De sphera activitatis*, in KUL, Ms. 354, *Physica*, Norbertus Josephus Ligiers, 1694–1695, fol. 484v.

which a body can act around itself (Figure 6.3).[25] The example given is that of fire that cannot heat up distant objects when they are out of the sphere of activity. In this context, the moth-to-the-flame motif was transformed into a visual mnemonic that resonated with a didactic message about proximity relative to affect.

Another case testifying to shifts from moral or religious significance toward one befitting the philosophical subject is that of the motif of the telescopic observation of the stars, which appears frequently in seventeenth-century texts. Optical instruments could be ascribed a range of meanings in period symbolic images.[26] In *Lux Evangelica* (1652, p. 397), the profane motto, taken from the story of Apollo and Daphne in Ovid's *Metamorphoses*

25 Ligiers, a member of the *Paedagogium Castri* (pedagogy of the Castle), studied at the Faculty of Arts during 1694–1695.

26 Blanchard (*L'optique du discours*, pp. 105–106) gives examples of emblems of *imprese* positively devised around the telescope. See also Hallyn, "Rhétorique."

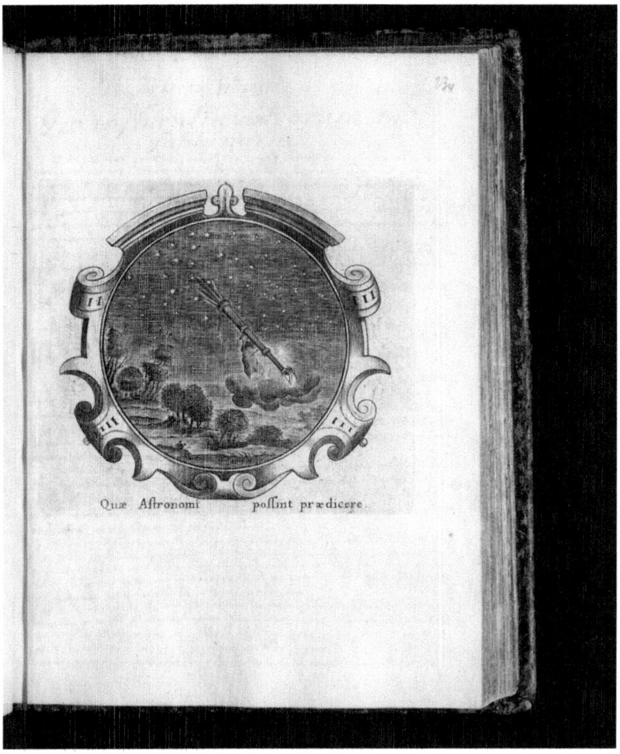

Figure 6.4 *Quae Astronomi possint praedicere*, engraving inserted in KUL, Ms. 213, *Physica*, Franciscus Josephus Gonzalez, 1739, fol. 234r.

(I:502), states: "whatever is hidden, he imagines even more beautiful." The biblical motto is a passage from the Song of Songs (4:1): "[How beautiful you are, my darling! Oh, how beautiful!] Your eyes are like doves behind your veil." The text below explains: "Celestial things are to be observed quite carefully by the eye of the mind, in order that all terrestrial matters are despised," gesturing to the hierarchy of spiritual over physical vision.[27] In the college notebook of Franciscus Josephus Gonzalez (*Physica*, 1739), the engraving from *Lux Evangelica* is accompanied by the title *Quae Astronomi possint praedicere* ("What astronomers might predict"), affirming that studying the celestial phenomena can evince consequences on earthly matters (Figure 6.4). Elsewhere in Gonzalez's manuscript, an engraving from *Firmamentum symbolicum* depicts the observation of stars using a

27 Si qua latent, meliora putat; [Quam pulchra es, amica mea! quam pulchra es! Oculi tui columbarum,] absque eo quod intrinsecus latet; and Oculo mentis, caelestia attentiùs contemplanda, quo terrena omnia despiciantur.

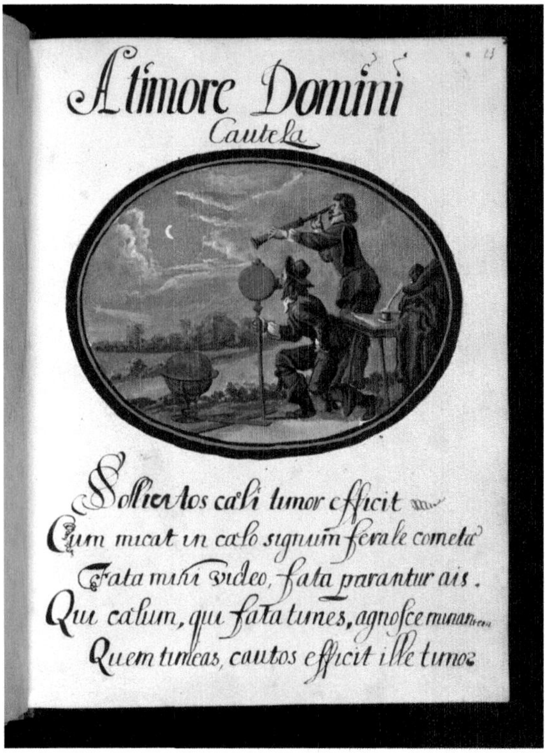

Figure 6.5 "A timore Domini cautela," in *Emblemata de timore Domini*. KBr, Ms. 20.306, 1647, fol. 23r.

telescope, with the caption *De aere* ("On air"). In Sebastianus a Matre Dei's book, the telescope was associated with the acquisition of knowledge, through an assimilation of the instrument's capacity to reveal things about the stars hidden to the eye with Mary's power of revelation of hidden truths to her devotees.[28] Similarly, an image in a 1647 *affixio* displays two men observing a comet on the horizon (Figure 6.5).[29] The *titulus* reads: *A timore Domini cautela* ("From the fear of the Lord comes vigilance"); the epigraph: *Sollicitos caeli timor efficit* ("Fear of Heaven makes them worried"); and the *carmen* or poem:

28 The captions read *Fistula dioptrica. Maria Erudiens* ("Perspective cylinder. Maria educating") and *Hac assumpta sciam* ("By taking this one with me, I will know"). I am very grateful to Aline Smeesters (UCLouvain) for her help in translating several Latin sentences in this essay.

29 *Emblemata de timore Domini affixa a rhetoribus gymnasii Societatis Jesu Bruxellis anno 1647. In timore Domini fiducia serenissimo Leopoldo archiduci* ("Emblems on the fear of the Lord exposed by the rhetoricians of the College of the Society of Jesus in Brussels, in the year 1647. In the fear of the Lord, there is trust for the most serene Archduke Leopold").

While the deadly omen of the comet shines in the sky,
you say: "I see the fates that are meant for me."
You who fear heaven and destiny, recognize the threat
of the one you must fear: this fear makes people vigilant.[30]

This ensemble foregrounds the personal device *Timore Domini* of the new governor of the Spanish Netherlands Archduke Leopold Wilhelm of Austria, while yoking celestial observation with admonitions to vigilance. An analogous *pictura*, but with a slightly different meaning, was used for the iconographic program of a printed booklet of poems. *Mathematicae scientiae emblematice expressae* was issued on the occasion of a public thesis defense at the Jesuit college of mathematics in Louvain in 1652, under the auspices of Leopold Wilhelm. The libretto was dedicated to the candidate, Theodore d'Immerseel, by his classmates. It contains four emblematic vignettes. In addition to the *pictura*, the page devoted to astronomy is provided with the motto *Mathematicè arcana detegenti* ("to unlock the secrets of mathematics") and the slogan *Industria pandit* ("laborious industry opens the way"), an allusion to the telescope (Figure 6.6).[31] The emblematic layout was inspired by the *Imago primi saeculi*, published twelve years earlier in 1640. This widely circulated book was published by Balthasar Moretus to celebrate the hundredth anniversary of the Society of Jesus.[32] It gathers 126 emblems engraved by Cornelis Galle the Elder, which were an important visual source for subsequent publications. Framing each image in *Mathematicae scientiae* is a cartouche adapted to the student and his patron, composed of two eagles holding a laurel leaf branch and a caduceus and Immerseel's coat of arms. The four compositions convey a double meaning, at the same time pedagogic and encomiastic, as they refer to the defended subjects in mathematics and offer praise toward the candidate, a potential future member of the socio-political elite.

All these examples testify to the communicative strength of the symbolic language of images in student notebooks that constituted a form of "applied

30 Cum micat in caelo signum ferale cometae / "Fata mihi video, fata parantur," ais. / Qui caelum, qui fata times, agnosce minantem / Quem timeas: cautos efficit ille timor. Transl. by Grégory Ems in *L'emblématique au service du pouvoir*, vol. 2, 25. On the comet as an omen of misfortune, see Le Boeuffle, *Astronomie*, pp. 97–98, *s.v. Cometa*.

31 Bousquet-Bressolier, "Pédagogie," pp. 164–166. See also, about the printed material related to this defense, de Mûelenaere, *Thesis prints*, cat. 13.1–13.6.

32 De Tollenaer, *Imago primi saeculi*.

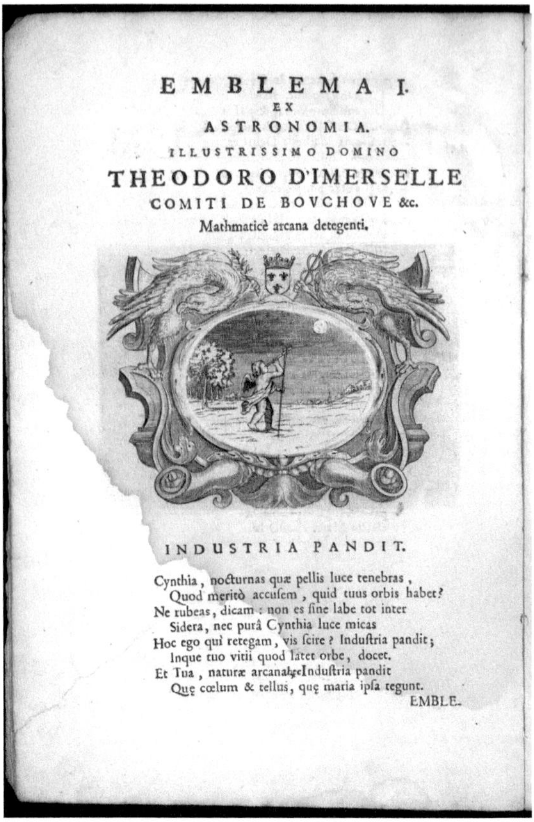
EMBLEMA I.
EX
ASTRONOMIA.
ILLUSTRISSIMO DOMINO
THEODORO D'IMERSELLE
COMITI DE BOVCHOVE &c.
Mathmaticè arcana detegenti.

INDUSTRIA PANDIT.

Cynthia, nocturnas quæ pellis luce tenebras,
Quod meritò accuſem, quid tuus orbis habet?
Ne rubeas, dicam: non es ſine labe tot inter
Sidera, nec purâ Cynthia luce micas
Hoc ego quì retegam, vis ſcire? Induſtria pandit;
Inque tuo vitii quod latet orbe, docet.
Et Tua, naturæ arcana hæc Induſtria pandit
Quę cœlum & tellus, quę maria ipſa tegunt.
EMBLE-

Figure 6.6 "Emblema I. Ex astronomia," in *Mathematicae scientiae emblematice expressae, in-folio* booklet of 20 pp. dedicated to Theodore d'Immerseel. Louvain: André Bouvet, 1652. UGent Library, inv. BIB.191G038.

emblematics" marshaled in service of university education.[33] The varied uses of the emblem outlined here are consistent with the emblem as an ornament whose main characteristic was its "plasticity" and "removability," even before its rise as a literary genre.[34] Perhaps more surprising, however, is the use of an old visual mode of representation to depict the new sciences, based on a visual account of reality. Symbolism, however, often inhabits

33 Spica, "Les jésuites et l'emblématique," p. 636. Similarly, Peter Daly divides emblem and impresa literature into six groups, among which illustrated emblem books in the strict sense, but also printed evidence of the use of emblem and impresa in the material culture. This last category deals with publications that were emblematically structured or that recorded emblematically devised events: ephemeral architecture of entries, consecration of bishops, plays, marriage and funeral solemnities, academic theses, etc. (Daly, "Emblematic Publications," pp. 249–250).

34 Ems, "Mises en scènes," p. 711; Châtelain, *Livres d'emblèmes*, pp. 17–19.

scientific imagery produced in the Southern Netherlands.[35] To a large extent, this can be explained by the influence Jesuits exerted in the visual culture of knowledge in this region. The Society of Jesus set up a pedagogy in their schools that exploited the creativity and enthusiasm of the youth to inculcate in them values and convictions, formulated through a playful and pleasant but also extremely powerful and effective mode of expression: emblematics.[36] The order promoted emblematic practice as an exercise in rhetoric and mnemonics, as education was more effective when interesting and amusing. Wit and refinement were more over important social graces to be taught in Jesuit colleges.[37] More generally, the Jesuits made a wide use of emblematics as a vehicle for knowledge.[38]

IV. Cut and Paste Practice: The Materiality of Prints

The use of emblematic devices adapted to a university context shows both the circulation of such printed material and attending shifts in meaning. In the examples cited above students' creativity was limited by the choice of prints offered for sale by the booksellers. More exceptionally, students could also cut engravings from personal printed books or booklets to ornament their lecture notes, affording them more inventiveness in arranging their manuscripts. This is the case of the notebook in physics of student Norbertus Ligiers, who reused engravings in medallions cut from two printed sources.[39] The first, *Laudatio funebris Ferdinandi III Austriaci* produced by the Louvain Jesuit College, contains the funeral oration for Ferdinand III of Austria (d. 1657), followed by verses and thirteen imperial laudatory emblems by Frederik Bouttats the younger.[40] Ten engravings are pasted on the pages of Ligiers's manuscript with a title written in pen under each image, such that the viewer can lift each printed medallion and discover its verso and the manuscript folio underneath.[41] The titles are

35 Dekoninck and Guiderdoni, "Knowledge in Transition," pp. 279–280.

36 Ems, "Mises en scènes," p. 711. See i.a. Porteman, "Jesuit Teaching and Education"; Loach, "Teaching of Emblematics"; Saunders, "Make Pupils Do Themselves."

37 Saunders, "Make Pupils Do Themselves," p. 188.

38 Dekoninck and Guiderdoni, "Knowledge in Transition," pp. 282–284.

39 Norbertus Ligiers, *Physica*, 1694–1695: http://lectio.ghum.kuleuven.be/lectio/items/show/631.

40 De Trauttmansdorff, *Laudatio funebris Ferdinandi III Austriaci* (Antwerp: s.n., 1657). See Daly, "Emblematic Publications," pp. 260, 277, n33–34; Daly and Dimler, *Jesuit Series*, p. 158, no. J.1358; De Schepper, "Amblemata," p. 1091 (no. 7); Sommervogel, *Bibliothèque*, col. 1438.

41 Emblems I, II and V (p. 49) of the *Laudatio* are not included in Ms. 354.

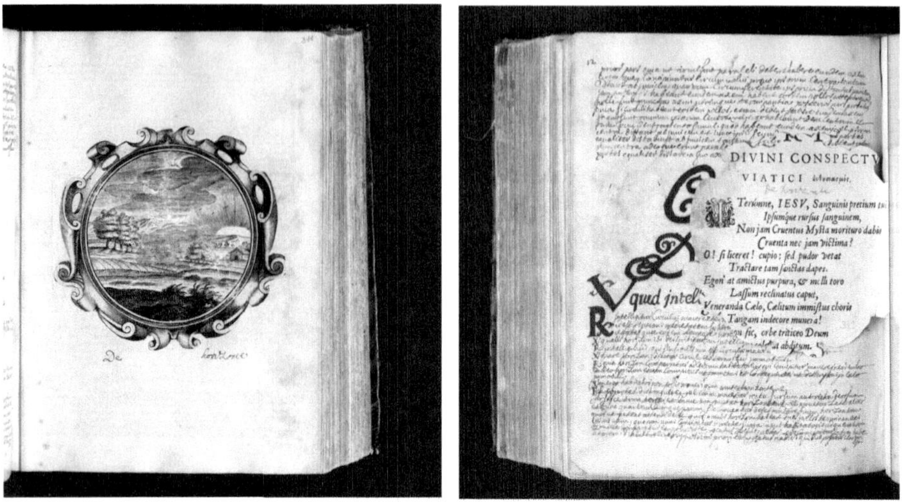

Figure 6.7 *De horizonte*, in KUL, Ms. 354, *Physica*, Norbertus Josephus Ligiers, 1694–1695, fol. 311r (LEFT) and 310v (RIGHT).

repeated in pencil on the back of the engravings, probably to aid the student in placing each emblem in his manuscript. The engraving with the caption *De horizonte* (f. 311r) taken from emblem VIII (p. 55) of the *Laudatio* and excised from its original encomiastic context, here illustrates the chapter of a treatise on the sphere (Figure 6.7).[42] Elsewhere, some of these vignettes are glued one against the other.[43] For instance, folio 329r contains three engravings pasted one on top of the other above the caption *De Zonis* in handwriting. A first composition, depicting the sun with zodiacal circle, is taken from emblem IV (p. 48). The engraving, whose reverse side is blank, can be lifted, revealing a second illustrated cartouche, the sun presented by a putto in a landscape (from emblem III, p. 47). The eagles and crown surrounding the image are partially cut to match the shape of the first cartouche. On the verso of this second image is pasted a third engraving, the same as the first composition with the zodiacal circle. This complex layout bears witness to the students' handiwork to train their minds. It

42 The engraving *De sensu interio* (f. 222r) comes from emblema XII (p. 65) of the *Laudatio*; *De meridiano* (f. 314v), from emblema V (p. 52); *De occasu astrorum* (f. 336v), from emblema IX (entitled "Major in occasu", p. 62).

43 On the back of *De tropicis* (f. 323v), taken from emblema VII (p. 54), is glued another medallion (emblema VI, p. 53). Similarly, two engravings are glued one against the other on f. 365v. In the first composition, the sun and full moon send their rays toward the terrestrial orb (emblema XI, p. 64); and in the second image, the sun with a zodiacal circle illuminates a landscape (emblema X, p. 63). The title *De Cœlis* is added in handwriting.

can be compared to the fugitive prints process, an innovative assemblage of printed cut-outs and paper flaps that changed the two dimensional image to an animated one.[44] It first appeared in Germany around 1538 and was widely copied throughout Europe, including in student notebooks.[45] By superimposing engravings that could be manipulated to make other visual devices visible, the same effect of surprise and discovery was sought in Ligiers's manuscript.

The second publication sourced for images for Norbertus Ligiers's physics notebook, *Argumentum salutationis gratulatoriae dicatae*, records the visit of Francisco de Mello (1597–1651), governor of the Southern Netherlands and Burgundy, to the Brussels Jesuit College on 10 February 1642.[46] This tribute to the governor, featuring De Mello's coat of arms and five engraved emblems, was signed by the "nobilitas studiosa" (fol. A2r), testifying to the close ties between the College and the Brussels court.[47] The five engravings pasted on the pages of the manuscript are accompanied by the captions: *De causis* ("on causes," f. 3r and f. 52v), *De causa materiali* ("on material cause," f. 59r), *An omnis sensatio fiat in cerebro?* ("But is all knowledge in the brain or not?," f. 144r), *De memoria* ("on memory," f. 224r) (Figure 6.8). In the *Argumentum salutationis*, the latter composition, which features a putto proffering crowns toward a heavenly glory, marshaled by Mercury, is accompanied by a laudatory text memorializing De Mello's fame that cloaks dynastic encomium in celestial metaphor: "You go as a legate to the pope and the Italian tetrarchs, Caesar sees you arrive at his feet. What a great light shines on your family! 'Because I announce great things, I shine': this sentence is yours, Mello, and not only yours, Star."[48] Perhaps inflecting this message and guided by the inclusion of Mercury – a figure associated

44 Moore, "Paper Cuts," p. 54.

45 It is, for instance, the case of *Cor Humanum* ("Human heart") edited by Michael Hayé, in the same manuscript (f. 113r). This engraving copied from a Cartesian textbook was frequently used in Louvain notebooks. Flaps could be lifted to let appear the internal working of the heart. About this print, see Berger, *The Art of Philosophy*, pp. 134–135; de Mûelenaere, "Illustrated Lecture Notebooks."

46 De Mello, *Argumentum salutationis gratulatoriae dicatae* (Brussels: Jan Mommaert, 1642). See Vermeir, *In staat van oorlog*, pp. 237–276.

47 These relationships were amply highlighted by Porteman, *Emblematic Exhibitions*. On the work's emblems see Daly, "Emblematic Publications," pp. 259–260; De Schepper, "Amblemata," p. 1091 (no. 8).

48 The motto "Quia nuncio magna, splendeo" printed under the *pictura* is repeated on the facing page of the libretto, with the text: "Pontifice Legatus adis, Italosque Tetrarchas, Te vidit pedibus Caesar adesse suis. Hinc generi lux quanta tuo! *Quia nuntio magna, Splendeo, Mello,* tuum est, non modò, *Stella,* tuum."

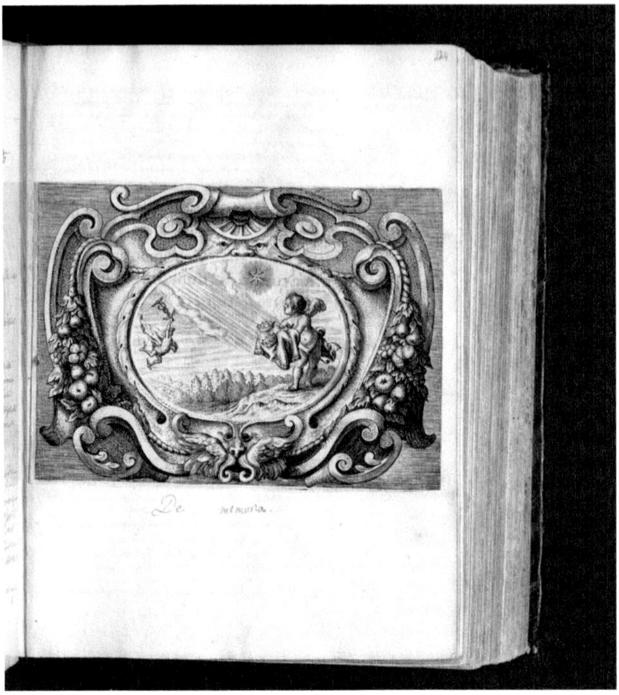

Figure 6.8 *De memoria*, in KUL, Ms. 354, *Physica*, Norbertus Josephus Ligiers, 1694–1695, fol. 224r.

with eloquence, reason, and learning, and sometimes with memory – the student placed this composition in a chapter on memory.[49]

Interpolations of images in lecture notebooks fulfilled manifold ornamental, ludic, and didactic functions. Ordering and manipulating prints gave students creative free rein: they could color and paint prints; draw after engravings; create dynamic layouts; or personalize existing compositions, stimulating the imagination and providing a fertile ground for memorization of philosophical content. These practices of inserting or trimming and pasting prints, putting them in dialog with both text and other images, were common in fifteenth- and sixteenth-century manuscripts. The religious devotional functions of the print medium gave rise to a tactile interplay between early prints and their collectors, who often altered their prints' format or coloring, and grouped them within albums, manuscripts, or printed books to suit

49 Hall, *Subjects and Symbols*, p. 207. According to Cesare Ripa, who associates Mercury with the firmness of speech, "although the words are light and seem to fly, if spoken with weight & judiciously, they are sure to make a strong impression in the memory" (*Iconologie*, p. 72).

their own pious mnemonic purposes.[50] The proliferation of interactive prints and illustrated single sheets in the sixteenth and seventeenth centuries appealed to an active personal, tactile, and auto-didactic viewership.[51] It also played a role in knowledge communication processes, as handling images and their dynamic interaction in a pedagogical context concretized course subject into a more comprehensible form.[52] By manipulating images when compiling their manuscripts, Louvain students could play with the multiple meanings conveyed by visual materials and experiment with new possibilities; the printed images' plasticity enabled their adaptation to accommodate different users, with the addition of new titles or details drawn in pen, application of color, association with other visual material.[53]

V. Conclusion

The various academic documents analyzed here, all of which belong to the same sphere of learning, demonstrate the extent to which student life constituted an important source for the creation of images. They were designed along the lines of the combinatorial art developed in early modern European visual culture, by reusing existent forms or by adapting visual patterns and conventional iconography. Illustrated college *dictata* were compiled during or after lecture courses, *affixiones* were an outcome of the school exercises (*exercitationes*), laudatory poems printed on decorated broadsheets and booklets were issued on the occasion of public *disputationes*. Designed to promote an excellent command of *eloquentia*, the *exercitationes* were an important educational tool in the *humaniora* to drill the memory and to put the theories of the *ars rhetorica* and the *ars poetica* into practice.[54] A characteristic of Jesuit teaching was the combination of these exercises with visual genres like allegories, emblems, hieroglyphs, devices, and riddles. College *dictata* produced at university were, in turn, steeped in Jesuit culture

50 Karr Schmidt, *Interactive Printmaking*, p. 11. See also Areford, *Viewer and the Printed Image* and Rudy, *Image, Knife and Gluepot*, pp. 300–301.

51 Karr Schmidt, *Interactive Printmaking*, pp. 1 and 15. For a state of the art on the materiality of printed words and images, see Larkin and Pon, "Materiality," pp. 1–6.

52 For instance, D. Baird highlighted the value of instruments as "crafted artifacts" in which "visual and tactile thinking and communication are central to their development and use." He insisted on the importance of the hands-on creation process to communicate meaning (*Thing Knowledge*, pp. XV–XVI).

53 Dackerman ed., *Pursuit of Knowledge*, p. 22.

54 Porteman, *Emblematic Exhibitions*, p. 11.

that perpetuated in higher education. They testify to the creativity of the students who integrated visual representations in their volumes for aesthetic, didactic, and mnemonic purposes, and contribute to a nuancing of the notional "epistemic image" increasingly invoked in early modern studies.[55] The concept can be capacious ("any image that was made with the intention of expressing, demonstrating or illustrating a theory"),[56] or more restrictive ("an image made with the intent not only of depicting the object of scientific inquiry but also of replacing it [...], a working object of science, a stand-in for the too-plentiful and too-various objects of nature").[57] While the engravings added in Louvain college notebooks do not correspond to the latter more narrow definition, they nonetheless transmitted philosophical knowledge through ancient allegorical and emblematic modes of representation and through handiwork experience.[58]

Bibliography

A Matre Dei, Sebastianus. *Firmamentum Symbolicum*. Lublin [=Antwerp]: Georg Forster, 1652.

Areford, David S. *The Viewer and the Printed Image in Late Medieval Europe*. Farnham: Ashgate, 2010.

Baird, Davis. *Thing Knowledge: A Philosophy of Scientific Instruments*. Berkeley, CA: University of California Press, 2004.

Baskins, Cristelle and Lisa Rosenthal, eds. *Early Modern Visual Allegory: Embodying Meaning*. Aldershot: 2007.

Berger, Susanna. *The Art of Philosophy: Visual Thinking in Europe from the Late Renaissance to the Early Enlightenment*. Princeton, NJ: Princeton University Press, 2017.

Blair, Ann M. "The Rise of Note-Taking in Early Modern Europe." *Intellectual History Review* 20, no. 3 (2010): 303–316.

Blanchard, Jean-Vincent, *L'Optique du Discours au XVIIe Siècle. De la Rhétorique des Jésuites au Style de la Raison Moderne* (*Descartes, Pascal*). Sainte Foy: Presses de l'Université Laval, 2005.

Bousquet-Bressolier, Catherine. "Pédagogie de l'Image Jésuite. De l'Image Emblématique Spirituelle aux *Emblemata* Mathématiques." In *François de Dainville*

55 Marr, "Knowing Images," pp. 1005–1006.

56 Lüthy and Smets, "History of Scientific Imagery," p. 399, n2.

57 Daston, "Epistemic Images," pp. 17–18.

58 For the "epistemology of handwork" see Smith, *The Body of the Artisan*, p. 28.

S.J. (1909–1971). Pionnier de l'Histoire de la Cartographie et de l'Éducation, edited by Catherine Bousquet-Bressolier, pp. 143–166. Paris/Geneva: Publications de l'École Nationale des Chartes, 2004.

Châtelain, Jean-Marc. *Livres d'Emblèmes et de Devises. Une Anthologie (1531–1735)*. Paris: Klincksieck, 1993.

Coleman, Dorothy Gabe. *An Illustrated Love "Canzoniere": The* Délie *of Maurice Scève*. Geneva: Slatkine, 1981.

Dackerman, Susan, ed. *Prints and the Pursuit of Knowledge in Early Modern Europe*. Cambridge, MA: Harvard Art Museum, 2011.

Daly, Peter M., and G. Richard Dimler, eds. *The Jesuit Series. Part Five (P–Z)*. Toronto: University of Toronto Press, 2007.

Daly, Peter. "Emblematic Publications by the Jesuits of the Flanders Belgium Province to the Year 1700." In *The Jesuits and the Emblem Tradition*, edited by John Manning and Marc Van Vaeck, pp. 249–278. Turnhout: Brepols, 1999.

Daston, Lorraine. "Epistemic Images." In *Vision and Its Instruments: Art, Science, and Technology in Early Modern Europe*, edited by Alina Payne, pp. 13–35. University Park, PA: Pennsylvania State University Press, 2015.

De Mello, Francisco. *Argumentum Salutationis Gratulatoriae Dicatae Excellentissimo Domino D. Francisco de Mello....* Brussels: Jan Mommaert, 1642.

De Mûelenaere, Gwendoline. *Early Modern Thesis Prints in the Southern Netherlands: An Iconological Analysis of the Relationships between Art, Science and Power.* Leiden: Brill, 2022.

De Mûelenaere, Gwendoline. "The Art of Learning: Illustrated Lecture Notebooks at the Old University of Louvain." In *The Epistemic Function of Vision in Science*, edited by Giulia Giannini, Enrico Giannetto, and Matteo Valleriani. Cham: Springer, forthcoming.

De Ridder-Symoens, Hilde. "De Praktijk van Kennisoverdracht aan de Europese Universiteiten voor 1800." In *Ex Cathedra. Leuvense Collegedictaten van de 16de tot de 18de Eeuw*, edited by Geert Vanpaemel *et al.*, pp. 7–21. Louvain: Universiteitsbibliotheek, 2012.

De Schepper, Marcus. "Amblemata voor de Uldinge. Een Zestigtal 'Onbekende' Zuid-Nederlandse Embleemdrukken in de Brusselse Koninklijke Bibliotheek." In *The Stone of Alciato. Literature and Visual Culture in the Low Countries. Essays in Honour of Karel Porteman*, edited by Marc van Vaeck, Hugo Brems, and Geert H.M. Claassens, pp. 1085–1118. Leuven: Peeters, 2003.

De Tollenaer, Jean, Jean Bolland, Sidronius de Hossche, and Godefroid Henschen. *Imago Primi Saeculi Societatis Iesu*. Antwerp: B. Moretus, 1640.

De Trauttmansdorff, Ferdinandus Ernestus. *Laudatio Funebris Ferdinandi III Austriaci....* S.l.: s.n., 1657.

Dekoninck, Ralph and Agnès Guiderdoni. "Knowledge in Transition: A Case of 'Scientific Emblematics' (Ciermans and Vaenius) at the Turn of the Seventeenth Century." In *Embattled Territory. The Circulation of Knowledge in the Spanish Netherlands*, edited by Sven Dupré *et al.*, pp. 279–297. Ghent: Academia Press, 2015.

Ems, Grégory. "Variété des 'Mises en Scène' Emblématiques dans la Province Jésuite Flandro-Belge au XVIIe Siècle." *Dix-Septième Siècle* 269, no. 4 (2015): 705–734.

Ems, Grégory. *L'Emblématique au Service du Pouvoir. La Symbolique du Prince Chrétien dans les Expositions Emblématiques du Collège des Jésuites de Bruxelles sous le Gouvernorat de Léopold-Guillaume (1647–1656)*, 2 vols. Louvain-la-Neuve: Presses Universitaires de Louvain, 2016.

Engelgrave, Hendrik. *Lux Evangelica sub Velum Sacrorum Emblematum Recondita in Anni Dominicas. Selecta Historia et Morali Doctrina*. Antwerp: Widow of Jan Cnobbaert, 1648.

Erasmus, Desiderius. *Dulce Bellum Inexpertis*. Text edited and transl. by Yvonne Remy and René Dunil-Marquebreucq. Brussels: Latomus, 1953.

Hall, James. *Dictionary of Subjects and Symbols in Art*. London: John Murray, 1996.

Hallyn, Fernand. "Rhétorique de la lunette." *Littératures classiques* 11 (1989): 13–23.

Karr Schmidt, Suzanne. *Interactive and Sculptural Printmaking in the Renaissance*. Leiden: Brill, 2018.

Karr Schmidt, Suzanne. *Altered and Adorned: Using Renaissance Prints in Daily Life*. New Haven, CT: Yale University Press, 2011.

Larkin, Graham, and Lisa Pon. "Introduction: the Materiality of Printed Words and Images." *Word & Image* 17, no. 1–2 (2001): 1–6.

Le Boeuffle, André. *Astronomie. Astrologie. Lexique latin*. Paris: Picard, 1987.

Loach, Judy. "The Teaching of Emblematics and Other Symbolic Imagery by Jesuits within Town Colleges in Seventeenth and Eighteenth-Century France." In *The Jesuits and the Emblem Tradition*, edited by John Manning and Marc van Vaeck, pp. 161–186. Turnhout: Brepols, 1999.

Lüthy, Christoph, and Alexis Smets. "Words, Lines, Diagrams, Images: Towards a History of Scientific Imagery." *Early Science and Medicine* 14 (2009): 398–439.

Marr, Alexander. "Knowing Images." *Renaissance Quarterly* 69, no. 3 (2016): 1000–1013.

Melion, Walter S., and Bart Ramakers, eds. *Personification. Embodying Meaning and Emotion*. Leiden: Brill, 2016.

Moore, Rosemary. "Paper Cuts. The Early Modern Fugitive Print." *Object* 17 (2015): 54–76.

Park, Katharine. "Allegories of Knowledge." In *Prints and the Pursuit of Knowledge in Early Modern Europe*, edited by Susan Dackerman, pp. 358–409. Cambridge: Harvard Art Museum, 2011.

Porteman, Karel. *Emblematic Exhibitions (Affixiones) at the Brussels Jesuit College (1630–1685): A Study of the Commemorative Manuscripts (Royal Library, Brussels).* Turnhout: Brepols, 1996.

Porteman, Karel. "The Use of the Visual in Classical Jesuit Teaching and Education." *Paedagogica Historica* 36, 1 (2000): 179–196.

Remmert, Volker R. *Picturing the Scientific Revolution: Title Engravings in Early Modern Scientific Publications.* Philadelphia, PA: Saint Joseph's University Press, 2011.

Ripa, Cesare. *Iconologie où les Principales Choses qui Peuvent Tomber dans la Pensée Touchant les Vices sont Représentées*, edited by Jean Baudouin. Paris: s.n., 1643.

Rudy, Kathryn. *Image, Knife and Gluepot. Early Assemblage in Manuscript and Print.* Cambridge: Open Book Publishers, 2019.

Russel, Daniel. "The Emblem and Authority." *Word & Image* 4, 1 (1988): 81–87.

Russel, Daniel. *Emblematic Structures in Renaissance French Culture.* Toronto: University of Toronto Press, 1995.

Saunders, Alison. "Make Pupils Do Themselves: Emblems, Plays and Performances in Jesuit Colleges Seventeenth Century." In *The Jesuits and the Emblem Tradition*, edited by John Manning and Marc van Vaeck, pp. 187–206. Turnhout: Brepols, 1999.

Smeyers, Maurits. "Een Collegeschrift van de Oude Leuvense Universiteit (1481–1482). Een Codicologisch en Iconografisch Onderzoek. Bijdrage tot de Studie van het Universitaire Onderricht Tjdens de Middeleeuwen." *Arca Lovaniensis* 4 (1975): 243–303.

Smith, Pamela H. *The Body of the Artisan: Art and Experience in the Scientific Revolution.* Chicago, IL: University of Chicago Press, 2004.

Sommervogel, Charles. *Bibliothèque des Écrivains de la Compagnie de Jésus.* Vol. 7 (Roeder-Thonhauser). Brussels/Paris: Schepens/Picard, 1896.

Spica, Anne-Élisabeth. "Les Jésuites et l'Emblématique." *Dix-Septième Siècle* 237, no. 4 (2007): 633–651.

Van Even, Edward. "Blendeff (Lambert)." In *Biographie Nationale vol. 2*, col. 470–471. Brussels: Thiry, 1868.

Van Vaeck, Marc en Johan Verberkmoes. "Humor in Collegedictaten." In *Ex Cathedra. Leuvense Collegedictaten van de 16de tot de 18de Eeuw*, edited by. Geert Vanpaemel *et al.*, pp. 187–121. Louvain: Universiteitsbibliotheek, 2012.

Van Vaeck, Marc. "Printjes [...] tot Ciraet van de Dictata Reeksen van Embleemgravures in de Leuvense Collegedictaten." In *Ex Cathedra. Leuvense Collegedictaten van de 16de tot de 18de Eeuw*, edited by Geert Vanpaemel *et al.*, pp. 151–168. Louvain: Universiteitsbibliotheek, 2012.

Vanpaemel, Geert. "The Louvain Printers and the Establishment of the Cartesian Curriculum." *Studium* 4, no. 4 (2011): 241–254.

Vermeir, René. *In Staat van Oorlog. Filips IV en de Zuidelijke Nederlanden, 1629–1648.* Maastricht: Shaker Publishing, 2001.

Wouk, Edward H. "Toward an Anthropology of Prints." In *Prints in Translation, 1450–1750. Image, Materiality, Space*, edited by Suzanne Karr Schmidt and Edward H. Wouk, pp. 1–18. New York: Routledge, 2017.

About the Author

Gwendoline de Mûelenaere is a postdoctoral researcher in history of art at Ghent University. Her current project focuses on illustrated lecture notebooks from the Old University of Louvain. Her book *Early Modern Thesis Prints in the Southern Netherlands* was published by Brill in 2022.

7. Visual Worlds on Early Modern Scientific Instruments: Types and Messages

Julia Ellinghaus and Volker R. Remmert[1]

Abstract

This chapter surveys early modern scientific instruments adorned with images. These images *per se* have no relevance for the instruments' use. To date, such "instrumental imagery" and its contexts have only sporadically been analyzed. This paper presents methods aiming at a systematic analysis of this visual material to inquire after its role in the various contexts of the adorned instruments (genesis, function, use) and importance for crafting histories of success and relevance within the emerging field of the sciences. The iconography points to quite a few significant topics: statements of specific positions in theoretical debates; mediation and illustration of knowledge, in particular by picturing the usability of the instruments; or the role of instruments as patronage artefacts.

Keywords: iconography, images, scientific instruments, scientific revolution, visual strategies

I. Introduction

Historical accounts conventionally highlight two central aspects of the *Scientific Revolution* of the sixteenth and seventeenth centuries: on the one hand, that this is a period of, arguably, never before imagined dynamics in the "sciences," embodied in such iconic figures as Copernicus, Galileo, and Newton, and giving rise to countless myths about progress. On the other

1 This essay draws on Remmert, "Analyse des imageries."

Noyes, R.S. (ed.), *Reassessing Epistemic Images in the Early Modern World.*
Taylor & Francis Group, 2023
DOI 10.5117/9789463723350_CH07

hand, that it signifies an essential step in the development of present-day scientific disciplines and a milestone in European history, marking the beginning of the scientization of modern societies. A fundamental trait of the *Scientific Revolution* was the growing epistemic value of practical knowledge in general and of scientific instruments in particular, allowing for a significant extension of the realm of experience.[2] Whereas astrolabes, armillary spheres, air pumps, microscopes, telescopes, and many other devices became increasingly important for the study of nature, they had not yet become standardized and impersonal tools.

That these usually unique items could, by means of function and design, mediate between scholars and social elites means that imagery incised in or imprinted on the surface of such instruments played a crucial role.[3] In fact, a great number of early modern instruments are adorned with such images that *per se* have no relevance to the object's practical use: for instance, the depiction of Atlas and Hercules on an astrolabe by Johannes Praetorius (1568, Dresden) or that of tradition in astronomy and geometry from the patriarchs to Copernicus on Jost Bürgi's astronomical equation clock (1591, Kassel). Despite the relative ubiquity of such imagery on instruments, this phenomenon and its context have only sporadically been analyzed. This chapter outlines a research project that aims at the first systematic analysis of the multifaceted visual material on instruments, interrogating its role in various contexts (genesis, function, use) and its importance for establishing or supporting narratives of success and relevance within the emerging field of the sciences.[4]

To begin with, we address the issues of what constitutes an early modern "scientific instrument" – a generic terminology that emerged only in the nineteenth century – and what can be gathered from studying their visual worlds. What follows expands on Gerard Turner's conceptualization of [metal] scientific instruments as *ideas made of brass* to include instruments irrespective of material as *materialized and shaped ideas* (thus including ephemeral devices of wood or paper). In the early modern period, scientific instruments were instead referred to as mathematical, optical, or philosophical instruments,[5] and included a broad spectrum from sundials and compasses, telescopes and microscopes, to air pumps and barometers. Paolo

2 For the role of practical knowledge, see Valleriani, *The Structures of Practical Knowledge*.

3 Borrelli, Korey, and Remmert, *Iconography on Early Modern Scientific Instruments*.

4 The project is funded by the German Research Foundation (2018–2021): *Iconography on Early Modern Scientific Instruments*.

5 Taub, "On Scientific Instruments," p. 337 and *Idem*, "Introduction." See also Bennett, "Early Modern Mathematical Instruments."

Brenni has classified scientific instruments according to three functional categories pertaining to the role of scientific instruments in "technical" or "scientific" contexts (however defined): instruments for research and investigation; professional instruments; and didactic instruments.[6]

According to Anthony Turner, apart from such functions "intrinsic to [the instruments] themselves," scientific instruments generally also have "a social function that may be linked with but may also be independent of their technical purpose."[7] This social function is closely related to the fact that often scientific instruments gained the status of "an *objet d'art*, in which the extremely elegant decorations at times suffocated the original function."[8] That scientific instruments as ideas made of brass were extremely suitable and fashionable as "collectors' items and objects of conspicuous consumption" is also relevant to the emerging trade of instrument makers in the early modern period.[9] Thus, at least two more interrelated categories must be added to Brenni's classification: (4) instruments as patronage artefacts or status symbols, produced in and for courtly contexts in Europe and beyond; and (5) instruments as material demonstrations of manufacturing skill and expertise. The latter evinces a process of self-fashioning and marketing on the part of instruments makers to serve economic interests in the trade and design successful patronage strategies at courts or similar elite milieux.[10]

Albert van Helden and Thomas Hankins underscore that instruments, while conferring authority and being able to "act as bridges between natural science and popular culture", were also created for specific audiences, such that "patronage [...] shapes the design and intended use," making these "highly culture bound" objects often entangled in a dense web of manifold potential functions and uses understandable only by taking into account their attending imagery.[11] By the same token, considering instruments' functions and uses often elucidates their imagery.[12]

That early modern *ideas made of wood* or those of paper survive in fewer numbers and have less frequently found their way into collections and museums (more often found today in libraries) poses problems of analysis,

6 Brenni, "Art and Science," p. 122.

7 Turner, "From Mathematical Practice," p. 135.

8 Brenni, "Art and Science," p. 127.

9 Johnston, "Mathematical Practitioners," p. 320.

10 Bennett, "Shopping for Instruments"; Morrall, "Entrepreneurial Craftsmen"; Biagioli, "From Print to Patents."

11 Van Helden and Hankins, "Introduction," p. 5; Behrmann, "Metrics of Justice"; Turner, "From Mathematical Practice."

12 Borrelli, Korey, and Remmert, *Iconography on Early Modern Scientific Instruments*.

though such objects are increasingly studied.[13] Thus, the majority of the early modern instruments known to the present study fall under Turner's rather exclusive class of *ideas made of brass*, with the caveat that this very much concentrates on high culture objects and stands to privilege the study of objects whose "social and symbolic role far outruns their practical utility or their intrinsic character," although it will be shown that imagery is in many cases closely intertwined with such social and symbolic roles.[14]

The development of the sciences since the sixteenth century has largely been perceived and narrated as a success story, shaped by an excellent track record of the sciences or their applicability.[15] Systematic study of the multifaceted visual worlds on early modern scientific instruments is closely linked to analysis of legitimization strategies employed in and for the emerging sciences. It is also deeply rooted in the deliberate and often highly creative strategies and narratives that their practitioners – or put anachronistically: scientists – employed since the sixteenth century to produce meaning and relevance with the goal of fashioning a legitimizing context for the emerging sciences.[16] Such strategies and narratives made use of a variety of textual and visual media and served as global and essential paratexts for early modern science and its historiographical legacy. Rooted in tradition, these strategies produced and employed myths and legends that would shape the future historiography of science, providing success stories essential to the genesis, development, and institutionalization of specific scientific disciplines. All these aspects can be found reflected in the visual worlds on early modern scientific instruments.

II. Classification of Visual Worlds on Scientific Instruments

The conceptualization of scientific instruments as *materialized and shaped ideas* discloses a wide array of objects of potential interest ranging from small portable sundials over astrolabes (made from brass, paper, and wood) to monumental astronomical clocks (such as that in Strasbourg Cathedral).[17] A preliminary classification of the five types of illustrations on instruments

13 See e.g. Dackerman, *Prints and the Pursuit of Knowledge*; Karr Schmidt, *Interactive and Sculptural Printmaking*; Kremer, "Experimenting with Paper Instruments"; Gingerich, "Astronomical Paper Instruments."

14 Turner, "From Mathematical Practice," p. 138.

15 Daston, "Science Studies," p. 24.

16 Remmert, *Picturing the Scientific Revolution*.

17 Globes will be excluded from the present discussion, despite being specific and often highly representative objects, because their programs require their study as a kind of three-dimensional

includes, firstly, the communication or representation of knowledge, which may include depictions of the instruments' use (e.g. measurement activities) or of their general applicability and utility. Such images are closely connected to the specific instrument. The second, that of period motifs and topoi, comprises often fairly standard designs such as religious imagery (Madonna with child, angels[18]), heraldic signs (family coat of arms), traditional motifs (four seasons, Chronos and death), or ornamentation (geometric patterns, leafwork/foliage). A third type, depictions of or pertaining to the mathematical sciences, encompasses motifs such as figurations of the liberal arts or quadrivium, astronomers, personifications (Urania), and so forth. A fourth type, that of legitimization strategies, can be subdivided into three further categories, including myths and legends, which includes reference to biblical and classical myths and legends, e.g. Atlas and Hercules as the first teacher and student of astronomy. Another subcategory, invention and construction of tradition(s), comprises forging ancestry or pedigree, such as the lineage stretching from Abraham to Copernicus on Jost Bürgi's 1591 astronomical clock (Kassel). The last subcategory, self-fashioning, includes that of instrument designers and makers as well as their customers and patrons (or a combination thereof), e.g. Tycho Brahe positioning his own portrait next to that of Copernicus on his own 1584 equatorial armillary sphere (Uraniborg). Finally, the fifth type, the delineation of theoretical positions, may include visualization of the debate on the world systems, e.g. figures of Ptolemy or Copernicus, heliocentric or geocentric cosmological diagrams. Interconnecting threads outlined further below run across these categories: first, that of instruments and their role as patronage artefacts bearing specific pictorial programs targeting particular clients and patrons;[19] and second, that of illustrations drawing imagery from prints and printed material.

III. Communication or Representation of Knowledge

This category is characterized by two aspects. On the one hand, these depictions often gestured to the specific task for which the instrument served. On the other hand, as Anthony Turner indicated, instruments could represent utility *per se*, for example on display in the environment of the

atlas in close relation with iconographic traditions related to atlases or cartography, which is beyond the purview of the present study.

18 On angels, see King, *Astrolabes and Angels*.

19 See, for instance, Behrmann, "Metrics of Justice"; Gaulke, "Perfect in Every Sense."

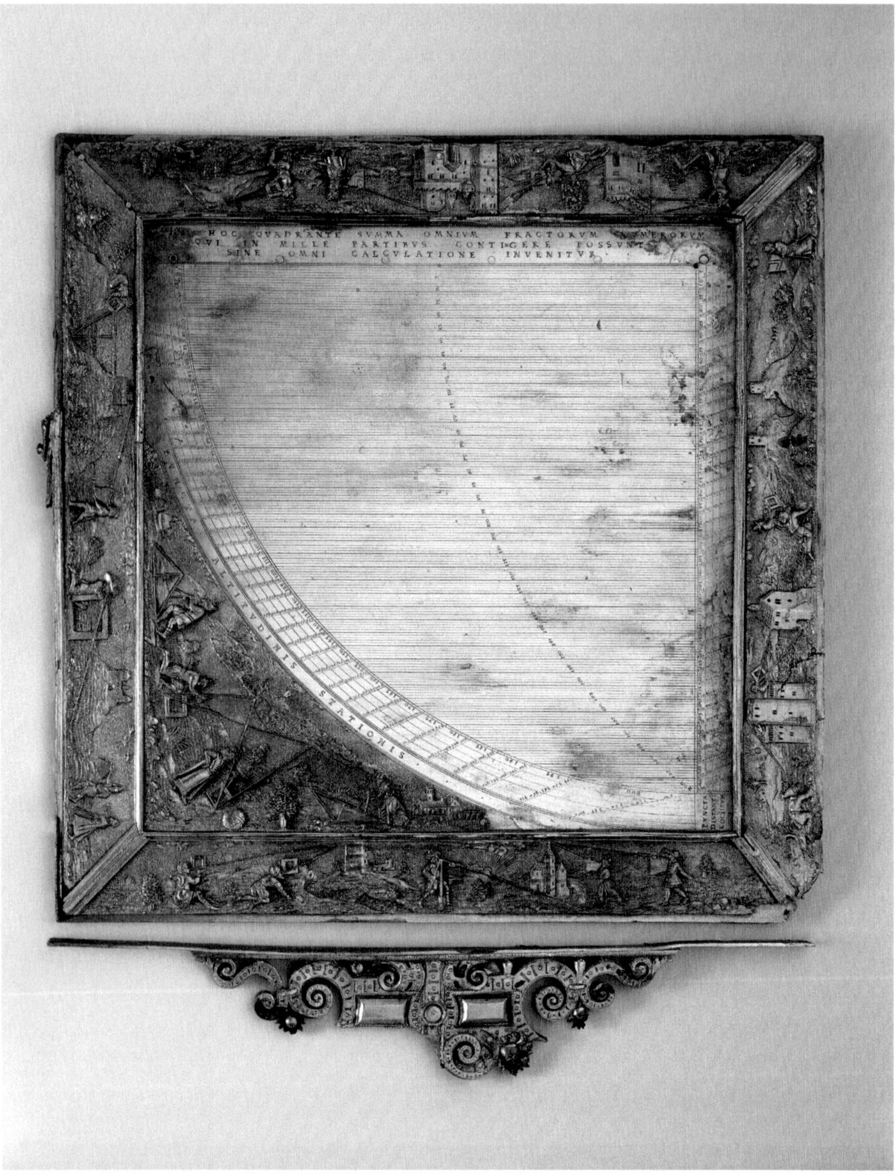

Figure 7.1 Christoph Schissler, geometric quadrant, 1569. Staatliche Kunstsammlungen Dresden, Mathematisch-Physikalischer Salon, Inv. No. C I 1. Staatliche Kunstsammlungen Dresden, Mathematisch-Physikalischer Salon / Photo: Michael Lange, Dresden.

early modern Kunstkammer like that in Dresden, where scientific instruments in the Elector's collection were closely linked to "utility, the part they played in carrying out some necessary activity".[20] From the wealth of

20 Turner, "From Mathematical Practice," p. 137.

relevant examples we may indicate similar geometric quadrants by Augsburg instrument maker Christoph Schissler (c. 1531–1608) in the Oxford History of Science Museum (c. 1579) and Mathematisch-Physikalischer Salon Dresden, the latter made in 1569 for August elector of Saxony[21] (Figure 7.1).

Both were showpieces highlighting the maker's talents, and both were decorated on their rim with reliefs indicating their use in surveying.[22] Some of the specific incised scenes have been traced to sixteenth-century booklets on the mathematical sciences by Walther Ryff (Rivius) and Oronce Finé.[23] While these illustrations may seem typical, they elucidate the intellectual context of the instruments' designers and patrons, e.g. adumbrating literature potentially consulted by maker and owner alike.[24]

IV. Period Motifs and Topoi

This category casts a wide net of examples, such as astrolabes adorned with coats of arms (astrolabe of Philis de Din, 1595, Oxford: bearing an IHS-sign: owner may have been Jesuit), maps (geographical astrolabe by Gillis Coignet, 1560, Oxford: map of the world), landscapes and rural scenes (astrolabe by Johannes Praetorius, 1591, Florence: rural scene), or portable sundials with religious imagery (inclining string-gnomon dial, 1585, Oxford: Judith and Holofernes, a frequent topic in Renaissance art). While largely conventional, certain of these depictions seem to carry a deeper meaning and merit closer examination. An example is a 1571 sun quadrant produced by the aforementioned Nuremberg mathematician and instrument maker Johannes Praetorius (1537–1616) for physician Melchior Ayrer (1520–1579) (Figure 7.2), a piece for which Praetorius (as he often did) worked with Nuremberg goldsmith Hans Epischofer (c. 1530–1585); instrument makers frequently collaborated with artists and artisans to produce high-end objects suitable as patronage artefacts.[25]

The bottom corner was engraved with a depiction of the biblical story of Jonah and the whale, itself copied from a 1566 series of engravings by Philipps Galle (1537–1612) after Maarten van Heemskerck (1498–1574).[26] It is not known why this topic was chosen for the precious and enormous

21 For the former, see Museum of the History of Science, Oxford, Inventory number 48659. For the latter, see Plaßmeyer, "Christoph Schissler."

22 Korey, *Die Geometrie der Macht*, p. 17.

23 Wunderlich, *Das Dresdner 'Quadratum geometricum'*, pp. 48–52.

24 On this, see Bobinger, *Christoph Schissler.*

25 See Hauschke, "Goldschmiede als Hersteller," p. 217.

26 See Ellinghaus, "Bilderwelten."

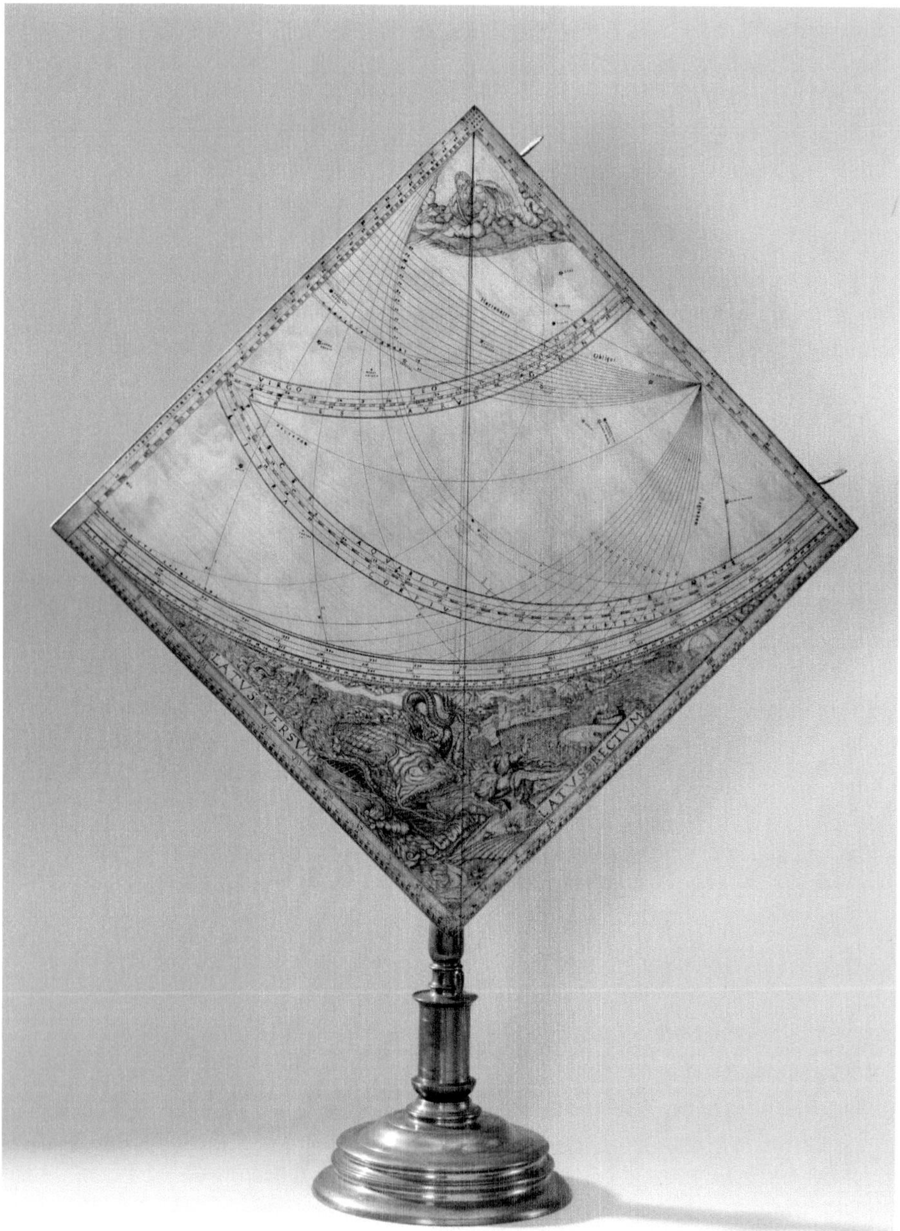

Figure 7.2 Johannes Praetorius, sun quadrant, 1571. Germanisches Nationalmuseum Nuremberg, Inv. No. WI 12. Germanisches Nationalmuseum Nuremburg / Photo: M. Runge.

instrument – with an overall hight of nearly one meter – for Ayrer. As a symbol of higher education, it surely served representative intentions more than practical usage. The topic might be connected to the general subject of fate or a more personal identification with the prophet Jonah, whom early moderns held to have been a physician himself.

V. Depictions of or Pertaining to the Mathematical Sciences

In the early modern period, the mathematical sciences were conceived of as a broad field of knowledge consisting of arithmetic and geometry (*mathematicae purae*) and many mixed fields (*mathematicae mixtae*) of a more practical character, ranging from the traditional disciplines of astronomy and music to architecture, fortification, geography, hydrology, navigation, etc. Most of these were suitable to representation as personifications often on instruments, especially the four of the quadrivium, as they were fairly standardized. Michael Scheffelt's 1705 circle instrument (Hamburg), for instance, is lavishly adorned with the seven liberal arts, inspired by a series of engravings by Jan Sadeler I (1550–1600) after Marten de Vos (1532–1603). Depictions of astronomers decorate various sorts of instruments, amongst them orreries or astronomical clocks, as Nicolaus Siebenhaar's 1651 tellurium (Frederiksborg Slot). Urania, the muse of astronomy, forms part of the pictorial program of the Strasbourg astronomical clock.[27] Geometry personified might typically be found on portable sundials, as on Christoph Trechsler's 1613 horizontal string-gnomon dial (Oxford); together with her pure sister Arithmetic, she was also featured on Tycho Brahe's great azimuth semicircle as described in his 1598 *Astronomiae Instauratae Mechanica*. In this book, Brahe proudly described the astronomical instruments he had designed for his Uraniborg observatory. It is particularly instructive insofar as he also dealt with the meaning of the imagery and embellishments, providing information that is rarely available. According to Brahe, the great azimuth semicircle was crowned by the figures of Urania flanked by Geometry and Arithmetic (Figure 7.3):

> On top it carries three figures cleverly and artfully carved out of strong wood. These are placed on bases, as appears from the figure, and their purpose is not only for ornament, but also that they should represent a symbolical meaning. For the figure that is placed highest is Urania

27 On this, see Remmert, "On Picturing the Past."

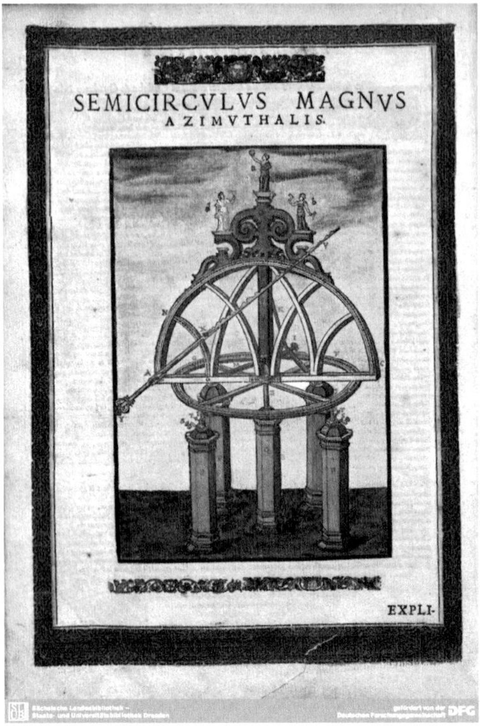

Figure 7.3 Tycho Brahe, *Astronomiae Instauratae Mechanica*, 1598, azimut semicircle. SLUB Dresden. Sächsische Landesbibliothek – Staats- und Universitätsbibliothek Dresden, http://digital.slub-dresden.de/id2767 Signatur S.B.14.

representing Astronomy herself. [...] With her right hand she is holding up the sphere of the celestial revolutions, while with her left hand she receives the objects extended towards her by the two women who stand lower, and who are to be imagined as being in her service. [...] The woman, standing below and to the left of Urania, is a symbol of Geometry. In her right hand she has a triangle which she extends towards Astronomy, and in the other a pair of compasses. By this she indicates that she is serving Astronomy by measurement and by mechanical construction, and also by the learned science of Trigonometry. She is looking up in the direction of Urania's face, and looks at her with awe. [...] On her head she wears a laurel wreath, thereby showing that she is eternal and able to grasp pure truth. The virgin standing on the other, or right, side represents Arithmetic. She too is looking up to Astronomy, showing awe and serving her. In one hand she is holding a piece of chalk, in the other a tablet, expressing hereby that she represents the numbers which Astronomy needs in order

> that it be subject to study, and that she analyzes into discrete quantities what Geometry first expressed in continuous quantities. She too wears a laurel wreath, since she is as eternal and true as Geometry. [...] The meaning of all this is partly that it should decorate the instrument, partly that it should show that Astronomy is the highest of the liberal sciences and, as it were, their queen, and that it has attached to itself as servants Geometry and Arithmetic in preference to all other sciences, although this does not mean that it looks down on the latter.[28]

The message here is twofold: on one level, Brahe took for granted that an appropriate instrument should be decorated. On another level, he sent a clear message with respect to the hierarchy of the sciences, wherein astronomy was attended by arithmetic and geometry serving as handmaidens.

VI. Legitimization Strategies: Myths and Legends, Invention and Construction of Tradition(s), Self-Fashioning

The field of legitimization strategies in early modern science is highly complex, especially in regard to the production of and reference to teleologies of success. Such legitimization strategies were crucial for the emerging sciences, and they could easily be depicted on early modern scientific instruments. References to myths and legends could play an important role in the legitimization of early modern astronomy. A classic example is the pair of Atlas and Hercules representing the first teacher of astronomy, Atlas (or King Atlas), and his first student, Hercules. Throughout the sixteenth and seventeenth centuries, we often find textual and visual references to the pair: Atlas and Hercules appear as honorary authorities for astronomy, and each was invoked to symbolise both the old and the new astronomy. Naturally, it would be difficult to come up with a line of tradition stretching further back.[29]

Sometimes, images of Atlas and Hercules would adorn scientific instruments such as the astrolabe made by Johannes Praetorius in 1568 (Dresden) mentioned at the outset where Atlas, wearing a crown, and Hercules can be seen striding together, eager to share their knowledge and place it at the service of astronomy and astronomers of the future.[30] Only a few years

28 Brahe, *Instruments*, 46. On the context, see Perkins, "Instruments of Authority."
29 On this see, Remmert, *Picturing the Scientific Revolution*.
30 *Ibid.*, pp. 139, 143; Ellinghaus, "Bilderwelten," pp. 118–120.

later, in 1578, Wentzel Jamnitzer put a similar scene on a richly decorated measuring instrument (Figure 7.5), and Brahe, too, referred to the story in his observatory where, in the summer of 1585, the great equatorial armillary sphere, the showpiece of Brahe's instruments, was installed. It was fixed on a globe of the heavens supported by the royal Atlas. Everyone entering the room came face-to-face with King Atlas, supporting the heavens on his shoulders as Brahe described in his *Astronomiae Instauratae Mechanica* of 1598.[31]

Apart from classical lore, biblical legends would be invoked on instruments, albeit less frequently, the most prominent example probably being the reference to three patriarchs on the astronomical equation clock built for Landgrave William IV of Hesse-Kassel in 1591 (Jost Bürgi, Kassel). As this figuration forms part of a series of eight famous astronomers, it leads directly to the next aspect discussed below.

In the case of the invention and construction of tradition(s), a typical strategy of legitimization was to make use of references to the past, particularly to biblical or classical antiquity.[32] Highlighting the long-accepted role of for example astronomy in the ancient world was part of broader discursive practices best understood as the invention of tradition(s) in early modern science. This could include reference to Atlas the first astronomer and Hercules his first disciple, the eminent mathematician and engineer Archimedes, who according to legend single-handedly defended Syracuse against the Romans,[33] and biblical patriarch Abraham as "planter of mathematics in Egypt".[34] While the role of the first two (particularly Archimedes) in legitimization strategies has been studied by historians of science, that of Abraham remains largely overlooked. He figures prominently in the pictorial program on a 1590–1591 astronomical equation clock (Figure 7.4) designed by court mathematician and clockmaker Jost Bürgi (1552–1632) with goldsmith Hans Jakob Emck for Landgrave William IV of Hesse-Kassel (1532–1592), the so-called Ptolemy of Kassel.[35]

Each of the object's four sides shows two episodes from the history of astronomy and geometry, tracing a line of tradition from ancient times to the sixteenth century: three patriarchs and Thales, Euclid, and Archimedes, Hipparchus and Ptolemy, and finally the medieval patron of astronomy, King

31 Remmert, *Picturing the Scientific Revolution*, pp. 149–151.

32 This paragraph is based on Remmert, "Inventing Tradition," p. 55.

33 Jaeger, *Archimedes and the Roman Imagination*.

34 On this aspect, see Popper, "'Abraham, Planter of Mathematics'."

35 Gaulke, "Perfect in Every Sense."

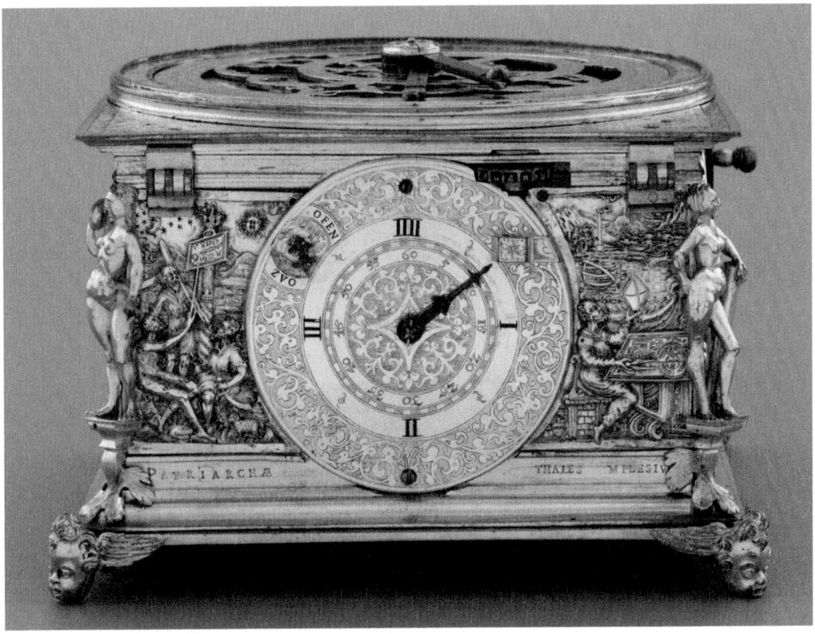

Figure 7.4 Jost Bürgi, equation clock, 1590/91. Museumslandschaft Hessen Kassel, Astronomisch-Physikalisches Kabinett, Inv. No. U 24. Museumslandschaft Hessen Kassel, Astronomisch-Physikalisches Kabinett / Foto: Mirja van Ijken.

Alfonso the Wise of Castile, and Copernicus. Thales, Euclid, and Archimedes stand for the geometrical tradition, while Hipparchus and Ptolemy, King Alfonso and Copernicus represent the astronomical, both lineages reaching back to Greek antiquity. The three patriarchs have been identified as Abraham, Adam, and Seth, to whom astronomical and geometrical knowledge had been attributed as of old.[36] The depiction here illustrates the well-known narrative of Abraham instructing the Egyptians in astronomy (disseminated through Flavius Josephus) and of the two pillars of Adam and Seth representing astronomical and geometrical knowledge. The double reference to the ancient twofold origins of astronomy and geometry in Greek antiquity and the Old Testament made a strong visual case for the relevance and nobility of both disciplines.

Regarding the issue of questions of self-fashioning, crucially, the iconography on Bürgi's clock for the landgrave extended an unbroken lineage from Abraham, pious scholar willing to sacrifice his son Isaac, through the pious astronomer Copernicus, to the devout William IV, who had turned Kassel into a major center of astronomical research, thus firmly situating

36 Remmert, "Inventing Tradition," p. 58.

William not only within a prominent ancestry of astronomical tradition, but also a royal line of leading elites. The pictorial program, pointing to and overlapping with the aspect of self-fashioning, served both the legitimization of astronomy and geometry and their practitioners and experts, and the religio-political dynastic ends of William IV, emblematizing the close connection between astrology and politics and alluding to his role as wise and pious prince. Among William IV of Hesse-Kassel's sixteenth-century contemporaries, Tycho Brahe – himself descended from a family of king makers – repeatedly proved himself a prince of self-fashioning; William, who numbered among Brahe's correspondents, collaborators, and, indeed, competitors, may have emulated him.[37] Portraits of astronomers – Ptolemy, al-Battānī, Copernicus, and Brahe himself – embodying a line of succession were to be found everywhere in Brahe's Palace of Uraniborg (named for Urania) on the island of Hven, while the base of Brahe's 1584 equatorial armillary sphere as described in *Astronomiae Instauratae Mechanica* was adorned by likenesses of Copernicus and Brahe, who was never shy to point to his own grandeur.[38]

The final category, delineations of theoretical positions, is tentative, insofar as it is exclusively identified with the period debate regarding the heliocentric or geocentric cosmological systems, supported prominently by sixteenth- and seventeenth-century visual media, which sometimes drew on easily depicted biblical passages from Joshua and the Second Book of Kings frequently cited as arguments against the earth's movement.[39] The Book of Joshua's account of the Israelites' conquest of the land of Canaan reported that, in the battle with the Amorites, Joshua stopped the Sun: "Sun, stand thou still upon Gibeon; and thou, Moon, in the valley of Ajalon. And the Sun stood still, and the Moon stayed, until the people had avenged themselves upon their enemies" (Joshua 10:12f). The Sun standing still over Gibeon, among the most spectacular miracles in the Bible, was ubiquitous in art throughout the Middle Ages and well into the eighteenth century.[40] Starting from the mid-sixteenth century, this passage and its pictorial representation constituted one of the standard biblical arguments against Copernicanism, given the explicit reference to the Sun's motion.

In the Second Book of Kings the prophet Isaiah announces to the dying King Hezekiah that he will recover and live a further fifteen years, to which

37 Christianson, *On Tycho's Island*.

38 On this, see Remmert, *Picturing the Scientific Revolution*, p. 172.

39 *Ibid.*

40 Kirschbaum, *Lexikon der christlichen Ikonographie*, 2: 436–442; Pigler, *Barockthemen*, 1: 113.

the dubious Hezekiah demands a sign from God: "And Isaiah the prophet cried unto the Lord: and he brought the shadow ten degrees backward, by which it had gone down in the [sun]dial of Ahas" (II Kings 20:9–11). That exegetical commentaries passage maintained that the measure of the Sun's retreat referred to a horological instrument – namely, a sundial installed by Hezekiah's father king Ahas – resulted in this passage's shorthand reference by the phrase *Horologium Ahas*. By the early seventeenth century, this passage had likewise become an oft-cited anti-Copernican argument, such that biblical exegetes often mentioned the two miracles together. References to the *Horologium Ahas* can be found in three hemispheric brass dials of Ahaz made 1547–1548 by Georg Hartmann, which restage the miracle in real time using an optical trick.[41] The subject was also depicted on instruments: Hartmann's 1561 printed paper sundial or compass, Christoph Schissler's 1578 refractive sundial (Philadelphia), and Praetorius's 1568 torquetum, also made for Ayrer (Nuremberg).[42] Hans Gruber's 1583 table clock (Stuttgart) depicts both miracles.[43] Despite the relative ubiquity of such imagery, it remains difficult to assess to what extent it was charged with geocentric undertones. A case study in this regard is the imagery on Wentzel Jamnitzer's magnificent 1578 measuring instrument (Paris) (Figure 7.5).[44]

The instrument's lavish pictorial program, outstanding for its dense and multifaceted celebration of astronomy, merits a study in of itself.[45] Suffice it to emphasize that the program carries distinct royal overtones, its incised portraits celebrating such "royal astronomers" as ancients King Atlas and Hercules – role models for rulers, including those of the Habsburg dynasty – as well as the biblical figures of King Hezekiah and Joshua stopping the sun. While Atlas and Hercules were inscribed on the side useful for astronomical, astrological, and horological purposes, Joshua and Hezekiah flank a now-lost compass on the side of the disk used for geodetical measurement. The disassociation of the figures of Joshua and Hezekiah from the context of

41 These objects are conserved today in Harvard, Toledo, and Madrid. See Mersmann, "Moving Shadows, Moving Sun," pp. 113–115. A lost paper Ahaz dial by Hartmann mentioned in a letter from 1528 may be preserved in a print by Georg Brentel the younger who copied many of Hartmann's paper instruments, see Dackerman, *Prints and the Pursuit of Knowledge*, pp. 296–297, cat. no. 69.

42 Lamprey, *Hartmann's Practika*, p. 297. On the refractive sundial in Philadelphia, see Plaßmeyer, "Christoph Schissler," p. 20; Mersmann, "Moving Shadows, Moving Sun," pp. 114–120; and the forthcoming paper by Andrew Morrall in Ellinghaus and Remmert, *Manipulating the Sun*.

43 Schaller, *Prunkuhren der Renaissance*, pp. 33–39.

44 On this instrument, see Savoi, "Le cadran solaire."

45 See our forthcoming chapter in Ellinghaus and Remmert, *Manipulating the Sun*.

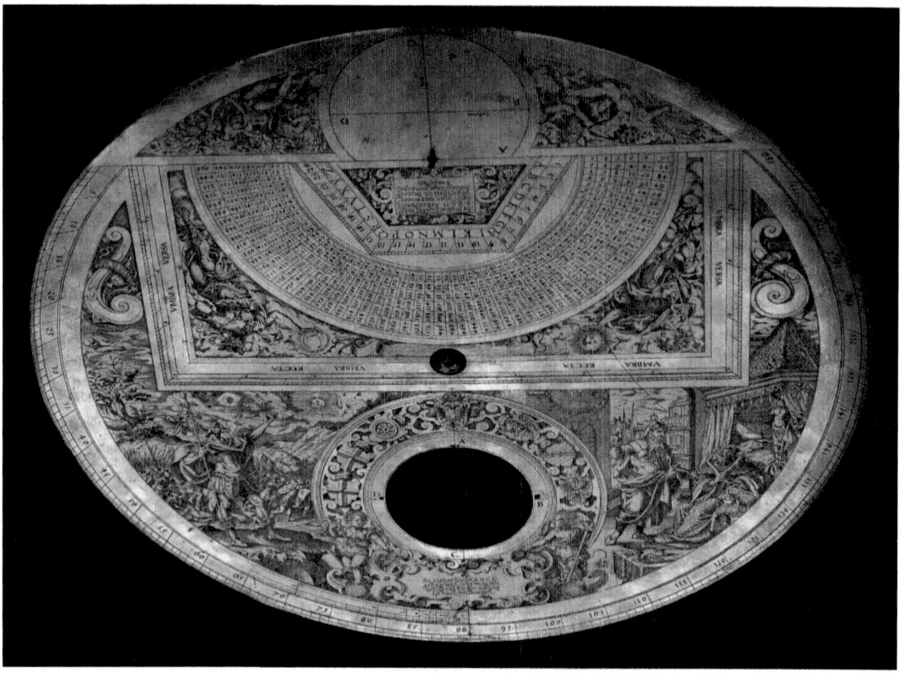

Figure 7.5 Wenzel Jamnitzer, mesauring instrument, 1578. Observatory Library Paris, Inv. No. 7. Photo: Dominique Monseigny / Bibliothèque de l'Observatoire de Paris.

the cosmos, their situation in the realm of geodesy, and the iconography of the depictions raise questions about these figures' meanings. That Joshua appears comparable to his representation as one of the virtuous leaders of the Nine worthies, and the richly attired and enthroned King Hezekiah adheres to the pictorial tradition whereby he embodies the divine right to rule, seems to foreground the theme of righteous rulership over any potential indication of cosmological preferences.

VII. Illustrations Drawing on Printed Material

Before concluding, the two interconnecting threads mentioned above warrant further observation. First, regarding patronage: thus far, quite a few examples of instruments and their role as patronage artefacts have emerged, each of which deserves an individual analysis as a case study *per se* accounting for specific respective contexts. This represents future research for most of the instruments mentioned (with the exception of Bürgi's astronomical clock). Second, pertaining to instruments with imagery drawing on printed material, a number of examples have also been cited here with motifs taken from a variety of printed matter used as a resource for templates.[46] In more than half our cases these consisted of burin-engraved and etched series representing subjects such as the four elements, the four temperaments, the twelve months, the seven planets, or the seven liberal arts, executed by famous Dutch, Flemish, French, or German artists of the sixteenth and seventeenth centuries – Hendrik Goltzius, Jan Sadeler, Maarten de Vos, Sebald Beham, Georg Pencz, Virgil Solis, and Etienne Delaune.

In contrast to the 1705 circle instrument by Michael Scheffelt (Hamburg) mentioned above, typically, a whole series was not transferred to the instrument, but only one or two figures, such as the personifications of Geometry and Astronomy from the liberal arts or the sun and the moon from the seven planets. Many templates were extracted from old as well as new books, including illustrations from Leonhard Zubler's *Novum instrumentum geometricum* (Basel 1614) and Hertel's *New ero[e]ffneter Geometrischer Schaw- und Ma[e]ss-Platz* (Brunswick 1675), used on several instruments to depict the applicability of certain measuring tools. Images of known and famous scholars were also commonly borrowed from title engravings; stripped of context and deprived of identification on the instrument, they often elude identification and thus were transformed into ideal personifications of astronomers, geometers, and so forth. The portrait of the Arabic astronomer Messahala (Mashallah ibn Athari) included in his *De Scientia Motus Orbis* published by Johann Stabius in 1504, was engraved on the lowest plate of Johannes Praetorius's 1568 torquetum (Nuremberg) in a completely new setting, combined with figures of an astronomer and surveyor from the title woodcut of Peter Apian's *Instrument Buch* (Ingolstadt 1533). A similar creativity with respect to printed models is demonstrated in the engravings on the side panels of a 1642 astronomical clock attributed to Nicolaus Radeloff made for the Kunstkammer of Schloss Gottorf in Schleswig-Holstein (Copenhagen). A portrait of Tycho Brahe accompanied by a diagram of his world system above him on one side is contrasted with Copernicus on the other, both figures adapted from the frontispiece engraving of Johannes Kepler's *Tabulae Rudolphinae* (Ulm 1627) (Figure 7.6). Whereas the figure of Brahe corresponds to his figure on the title page model, in representing Copernicus on the clock the artist copied the figure of Hipparchus from the book, possibly because Hipparchus

46 On graphic templates for scientific instruments, see Ellinghaus, "Bilderwelten."

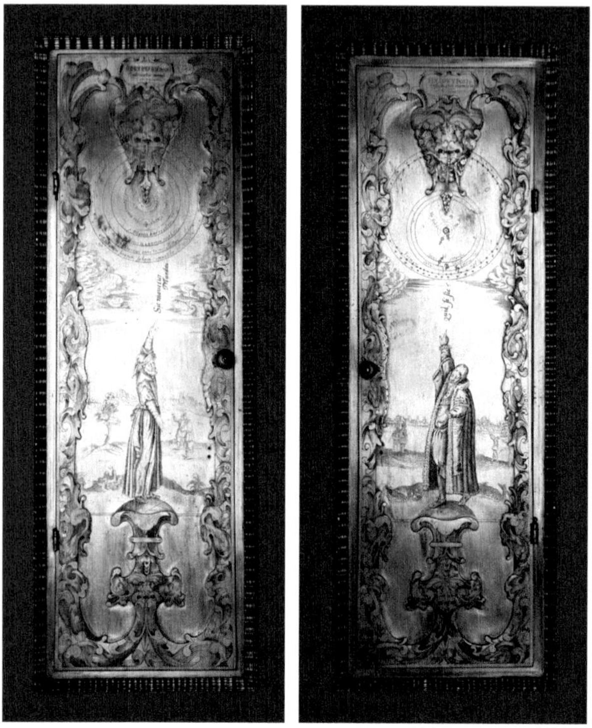

Figure 7.6 Nicolaus Radeloff, astronomical clock, 1642. National Museum of Denmark Copenhagen, Inv. No. N04 Inv D138. National Museum of Denmark Copenhagen / N04 Inv D138 / CC-BY-SA 4.0.

shown standing in the frontispiece could be adapted more easily to the clock's format.[47]

VIII. Conclusions and Perspectives

This chapter's modest goal, to make a case for the study of the visual worlds on early modern scientific instruments, suggests that more detailed analysis of this imagery is highly relevant to understanding the intellectual, cultural, and artistic contexts shaping and determining the production of instruments in the early modern period. It opens a window on the investigation of collaborative processes during the conception, design, and construction of instruments in the multi-layered field between instrument makers, artists, artisans, patrons, and scholars. While not necessarily exhaustive,

47 On this, see Remmert, *Picturing the Scientific Revolution*, pp. 49–52; Ellinghaus, "Bilderwelten," pp. 120–123.

the preliminary classification set forth here serves as a travel guide through the visual landscape of instrument imagery, illustrating this landscape's capacity to pose more questions for subsequent research.

Bibliography

Behrmann, Carolin. "Metrics of Justice. A Sundial's Nomological Figuration." *Nuncius.* 30, no.1 (2015): 161–194.

Bennett, Jim. "Early Modern Mathematical Instruments." *ISIS* 102, no.4 (December 2011): 697–705.

Bennett, Jim. "Shopping for Instruments in Paris and London." In *Merchants and Marvels: Commerce, Science and Art in Early Modern Europe,* edited by Paula Findeln and Pamela H. Smith, pp. 370–395. New York: Routledge, 2001.

Biagioli, Mario. "From Print to Patents: Living on Instruments in Early Modern Europe." *History of Science* 44, no. 2 (2006): 139–186.

Bobinger, Maximilian. *Christoph Schissler der Ältere und der Jüngere.* Augsburg: Verlag Die Brigg, 1954.

Borrelli, Arianna, Michael Korey, and Volker R. Remmert, eds. *Iconography on Early Modern Scientific Instruments,* special issue of *Nuncius. Journal of the Material and Visual History of Science* 30, no. 1 (2015).

Brahe, Tycho. *Astronomiae Instauratae Mechanica.* Hamburg: Philippi de Ohr chalcographi, 1598.

Brahe, Tycho. *Instruments of the Renewed Astronomy,* edited by Alena Hadravová, Petr Hadrava, and Jole R. Shackelford. Prague: Koniasch Latin Press, 1996.

Brenni, Paolo. "Art and Science: Design of Scientific Instruments." In *Creativity in the Arts and Science,* edited by William R. Shea and Antonio Spadafora, pp. 121–141. Canton: Science History Publications, 1990.

Christianson, John Robert. *On Tycho's Island: Tycho Brahe and His Assistants, 1570–1601.* New York: Cambridge University Press, 2000.

Dackerman, Susan, ed. *Prints and the Pursuit of Knowledge in Early Modern Europe.* Cambridge, MA: Harvard Art Museum, 2011.

Daston, Lorraine. "Science Studies and the History of Science." *Critical Inquiry* 35, no. 4 (2009): 798–813.

Ellinghaus, Julia. "Bilderwelten auf wissenschaftlichen Instrumenten der Frühen Neuzeit und ihre graphischen Vorlagen." *Imprimatur. Ein Jahrbuch für Bücherfreunde* 27 (2021): 107–130.

Ellinghaus, Julia, and Volker Remmert, eds. *Manipulating the Sun: Picturing Astronomical Miracles from the Bible in the Early Modern Era.* Leiden: Brill, forthcoming.

Gaulke, Karsten. "Perfect in Every Sense: Scientific Iconography on an Equation Clock by Jost Bürgi and the Self-Understanding of the Astronomers at the Kassel Court in the Late 1580s." *Nuncius* 30, no. 1 (2015): 37–74.

Gingerich, Owen. "Astronomical Paper Instruments with Moving Parts." In *Making Instruments Count: Essays on Historical Scientific Instruments Presented to Gerard L'Estrange Turner*, edited by R. G. W. Anderson, J. A. Bennett, and W. F. Ryan, pp. 63–74. Aldershot: Variorum, 1993.

Hauschke, Sven. "Goldschmiede als Hersteller wissenschaftlicher Instrumente und Geräte." In *Nürnberger Goldschmiedekunst 1541–1868, vol. 2, Goldglanz und Silberstrahl*, edited by Karin Tebbe, Ursula Timann, and Thomas Eser, pp. 216–332. Nuremburg: Verlag des Germanischen Nationalmuseums, 2007.

Jaeger, Mary. *Archimedes and the Roman Imagination.* Ann Arbor, MI: University of Michigan Press, 2008.

Johnston, Stephen. "Mathematical Practitioners and Instruments in Elizabethan England." *Annals of Science* 48, no. 4 (1991): 319–344.

Karr Schmidt, Suzanne. *Interactive and Sculptural Printmaking in the Renaissance.* Leiden: Brill, 2018.

King, David A. *Astrolabes and Angels, Epigrams and Enigmas: From Regiomontanus' Acrostic for Cardinal Bessarion to Piero della Francesca's Flagellation of Christ.* Stuttgart: Steiner Verlag, 2007.

Kirschbaum, Engelbert, ed. *Lexikon der christlichen Ikonographie.* Rome: Herder, 1970.

Korey, Michael. *Die Geometrie der Macht. Mathematische Instrumente und fürstliche Mechanik um 1600.* München: Deutscher Kunstverlag, 2007.

Kremer, Richard. "Experimenting with Paper Instruments in Fifteenth- and Sixteenth-Century Astronomy: Computing Syzygies with Isotemporal Lines and Salt Dishes." *Journal for the History of Astronomy* 42, no. 2 (2011): 223–258.

L'E Turner, Gerard. *Elizabethan Instrument Makers: The Origins of the London Trade in Precision Instrument Making.* Oxford: Oxford University Press, 2000.

Lamprey, John. *Hartmann's Practika: A Manual for Making Sundials and Astrolabes with the Compass and Rule.* Bellevue: J. Lamprey, 2002.

Mersmann, Jasmin. "Moving Shadows, Moving Sun: Early Modern Sundials Restaging Miracles." *Nuncius* 30, no. 1 (2015): 96–123.

Morrall, Andrew. "Entrepreneurial Craftsmen in Late Sixteenth-Century Augsburg." In *Mapping Markets for Paintings in Europe 1450–1750*, edited by Neil de Marchi and Hans J. van Miegroet, pp. 211–236. Turnhout: Brepols, 2006.

Nowotny, Helga. *Es ist So. Es Könnte auch Anders Sein.* Frankfurt a. M.: Suhrkamp Verlag, 1999.

Pechstein, Klaus. "Hans Epischofer – Der Monogrammist HsE." In *Anzeiger des Germanischen Nationalmuseums 1970*, pp. 96–102. Nürnberg: Verlag des Germanischen Nationalmuseums, 1970.

Perkins, Emma. "Instruments of Authority: Tycho Brahe's Technological Illustrations." *History and Technology* 34, no. 3–4 (2018): 259–272.

Pigler, Andor. *Barockthemen Vol. 1.* Budapest: Akadémia Kiadó, 1974.

Plaßmeyer, Peter. "Christoph Schissler: The Elector's Dealer." In *European Collections of Scientific Instruments, 1550–1750*, edited by Giorgio Strano, Stephen Johnston, Mara Miniate, and Alison Morrison-Low, pp. 15–25. Leiden: Brill, 2009.

Popper, Nicholas. "'Abraham, Planter of Mathematics': Histories of Mathematics and Astrology in Early Modern Europe." *Journal of the History of Ideas* 67, no. 1 (2006): 87–106.

Remmert, Volker R. "On Picturing the Past: Arithmetic and Geometry as Wings of the Mind." *The Mathematical Intelligencer* 31, no. 3 (2009): 42–47.

Remmert, Volker R. *Picturing the Scientific Revolution: Title Engravings in Early Modern Scientific Publications.* Philadelphia, PA: St. Joseph's University Press, 2011.

Remmert, Volker R. "Inventing Tradition in 16th- and 17th-Century Mathematical Sciences: Abraham as Teacher of Arithmetic and Astronomy." *The Mathematical Intelligencer* 37, no. 3 (2015): 55–59.

Remmert, Volker R. "Analyse des Imageries des Instruments Scientifiques de l'Époque Moderne (XVIe–XVIIe Siècle)." *Images des Mathématiques* (2019): https://images.math.cnrs.fr/Analyse-des-imageries-des-instruments-scientifiques-de-l-epoque-moderne-XVIe.html?lang=fr (accessed 3 July 2022).

Savoi, Denis. " Le Cadran Solaire de Hauteur de Wenzel Jamnitzer de l'Oberservatoire de Paris." *Commission des Cadrans Solaires – Société Astronomique de France, Cadran Info* 36 (2017): 136–151.

Schaller, Andrea. *Prunkuhren der Renaissance. Württembergisches Landesmuseum Stuttgart.* Stuttgart: Württembergisches Landesmuseum, 2001.

Taub, Liba. "On Scientific Instruments." *Studies in History and Philosophy of Science* 40, no. 4 (December 2009): 337–438.

Taub, Liba. "Introduction: Reengaging with Instruments." *ISIS* 102, no.4 (2011): 689–696.

Turner, Anthony. "From Mathematical Practice to the History of Science: The Pattern of Collecting Scientific Instruments." *Journal of the History of Collections* 7, no. 2 (1995): 135–150.

Valleriani, Matteo, ed. *The Structures of Practical Knowledge.* Dordrecht: Springer, 2017.

Van Helden, Albert, and Thomas L. Hankins. "Introduction: Instruments in the History of Science." *Osiris* 9 (1994): 1–6.

Wunderlich, Herbert. *Das Dresdner 'Quadratum geometricum' aus dem Jahre 1569 von Christoph Schißler d. A.* Berlin: VEB Deutscher Verlag der Wissenschaften, 1960.

About the Authors

Julia Ellinghaus is art historian and research assistant at the Interdisciplinary Center for Science and Technology Studies (IZWT) at Wuppertal University.

Volker Remmert is professor in the history of science at Wuppertal University and director of the Interdisciplinary Center for Science and Technology Studies (IZWT).

8. Visual Tools and Searchable Science in Early Modern Books

Britta-Juliane Kruse and Stephanie Leitch

Abstract

This chapter investigates visual tools in sixteenth-century manuscripts and early printed books. By presenting data in useable and searchable formats, printers of short, pithy vernacular pamphlets effectively rewired the emphasis of older genres. This trajectory by which ancient knowledge domains became how-to data can perhaps best be tracked in the *Liber Quodlibetarius,* a codex produced in southwestern Germany c. 1524, which curiously imported images from a variety of printed genres into a hand-produced manuscript form. The sovereignty of images derived from cosmography, hippiatria, chiromancy, and metoscopy can provide important clues about their reception as data, namely, how those genres were cross-referenced, collected, and shaped up for a new visually attentive vernacular viewer.

Keywords: visual tools, Liber Quodlibetarius, chiromancy, early prints, knowledge transfer, technical literature

I. Introduction

The sixteenth-century press codified disciplines of knowledge by distilling available information into visual tools. By presenting data in useful and easily legible formats, printmakers rewired the emphasis of older genres such as the *Book of Secrets* and repackaged those secrets in printed editions of the more marketable genres of complexion literature, physiognomy, and cosmography. While scholars of early modern prints are usually eager to search for prints' agency in the technologizing of the image, or in the prompting of the collective viewing practices that underwrote emerging

Noyes, R.S. (ed.), *Reassessing Epistemic Images in the Early Modern World.*
Taylor & Francis Group, 2023
DOI 10.5117/9789463723350_CH08

scientific consensus, a look at prints' intersection with manuscript culture proves a fruitful route to uncovering a path to early modern data development. The trajectory by which ancient knowledge domains became how-to skills can perhaps best be tracked in a manuscript likely assembled in Passau, c. 1524. The so-called *Liber Quodlibetarius*, which seamlessly imported images from a variety of printed genres, forms the focus of this chapter. The manuscript's compiler merged information from books in the humanistic tradition, such as cosmography and chiromancy, with pamphlets advertising designs for bridle bits of horses, tools for field surgery, and artists' manuals, on the strength of their visual kinship and the degree to which images were critical messengers of the books' program. In transposing the images from these printed books, the manuscript aimed to make that data more useful to the reader by deputizing images to stand in for text-based knowledge.[1] This chapter argues that the *Liber Quodlibetarius* allows us to see through to the pictorial apparatus of the printed books that had already condensed and visualized arcane knowledge associated with the Secrets tradition. The printing press distilled text-based knowledge into searchable visual units.

II. Creation of the Manuscript

The codex *Liber Quodlibetarius*, widely referred to as a pictorial encyclopedia, or *Bayerische Bild-Enzyklopädie* in the secondary literature, resides today in the University Library in Erlangen.[2] A closely related manuscript is preserved in the Jagiellonian Library in Cracow (Rkp. Przyb. 35/64 (49a.5.2.).[3] *Liber Quodlibetarius* was written in Bavarian dialect and

1 Henderson, *Images*, pp. 70–73.

2 Ferrari, *Le monde en images*, pp. 4–30. The manuscript is the topic of a research team (Prof. Michele C. Ferrari, Dr. Manuel Teget-Welz, and Dr. Stefan Weber) at the Friedrich-Alexander Universität Erlangen-Nürnberg that investigates early modern illustrated encyclopedias, *Bildenzyklopädien des frühen 16. Jahrhunderts*: https://www.mittellatein.phil.fau.de/forschung/#collapse_4 (accessed 3 July 2022).

3 For an overview of the content and partial identification of a selection of printed templates, see Rudolph, "Hausbücher. Bayerische Bild-Enzyklopädien," pp. 466–473 (Nr. 49a.5.1) in *Katalog der deutschsprachigen illustrierten Handschriften des Mittelalters* (*KdiH*), ed. Ulrike Bodemann, *et al.* (Munich: C.H. Beck, 2015), pp. 466–473. https://kdih.badw.de/datenbank/handschrift/49a/5/1. This entry is followed by a description of an associated manuscript ("Schwesterhandschrift") without explantory texts preserved in Kraków, Biblioteka Jagiellońska, Rkp. Przyb. 35/64 (49a.5.2.), https://kdih.badw.de/datenbank/handschrift/49a/5/2. See also https://handschriftencensus.de/24571 (accessed 3 July 2022).

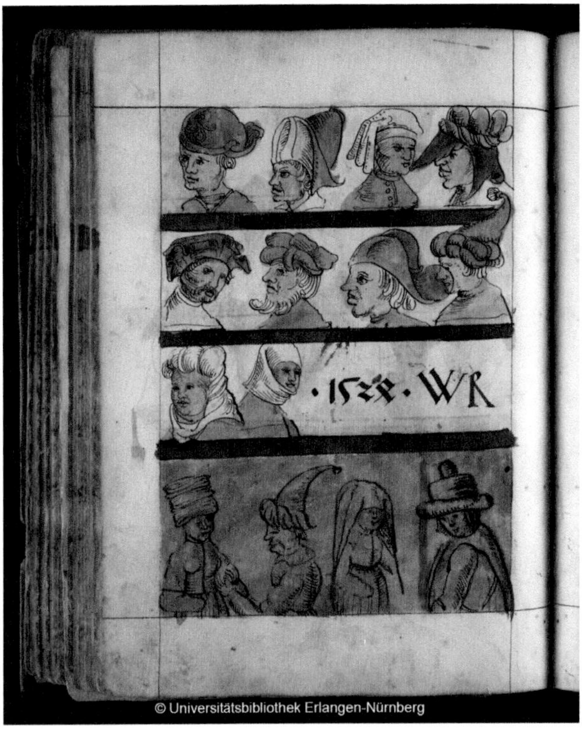

Figure 8.1 Master WR, Benedictus Rughalm, *Liber Quodlibetarius*. University Library Erlangen-Nuremberg, Ms. B 200, fol. 86v.

probably produced in Passau. Of the codex's 288 pages of images, many are sourced in contemporary circulating printed books. The manuscript twice bears the date 1524 (Figure 8.1). This was likely penned by the scribe who identifies himself as a cleric, "Benedictus Rughalm patavius pfaff" (f. 164r).[4] The Rughalm family can be traced to Passau, where, around 1500, Benedictus and Wolfgangus Rughalm (probably his brother) were enrolled at the University of Vienna.[5] The two illustrators involved in the manuscript's production are noted in the codex itself: Benedictus, mentioned above, and another, who can be identified by the monogram "WR" (Wolfgangus Rughalm?). The compilation demonstrates the high level of erudition of the compiler, who collected images from a variety of

4 A second hand wrote the collection of "Priamel," included with the *Liber Quodlibetarius* in this codex. Illustrator's monogram "WR" can be found on fol. 86v. Labels of some drawings are also from Benedictus Rughalm. UB Erlangen-Nürnberg, MS. B 200: http://digital.bib-bvb.de/view/bvbmets/viewer.0.6.4.jsp?folder_id=0&dvs=1605197036959~61&pid=5281832&locale=de&usePid1=true&usePid2=true (accessed 3 July 2022).

5 Identified in Bodemann, *Katalog der deutschsprachigen illustrierten Handschriften*, p. 470.

printed genres,[6] and imported them into a hand-produced manuscript form predating print.[7] The illustrators cribbed information from contemporary books like Hartmann Schedel's *Weltchronik*, Peter Apian's *Cosmographicus Liber*, Bartolommeo della Rocca Cocles's *Phisionomi vnd Chiromanci*, and Hans von Gersdorff's *Feldbůch der Wund-Artznei.*[8] In addition to these, the manuscript imported the pictorial content from books advertising designs for bridle bits for horses, manuals of musical instruments, and artists' model books. The inclusion of particular works seems to be based on two factors: the visual kinship of the respective images across their printed texts, and the degree to which images were critical to the books' argumentation. In transposing the images from these printed books, the manuscript aimed to make the collected data more searchable. The images populate visual fields of a new type of storage solution: we can think of the *Liber Quodlibetarius* as a type of database for much of the available information that could be visualized.

The idea that printed illustrations promoted empirical investigation and shaped disciplinary fields of study is not new. Prints as critical technology that built knowledge acquisition has been the subject of sustained inquiry in both art history and the history of science. Important studies by Sachiko Kusukawa, Susan Dackerman, Brian Ogilvie, Volker Remmert, William Eamon, and Deborah Krohn have scrutinized how prints fixed the approach and parameters of the disciplines of botany, anatomy, medicine, astronomy, natural history, and the culinary arts.[9] More recently, Suzanne Karr Schmidt has demonstrated the critical access prints provided to the how-to knowledge of scientific instruments.[10] Publications devoted to these topics featured prints that flickered between visual tools and data reserves. These prints

6 For identification of printed templates used for the illuminations, see Bodemann, *Katalog der deutschsprachigen illustrierten Handschriften des Mittelalters*, pp. 471–472.

7 For this practice in the sixteenth century, see Pettegree, "The Renaissance Library and the Challenge of Print," pp. 72–102, esp. p. 75. Jordanova, *The Look of the Past.*

8 See e.g. Hartmann Schedel, *Registrum huius operis libri cronicarum. Cu[m] figuris et ymagi[ni]-bus ab inicio mu[n]di* (Nuremberg: Anton Koberger for Sebald Schreyer and Sebastian Kammermeister, 1493). Woodcuts by Michael Wohlgemut, Wilhelm Pleydenwurff and Albrecht Dürer. See ISTC https://data.cerl.org/istc/is00307000, with links to several electronic facsimiles (accessed 3 July 2022). Gersdorff, *Feldbůch der Wund-Artzney,* provides several prints as templates, see http://mdz-nbn-resolving.de/urn:nbn:de:bvb:12-bsb00010085-8 (BSB Munich) (accessed 3 July 2022). See also Panse, *Hans von Gersdorffs Feldbuch.*

9 Kusukawa, *Picturing the Book of Nature*; *Idem*, "The Uses of Pictures"; Dackerman, *Prints and the Pursuit of Knowledge*; Ogilvie, *The Science of Describing*; Remmert, *Picturing the Scientific Revolution*; Eamon, *Science and the Secrets of Nature*, and Krohn, *Food and Knowledge in Renaissance Italy.*

10 Schmidt, *Interactive and Sculptural Printmaking.*

performed a range of functions: they documented, recorded, and served as mnemonic prompts. Importantly, many also cued first-hand engagement with the world for an agent whom we might call an amateur or vernacular viewer: a non-academic urban dweller who preferred to read in their native German over cranky medieval Latin, and preferred pictures even about that. From the way pictures organized items on the visual horizons of early moderns, we can infer new types of visual searches.

The succinct display of information contained in the manuscript suggests that a repertoire of images was developing around related scientific fields, some of which previously had been the province of Secrets.[11] The visual data embedded in the manuscript's display offered the reader a practical way to control the advanced knowledge unleashed by a massive outpouring of printed books in the previous half-century. The *Liber Quodlibetarius* can be thought of as a visual appendix in which the informational content of some images approached data from other fields of inquiry. A collective emphasis on the functionality of those images as tools characterized a number of disparate genres that featured diverse groups of pictures. Despite superficial differences, however, one thing united them: the images' desire to instruct, to be memorable, repeatable, and searchable. Together, these diverse illustrations function surprisingly well as a tool-kit. We can say that the metadata provided by their knowledge-making tendencies linked them and allows us to group them together as images with epistemological functions. In their profound distillation of knowledge disciplines, the manuscript's pictures exhibit a sovereignty that points to new knowledge-making practices and their reception.[12] This codex therefore sheds light on early modern cross-referencing, collecting practices, as well as the role of printed books in shaping the visual horizon of amateur viewers.

Short vernacular pamphlets ardently embraced the eye's role in their pictorial programs. In cosmographic and physiognomic texts of the 1520s and 1530s, we see the eye of the viewer being beckoned by images to discriminate among visual phenomenon. Cosmographic images compelled the eye to make visual judgments: for example, in Peter Apian's *Cosmographicus Liber*, active eyes peer out at topographic views, or sight the altitude of towers, or reside at the center of diagrams of lunar phases. The display of the paired faces in physiognomic texts also presented a surface whose format implied

11 For the *Liber Quodlibetarius* in the tradition of the Secretum Secretorum, see Leitch, "Getting to How-To," pp. 635–659.

12 Mulsow, "History of Knowledge," pp. 159–188, with comments by Lorraine Daston and response of the author.

an eye in the active scanning those images demanded. Without the texts in which they were originally embedded, such images accumulate in the manuscript as the residue of knowledge-making gestures in cosmography and physiognomy. The "knowing" or "epistemic" image has been invoked in the secondary scholarship to refer to a broad range of knowledge-based pictorial gestures that include organization, collecting, demonstrating theories, scientific tools, or direct reference to natural objects. These uses are succinctly cataloged in Alexander Marr's 2016 essay "Knowing Images," which also vets the relevance and limitation of this term.[13] To the current inventory of types of epistemic images, we can also add images of the elements of the heavens, faces in profile, or palms marked with lines for palm reading that built visual fluency or literacy.

Early modern printed books hailed readers as users, advertising themselves as trusty companions in do-it-yourself endeavors and presented visual interfaces that could be mined for specific information. Searchable interfaces became important to readers of books increasingly designed as how-to manuals.[14] A new functionality attends these images for everyday observers, specifically helping them measure, gauge, and assess things empirically. New sensory and cognitive exercises were inspired by these prints, and their designs were increasingly important for a user who would privilege easy call-up, condensed information, and tools to store and generate new data.[15] This essay will argue that the presentation of images in *Liber Quodlibetarius* as visualized knowledge sharpened practical skills already active in their printed source material. While the manuscript's images borrow this ontology from the printed images, they package that information in pictorial ways that could be easily referenced. The manuscript compiler mobilized the knowledge appearing in the printed books under the guiding principles of useful aims, inviting novel empirical activity and diversifying the differentiated cognitive potential of the reader. In some cases, the codex's formulaic reformatting of images even surpassed those printers' strategies to clarify information via images. Reflecting on the book as a repository of knowledge, this paper also proposes an epistemology for the printed images that comprise them. As technical shorthand for knowledge-based pursuits, those images spurred new investigations of the reader's local environment.

13 Marr, "Knowing Images." Marr discusses the range of use of this term by scholars Martina Heßler, Dieter Mersch, Christoph Lüthy, Alexis Smets, and Lorraine Daston.

14 Enabling knowledge-based queries as a proprietary feature of printed images underwrote the grouping of prints articulated by Dackerman, *Prints in the Pursuit of Knowledge*.

15 Karr Schmidt and Wouk, *Prints in Translation*, pp. 6–7.

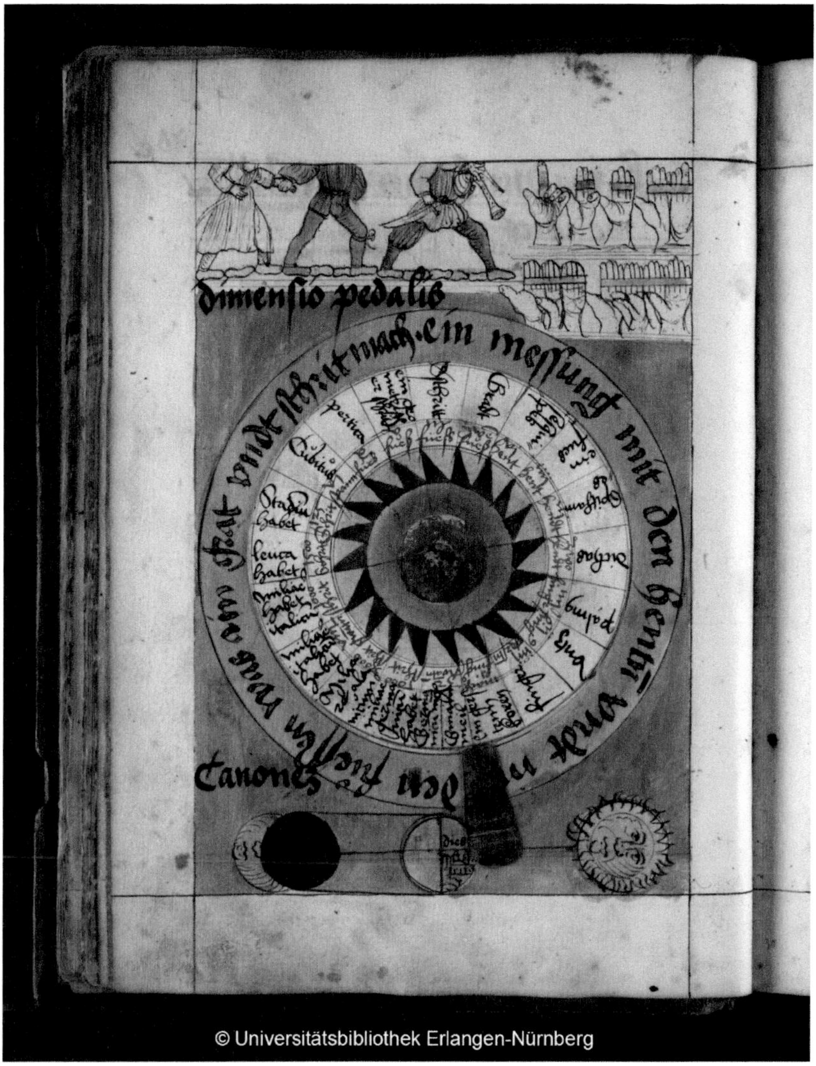

Figure 8.2 Master WR, Benedictus Rughalm, *Liber Quodlibetarius*. University Library Erlangen-Nuremberg, Ms. B 200, fol. 27v.

III. Cataloging Bodily Tools

Broken down into component pieces, the images that appear on a single page of the *Liber Quodlibetarius* (Figure 8.2) reveal themselves as computational tools. In the center floats a richly inscribed moveable dial. Images that demonstrate how to measure distances float above the dial; a proof that demonstrates the earth's sphericity hovers beneath it. Each of these images appear as free-standing data elements whose connection to each other might need elaboration. To fully grasp their meaning, or we might say their use, the reader must consult their common source: a printed cosmography from 1524. The manuscript's compiler culled these designs from different pages of the printed *Cosmographicus Liber* (Landshut: Weyssenburger, 1524), a cosmographical text that was the brainchild of mathematician and astronomer Peter Apian, who served on the local mathematics faculty and operated his own printing press in Ingolstadt.[16]

Apian's printed *Cosmographicus Liber* was a zippy, updated, pictorially rich quarto edition that served a demand for digestible cosmographic information. One of the first books to market scholarly astronomy to a wide amateur audience, the *Cosmographicus Liber* converted dense geographic and astronomic literature in the Ptolemaic tradition into clear and forceful images. Apian designed his text for the scientific laymen: its pages were rife with diagrams that centered man at the nexus of a world that, with the aid of a book, could be experienced and observed first-hand. Featuring visual demonstrations that animated and explained the celestial sphere, Apian's manual is a practical guide for the amateur mathematician and stargazer.[17] Apian's strategic use of illustrations and diagrams removes this book from the purely scholarly realm and tentatively situates it within the library of useful literature. According to Tom Conley, the way in which the *Cosmographicus Liber* activates the limbs of the reader in the service of an experiential understanding of the world reinforces Apian's cosmography as an art to be apprehended somatically.[18] To this, we would add that Apian's illustrations also insist on the body's agency in their appeal to the idea of visual knowing.

The *Liber Quodlibetarius* mobilizes Apian's images into independent tools for deriving data. The top line of fol. 27v was inspired by Apian's demonstration of how to measure distances via strides and hands (*Cosmographicus Liber,* fol. 34).[19] Hands and feet were critical common denominators of early modern empirical navigation of the world: here they are exploited for their

16 Röttel, *Peter Apian*; Grimm, "Geadelte deutsche Buchdrucker-Familien," pp. 257–271.

17 Gaida, "Reading Cosmographia," pp. 277–302.

18 Tom Conley, *The Self-Made Map*, pp. 62–72. Conley raises the intriguing possibility that the woodcuts of bodily parts in Apian might be related to tradition of anatomical blazons (69). To this we would add that they also evoke the focus on individual body parts featured in contemporary physiognomies.

19 https://archive.org/details/Cosmographicus00Apia/page/n45/mode/2up (accessed 3 July 2022).

epistemic agency.[20] The manuscript's distillation of images from Apian succinctly advocated for the body's agency in negotiating outdoor experience. They show how astronomy and mathematized geography could help readers find their terrestrial coordinates by using their hands, feet, strides, and eyes to measure their locations. The eye is invoked by the bottom fragment from a proof that demonstrates what kind of astronomical knowledge can be derived from faithful observation of the lunar cycle. Each of these images point to knowledge that can be obtained from empirical experience.

These bodily agents were chosen to frame the page's centerpiece: a decorative spiked volvelle with a tab to move it. This computational tool has no precedent in Apian's printed text, but it shows well how the manuscript modified data in the *Cosmographicus Liber* by foregrounding the searchability of that information. An inscription on the green exterior field of the central dial also points back to somatic investigation: ein messung mit den henten undt mit den *fůessen* was ain grat undt schrit mach[er]— "measuring by way of hands and feet/ which units comprise a degree and a stride." With this caption, the dial announces itself as a computational tool that converts theoretical measurements in Latin into useful vernacular ones. The wheel's two layers converts common units found in black in Latin on the circle's exterior into measurements based on hands and feet, penned in rubricated vernacular German in the interior. This data cued to body parts could be more easily deployed by the book's vernacular reader.

The wheel's formal resemblance to the nocturnal dial, a physical tool used by astronomers, reminds the viewer that the *eyes* were yet another organ that could be used to calibrate sightings. Instructions for producing such a device can be found in Apian's printed text, along with an explanation of how the reader could learn to tell the local time by determining the relative position of bodies in the night sky. For this, the compiler constructed a round dial that would establish continuity with the codex's graphic design around centralized roundels. More than just well-engineered design, however, the dial has become a theoretical database analogous to an actual tool: a calculator that stores and translates common units of measurement.

A reference to the nocturnal dial appears a few pages after this "fictive" dial (at fol. 33r), where it is held by a man pictured measuring the height of stars with such a device in the lower left corner of the page. The

20 Pamela Smith's work on the embodied experience that drove early modern artisanal pursuits can be usefully brought to bear on Apian's methods, and by extension, those at work in the *Liber Quodlibetarius*. Smith, *The Body of the Artisan*. See also Sullivan and Wear, "Materiality, Nature and the Body."

compiler elides images from two facing pages of the *Cosmographicus Liber*'s appendix.[21] The nocturnal appears adjacent to a diagram of lunar observations in the *Liber Quodlibetarius*; the compiler has connected lunar phase observations with the related pursuit of night-time data gathering: finding the hour of night. The manuscript's juxtaposition of these two programs of evening observation argues that these are related empirical acts. It showcases for the reader how to generate useful data from the night sky, or what data sets can be derived from a program of visual observation.

As early sixteenth-century printed books refocused the study of disciplines like cosmography onto pictures of the instruments themselves, the pressure of utility bore down on the images. Some tools withstood this force admirably and were even shaped by their aesthetic presentation of data.[22] Indeed, elegantly presented data sets such as the measurement dial could obscure the distinction between real and fake tools. A fake tool could introduce the reader/viewer to previously unknown fields of disciplinary inquiry, or even invent one. Volvelles showcase the way in which astronomical and astrological information was made user-friendly by synthesizing the material into interactive media. This data became searchable through the use of movable wheels that could help the reader derive the number of hours of sun and moonlight, for example. Other volvelles assisted searches that would help the reader schedule hygenic and curative activities of the months in order to balance bodily humors: the early modern user would spin the dial to determine if it was propitious to bathe in Pisces or to let blood in Taurus. Volvelles were themselves activated by a search for practical data. We can thus add the variety of empirical pursuits linked in the moveable diagrams to other epistemic instruments already cataloged by historians of science.

IV. Horse Hardware

Among the other tools the *Liber Quodlibetarius* collected was hardware used to domesticate horses. The manuscript culled designs for horse bits from printed literature that explained how the characters of horses could be conditioned by types of bits. While the earliest printed bit-books were

21 Apian, *Cosmographicus liber*. See appendix at https://archive.org/details/Cosmographicus00Apia/page/n119/mode/2up (accessed 3 July 2022).

22 Karr Schmidt, *Interactive and Sculptural Printmaking*, pp. 245–58; Marr, "Ingenuity in Nuremberg."

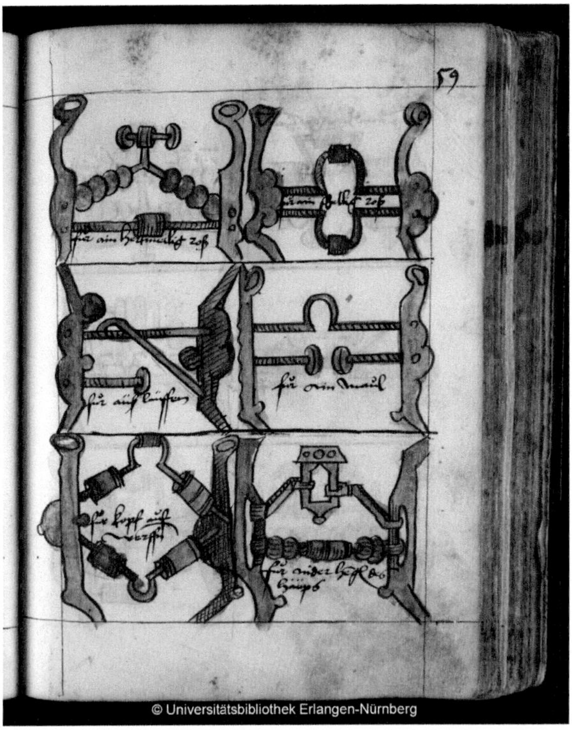

Figure 8.3 Master WR, Benedictus Rughalm, *Liber Quodlibetarius*. University Library Erlangen-Nuremberg, Ms. B 200, fol. 59r.

printed in Augsburg and Regensburg in the 1530s, the manuscript shares content with printed manuals that proliferated later in the sixteenth century.[23] Ten pages of the *Liber Quodlibetarius* (UER fos. 58r–62v) display designs of the elaborate ways contemporary equestrians fantasized about imprisoning the tongues of horses.[24] Unlike the single page designs readers would encounter in the printed sources, the manuscript shapes a collection of highly wrought bits into a serial display. Because each bridle bit sports slightly different furnishings, it seems that the compiler wanted to feature them in their individuality.[25] However, those bits presented in *Liber Quodlibetarius* radically epitomize the more expansive detail found in their printed sources (Figure 8.3). These images highlight their sovereignty

23 See Cuneo, "Just a Bit of Control," p. 144, locating seventeen editions of such texts printed between 1532 and 1639 in the collections of HAB and BSB Munich.

24 For this rhetorical flourish, I thank my colleague François Dupuigrenet Desroussilles.

25 This degree of particularity mirrors the highly differentiated forms of knowledge that we see in technical images of machines. See Popplow, "Why Draw Pictures of Machines?," p. 48.

as visual implements and demonstrate confidence on the part of compiler that they could distill the genre's function. Instead of combing through passages of texts to read about a design that would correct an unruly horse, the viewer of the *Quodlibetarius* could simply search for the bit that would tame that particular horse's nature.

Printed books of bridle bits can illuminate generally how hardware designs began to govern the repertoire of equestrian knowledge. The extent to which representations of tools take over the field of hippology is reflected in the variety and particularity of the designs of these items. Fantastic images like these catapulted instrumentation into the foreground of trades, occupations, and training processes, leaving the text to back up warrants made by the pictures themselves.[26] The bits, presented here as independent data, likely amplified the repertoire of designs already in circulation. The manuscript's group of designs could function as a model book for a blacksmith or a spur maker.[27] This need for variety was apparently strong enough that one early printer of bit books resorted to stamping a portion of the image simply in order to generate them expediently, a practice that produced somewhat sloppily registered designs.[28] The bits in these books seem to be prototypes for custom fabrication; certainly no blacksmith would have had such a wide assortment on hand for "ready to wear" purchases. They, like many images in other technical literature, are speculative.[29] They show how theoretical reflections on the behavior of horses translated into the *fantasia* of instruments.

None of the images of horse bits in this section are titled or captioned as such, rather they seem to function as sovereign images, and ones that viewers could apparently easily interpret: this suggests that early modern viewers of this book might have been better visually informed about these items than the average modern non-equestrian. So confident was the compiler of the images' sovereign ability to embed knowledge that only sparse text was needed as an accompaniment to the pictures. All that remains from

26 See Hall, "The Didactic and the Elegant," p. 36.

27 Cuneo, "Just a Bit of Control," p. 149. Mang Seuter, stable master to Marx Fugger, printed luxury editions of large format and highly wrought decorative bit books, beginning in 1584. In these later editions, the writers were the craftsmen and spur makers themselves who assert their authorial presence and boast that they had skills to bring the specialized bits to fruition.

28 Hier inn begriffne pis, zaigen clärlich an, wie ain yedess Roß, iüng und alt soll gezämbt, dadurch ime, angenomene pöse gewonhait ubelstandt, [...] benomen [...] werden mag. [Augsburg?], [c. 1530], VD 16 G 592. HAB, A: 108.14 Quod. (3). I am grateful to Abigail Youngblood and other members of a graduate seminar at Florida State University in the spring of 2019 for discovering this and pointing it out (Stephanie Leitch).

29 Popplow, "Why Draw Pictures of Machines?," p. 43.

the older manuscript tradition of *Hippiatria*, a genre devoted to the care and medical treatment of horses, is now the character of the horse that the bits were meant to condition. The serial nature of these images suggest that they are technical trappings inspired by another knowledge regime: the profiling of the horse's character. The page itself guides a search for exactly the restraint that should be used for controlling horses of different stripes, for example: the bit in the top left corner is to be used for a *hertmeilig* or an aggressive horse. The bit on the upper right will tame a *schellig* or wild and excited horse. Desire to meet the demand to correct for a broad range of character diagnoses seems to have spurred the fabrication of pictorial tools so marvelous that they began to eclipse medical care. The *Quodlibetarius* presents *hippiatria* as a character-driven science that the book makes searchable, bit by bit.

V. Chiromantic Analysis in Print and Handwriting

The idea that the surface of the body could serve as a map of decipherable information – both for empirical pursuits like measuring, and also as a clue to human nature – was popular in the epistemic literature. The concept of a person's nature (or *complexion*) gradually shifted in early modernity, from the interiority of humoral disposition and the balance of bodily fluids to externally searchable traits that gave rise to a number of genres dealing with identity politics, such as chiromancy and physiognomy.[30] Since ancient times, the prognostics of chiromancy (palm reading) tried to determine the mental and physical characteristics of humans, concentrating on the hands, considering the form, texture, color, and length of fingers, as well as the palm's lines and features. To judge from the proliferation of illustrated printed volumes in the sixteenth century, the art of reading marks on the palm became an accessible and popular type of diagnostic skill enabled by the print market. A spate of printed volumes, variously titled *Die Kunst der Chiromantzey* and *Phisionomi und Chiromanci*, circulated post-1520s in Strasbourg, roughly presenting content first introduced in print by Johannes Hartlieb's blockbook *Die Kunst Ciromantia* (Augsburg: Jörg Schapf c. 1485/95). These can be considered thematic precursors of the material that appear in *Liber Quodlibetarius*.

The collection of hands in the manuscript (fos. 81r–85v) (Figure 8.4) reduces information available in a number of popular printed contemporary

30 Groebner, "Complexio/Complexion."

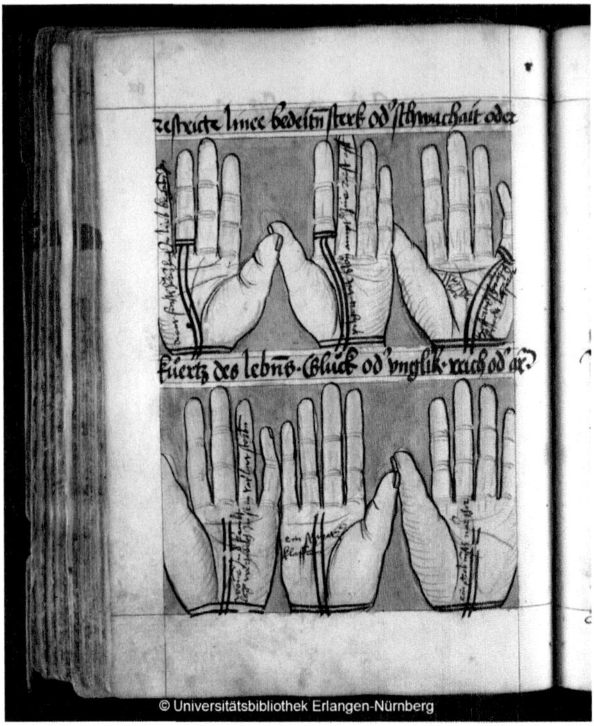

Figure 8.4 Master WR, Benedictus Rughalm, *Liber Quodlibetarius.* University Library Erlangen-Nuremberg, Ms. B 200, fol. 82v.

palm reading manuals, such as the *Phisionomi und Chiromanci* printed c. 1540 in Strasbourg by Jakob Cammerlander.[31] The ten-page image sequence of hands omitted the extensive diagnoses spelled out by the printed texts, stressing instead how those images could visually cue palm readings. The captions that snake around the images synopsize the major palm axes and remind readers to alternate hands when conducting their analyses on either men or women: "in men, judgments are made on the right hand; for females, the left hand is used" (fos. 81v–82r). These images condense textual descriptions from the printed editions with the intent to summarize chiromantic content via the visual shorthand of pictures.

In a page that is typical of printed chiromancies (Figure 8.5), we can see that images of hands reside near text passages that articulate more extensive diagnoses. But a different tactic is active in the manuscript where the individual hands are collapsed with short explanatory captions in a single visual frame, connecting closely the expressions of text and image.

31 See for example, Cocles, *Phisionomi vnd Chiromanci.*

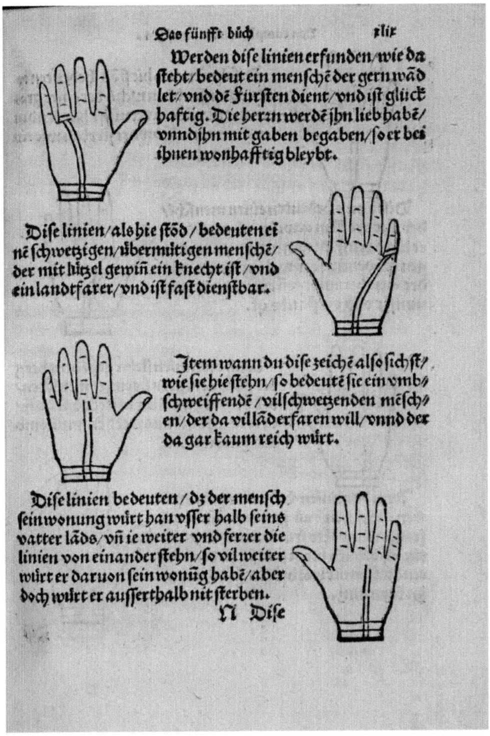

Das fünfft bůch xlix

Werden diſe linien erfunden/wie da ſteht/bedeut ein menſchē der gern wād let/vnd dē Fürſten dient/vnd iſt glück haftig. Die herꝛn werdē jhn lieb habē/ vnnd jhn mit gaben begaben/ſo er bei ihnen wonhafftig bleybt.

Diſe linien/als hie ſtōd/bedeuten ei nē ſchwetzigen/übermütigen menſchē/ der mit lützel gewiñ ein knecht iſt/vnd ein landtfarer/vnd iſt faſt dienſtbar.

Item wann du diſe zeichē alſo ſichſt/ wie ſie hie ſtehn/ſo bedeutē ſie ein vmb-ſchweiffendē/vilſchwetzenden mēſch-en/der da vil lāder faren will/vnnd der da gar kaum reich würt.

Diſe linien bedeuten/dz der menſch ſein wonung würt han vſſer halb ſeins vatter lāds/vñ ie weiter vnd ferꝛer die linien von einander ſtehn/ſo vil weiter würt er daruon ſein wonūg habē/aber doch würt er auſſerthalb nit ſterben.

N Diſe

Figure 8.5 Bartolommeo della Rocca Cocles: *Phisionomi vnd Chiromanci*. Straßburg: Jakob Cammerlander 1540. Herzog August Bibliothek Wolfenbüttel: H: N 97.4° Helmst. (7), Xlix.

Interpretations of the palm lines are inscribed and summarized within the manuscript's hands like keywords on top of the image. Information in headings present the palm lines that reflect opposition: e.g. which axes of the palm indicate strength, weakness, brevity of life, happiness or misfortune, poverty or wealth. Also included are the essential conditions of human existence, and the success or failure of life plans – these are all expressed in short and succinct ways. The scribe avoided the superfluous text of the source edition, retaining only a brief caption and the image of the hand on which the diagnosis was noted (see Figure 8.4). The text that runs between the images announces the types of predictions that can be made by tracking certain axes on the palm. A palm reader would consult this page when they encountered creases in a subject's palm that presented as double-lines: *zestricte linee bedeiten sterk oder schwachait oder kůertz des lebens/Glůck oder vnglik/reich oder arm*, or "double lines mean strength or weakness or a short life. Happiness or misfortune, rich or poor."

To distinguish among palms with a variety of double lines, the reader now refers to the images themselves. The way the scribe notes information on the hands themselves points to a shift in focus to particular qualities. For instance, the inscription on the top leftmost hand (see Figure 8.4) tells us that two continuous lines descending from the left-hand ring finger signal an inveterate pleaser: someone who can cultivate the loyalty of princes and serve them admirably: *dient fursth herren / hant jn lieb / begaben jn*, or "serves princes, gentlemen. They value him (appreciate him), shower him with gifts." Here, the images of hands function as touchstones for diagnostic practices, moving assessments to the surface of the palm itself. On the palm in the center of the top row, the inscription on the unbroken double lines ascending to the ring finger draws the reader's attention to the subject's proclivity for travel: *pilgert vil / ston ampter züe / reist offt*, or "undertakes many pilgrimages / is suitable for various appointments / travels often."

Unbroken lines at the palm's center (as seen on the center hand in the bottom row) (see Figure 8.4) indicate someone with the proverbial gift of the gab: *ein schwatzer, klaffer*, or "a chatterbox, yapper." The perforated lines on the hand at the bottom right indicate *Ein grob mensch natur thor* or "a rude man, with the nature of a fool." These captions aphorize the longer diagnoses spelled out in printed chiromancies. But the *Quodlibetarius*'s systematic presentation of hands enable a quick search: it allows the images to be easily scanned for the thickness and length of lines. These particulars inscribed on the images themselves provide quick reference material for the amateur reader of palms, almost obviating the need to consult the printed text for more indepth diagnoses. Like the chiromantic approach to horses unpacked by the images of bridle bits, these hands also embed diagnostic data that specifically signal a field of knowledge-making. Many such aspects of judicial astrology are featured in *Quodlibetarius*, represented mainly by the images that were to assist such diagnoses.

VI. Physiognomic Analysis

The repertoire of heads found in the *Quodlibetarius* also codify a set of visual tools; these heads are for the practitioner of facial readings. The art of physiognomy, or the reading of character from profiles and facial characteristics, is synopsized by images of faces found on three successive pages (Figure 8.6). While the faces are copied from a printed source, little attempt is made to transfer the specific diagnoses of the physiognomy edition from which these images come, Cocles's *Ein kurtzer Bericht*...(Straßburg: Albrecht, 1534), except

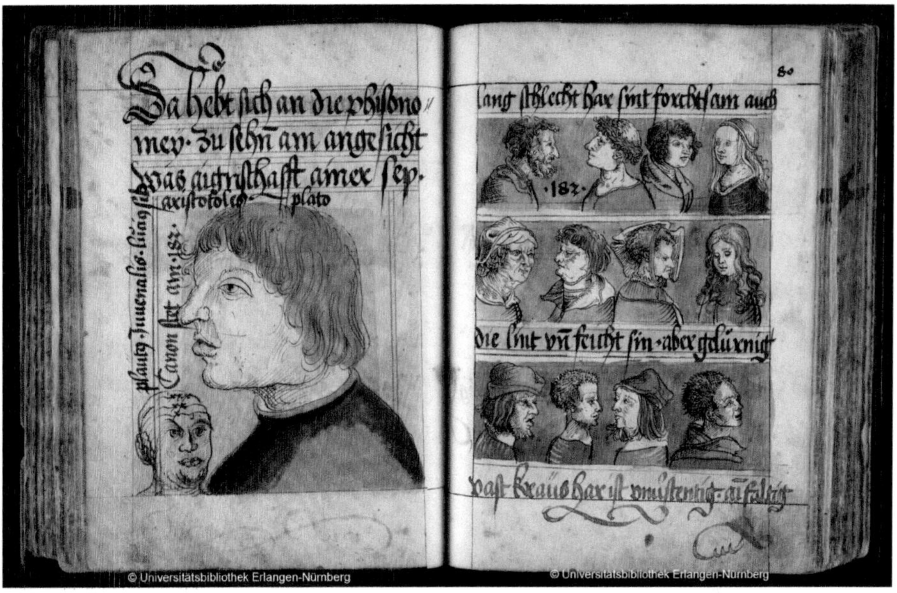

Figure 8.6 Master WR, Benedictus Rughalm, *Liber Quodlibetarius*. University Library Erlangen-Nuremberg, Ms. B 200, fos. 79v–80r.

to synopsize it by its images. The heads are presented in neat rows of mostly paired sets that communicate the principles and practice of physiognomy.[32] Paired sets of heads form the basis of diagnostic practice in period printed manuals. In the latter, each of the heads represent poles between which facial characteristics such as foreheads, noses, mouths and chin are to be assessed (Figure 8.7). Instead of expounding on the meanings linked to these profiles in the printed text, however, the manuscript's inscription alludes very loosely to the art of physiognomy: "This section pertains to physiognomy, by which one can tell character of the subject from the face."[33] The captions that snake around the following page describe only very generally symptoms that can be diagnosed by physiognomic readings. If the reader needed more information on the fundamentals of physiognomic analysis as a discipline, they were referred on fol. 79r to consult the Canons found on fol. 182v. On that page of text, one could find the crowded catalog of traits that once comprised the complexion literature (with wide-ranging remarks on voices, eyebrows, shoulders, shins, and chest hair) before it solidified around a standardized repertoire of heads.[34]

32 Leitch, "Visual Acuity."

33 "Das hebt sich an die phisonomey. Zu sehen am angesicht, was aigenschafft ainer sey."

34 http://digital.bib-bvb.de/webclient/DeliveryManager?custom_att_2=simple_viewer&pid=5281832&childpid=5282632 (accessed 3 July 2022).

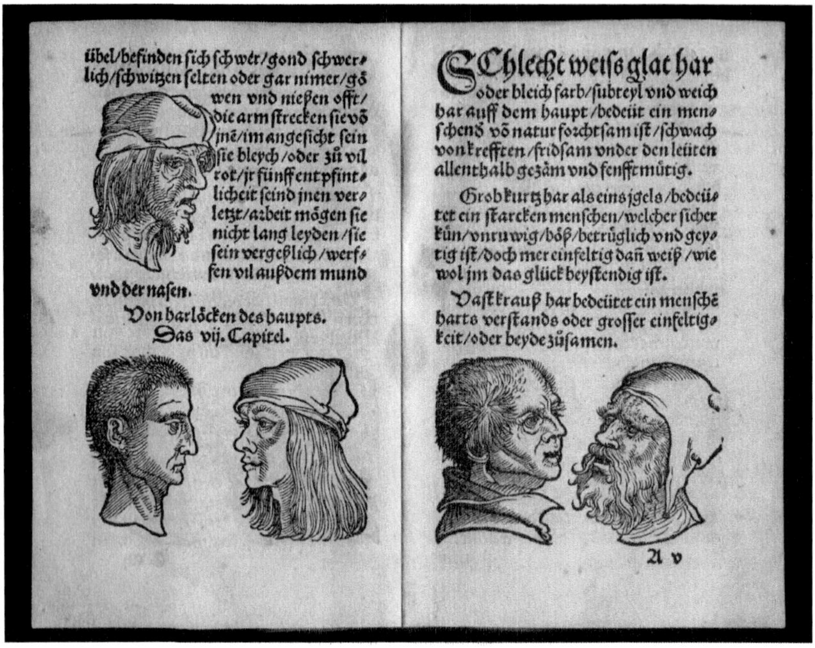
übel/befinden sich schwer/gond schwer-
lich/schwitzen selten oder gar nimer/gö-
wen vnd nießen offt/
die arm strecken sie vō
jnē/im angesicht sein
sie bleych/oder zů vil
rot/jr fünff entpfint-
licheit seind jnen ver-
letzt/arbeit mögen sie
nicht lang leyden/sie
sein vergeßlich/werf-
fen vil auß dem mund
vnd der nasen.

Von harlöcken des haupts.
Das vij. Capitel.

SChlecht weiss glat har
oder bleich farb/subteyl vnd weich
har auff dem haupt/bedeüt ein men-
schen d vō natur forchtsam ist/schwach
von krefften/fridsam vnder den leüten
allenthalb gezäm vnd senfftmůtig.

Grob kurtz har als eins jgels/bedeü-
tet ein starcken menschen/welcher sicher
kůn/vnruwig/böß/betrüglich vnd gey-
tig ist/doch mer einfeltig dañ weiß/wie
wol jm das glück beystendig ist.

Vast krauß har bedeüt et ein menschē
harts verstands oder grosser einfeltig-
keit/oder beyde zůsamen.

A v

Figure 8.7 Bartolommeo della Rocca Cocles: *Ein kurtzer Bericht der gantzen Phisionomey vnnd Chiromancy*. Straßburg: Johann Albrecht 1534. Herzog August Bibliothek Wolfenbüttel: A: 565.2 Quod. (1), Aivv–Av.

That an impulse to collect groups of related images was the priority of the manuscript compiler can be seen in the fact that when an important field was neglected, the compiler or scribe sought to repair that omission (see Figure 8.6). In these images, we can measure the growth of printed images' sovereignty on two levels. First, we see the efficacy with which pictures distilled prescriptions embedded in narrative passages, and thus encouraged empirical processes. And secondly, we witness here the degree to which images could link to other fields of judicial astrology. For example, the cross-reference to the art of metoscopy appearing at the base of the physiognomy page connects the two fields of diagnostic inquiry. A drawing of a head with an articulated brow appears at the bottom left of the introductory page of the physiognomy section, almost as if it were a footnote. Indeed, the marked forehead is drawn by a later hand, so perhaps thinking of this sketch as a scholarly footnote is not too far from the mark. This forehead inscribed with lines and symbols is of the type that we see in manuals relating to metoscopy, or the art of reading lines in the forehead, on which planetary symbols are engraved in order to show that planet's influence over the individual subject. Later printed manuals, such as *Aphorismorum metoposcopicorum* published in Frankfurt in the 1580s, instruct the user exactly how to decode such lines.[35] The drawing in *Quodlibetarius*, inserted by a later scribal hand, linked the diagnostic practices of chiromancy and physiognomy to forehead reading. This edit seeks to repair the omission of metoscopy, a related method of reading facial surfaces by decoding sets of brow lines. The second scribe's link to metoscopy on the physiognomy page completes a set of epistemic gestures related to deciphering marks on the human face.

The *Liber Quodlibetarius* shows how each representative field of epistemic practice was cued by images. It leaves no doubt that the how-to factor was the driving force that included each of these printed genres, a point perhaps best made by a final example for which no printed precedent has yet been identified. Its sovereignty, nonetheless, has clearly been inflected by the printed how-to literature. The central image of the page in question purports to demonstrate a common procedure for obtaining data, in this case, the volume of a jar of a staple of early modern Bavarian home cooking: *Schmalz*, or lard (Figure 8.8.).[36] This image of a man holding a ceramic pot with a measuring tape cascading over it explains how to calculate the container's volume. The compiler recommends a method for doing this via the creation of a paper instrument: *How to measure the weight of lard in a pot: make a note* [*on*] *parchment, and find* [*the measurement*] *against the red numbers* [*provided*] *below*.[37] The scale of units to which the reader's DIY tool should be calibrated spans the next few folios at the bottom (UER, fos. 55r–56r), extending through images that depict playing cards, juggling instruments, games of chance, and nature prints. This image attempts to elevate the calculation of a volume of lard to the level of an epistemic act. Certainly, the didactic weight of the other images surrounding it helped shore up the scientific veneer of this experiment.[38] Even while the scale looks irregular to present-day viewers (the units are randomly subdivided until they reach 62), the force of the book's other technical images unequivocally recommends this device as a useable tool. Passing through the bottom of pages devoted to warnings about games of chance and quantifiable juggling tools, the red ruler passes through the nature prints taken from herbs and plants, so-called *Naturselbstdrucke*. These prints are epistemic images of a related order, inked indexical traces of actual leaves. The measuring tape at

35 Hájek z Hájku, *Aphorismorum metoposcopicorum libellus unus*.

36 http://digital.bib-bvb.de/webclient/DeliveryManager?custom_att_2=simple_viewer&pid=5281832&childpid=5282390 (accessed 3 July 2022).

37 *Zü messen wie vil pfunt schmaltz jn aim hefen sint: mach ein zetl* (*zettel*) *pergamenen/ suchs ab gegen den untern roten zifferen.*

38 These pages are cinched between images of star charts and a map with coordinate markings.

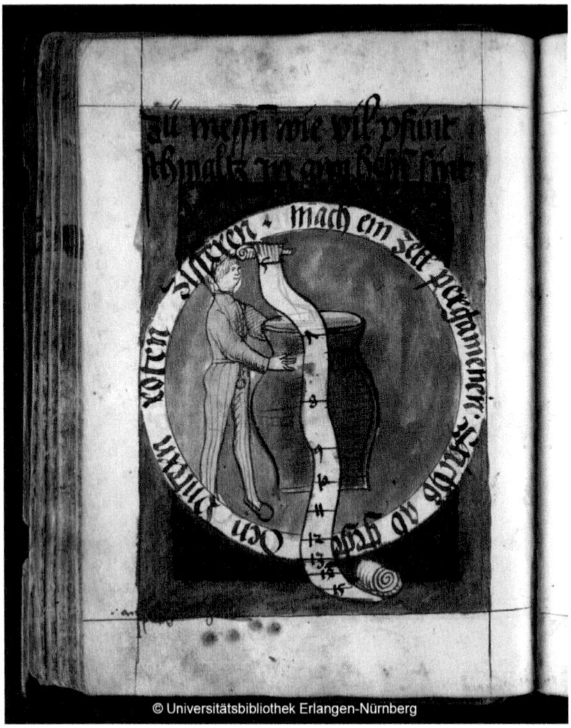

Figure 8.8 Master WR, Benedictus Rughalm, *Liber Quodlibetarius*. University Library Erlangen-Nuremberg, Ms. B 200, fol. 54v.

the edge of the image serendipitously provides a scale for their dimensions. Epistemic warrants served by the manuscripts' images from other printed texts cast a technical veneer over content more difficult to quantify like lard, but seemingly also in need of reconciliation. The confidence that lard could be measured by this method was constructed by the expectations of surrounding technical images.

VII. Conclusion

In conclusion, the *Liber Quodlibetarius* provides us with an encyclopedic view of the kinds of authoritative information, or data, which was circulating in printed sources in early modernity. The period of the compilation of this codex coincided with the intense visualization of most early modern how-to genres in print. The manuscript's concentration on the way in which such books structured information through illustrations and making knowledge verifiable through visual models illuminates the commonality amongst diverse genres like cosmography, physiognomy, chiromancy, and the science of gauging a volume of lard. In transposing and reworking mostly the images from such printed manuals, the manuscript's compiler deputized that data to speak for those genres in the shape of searchable and arguably more useful, formats. Data from these books were included on the strength of their visual formatting and the degree to which they invited observations that could be checked against images. Insofar as they did this, as this essay has argued, they encouraged and shaped the contours of early modern searchable science. Is "epistemic" a fitting modifier for such diverse tools that communicate knowledge-making claims? It is already a fraught term in the scholarship, one whose capaciousness causes it to snag between an image that demonstrates a theory or a method of obtaining knowledge on the one hand, or a "working object of science" on the other. Most of the images in the manuscript tread a tightrope between these ontologies. Some show the compiler's intention to cram images of epistemic inquiry into a communal storehouse; others express the functionality of existing tools known in part from the repertoire of data established by prints. Others still forged novel epistemic commerce amongst the images themselves. Almost all of the images serve to populate the visual field of the compendia, "collecting" representative samples from most known sources of printed knowledge.

Bibliography

Apian, Peter. *Cosmographicus liber Petri Apiani mathematici studiose collectus.* Landshutae: impensis P. Apiani, 1524.

Baigrie, Brian Scott. *Picturing Knowledge: Historical and Philosophical Problems Concerning the Use of Art in Science.* Toronto: University of Toronto Press, 1996.

Cocles, Bartolommeo della Rocca. *Phisionomi Vnd || Chiromanci.|| Eyn Newß Complexion Büch=||lein/ Der Menschen Geburt/ Sitten/ Geberden Vnd Neyg=||ligkeyten/ Auß Der Phisionomi/ Chiromanci/ Den Siben || Planeten/ Zwölff Zeichen/ Vnnd Den XXXvj. Bildern || Deß Himels/ Auch Nach Den Zwölff Monaten/ Leichtlich || Vnd Grundtlich Zulernen/ Auß Platone/ Aristotele/ Pto||lomeo/ Hali/ Albumasar/ Vnnd Johanne Künigsper=||ger #c. in Fünff Büchlin Kurtzlich Gezogen/ Vnd Im Sech||sten Werden Erzelt Wunder Gestalt Lewt Mit Jren Siten.||*. Strasbourg: Cammerlander, 1540.

Conley, Tom. *The Self-Made Map: Cartographic Writing in Early Modern France.* Minneapolis, MN: University of Minnesota Press, 1996.

Cuneo, Pia F. "Just a Bit of Control: The Historical Significance of Sixteenth- and Seventeenth-Century German Bit-Books." In *The Culture of the Horse: Status,*

Discipline, and Identity in the Early Modern World, edited by Karen Raber, Treva J. Tucker, and Pia F. Cuneo, pp. 141–173. New York: Palgrave Macmillan, 2005.

Dackerman, Susan, ed. *Prints and the Pursuit of Knowledge in Early Modern Europe*. Cambridge, MA: Harvard University Art Museums, 2011.

Eamon, William. *Science and the Secrets of Nature: Books of Secrets in Medieval and Early Modern Culture*. Princeton, NJ: Princeton University Press, 1996.

Ferrari, Michele C. *Le monde en images. Une encyclopédie bavaroise au début du XVIe siècle*. Dijon: Éditions Faton, 2021.

Gaida, Margaret. "Reading Cosmographia: Peter Apian's Book-Instrument Hybrid and the Rise of the Mathematical Amateur in the Sixteenth Century." *Early Science and Medicine* 21, no. 4 (2016): 277–302.

Gersdorff, Hans von. *Feldbůch der Wund-Artzney, sampt vilen Instrumenten der Chirurgey*. Strasbourg: Johann Schott, 1517.

Grimm, Heinrich. "Geadelte deutsche Buchdrucker-Familien im 16. und 17. Jahrhundert." *Gutenberg-Jahrbuch* 36 (1961): 257–271.

Groebner, Valentin. "Complexio/Complexion: Categorizing Individual Natures, 1250–1600." In *The Moral Authority of Nature*, edited by Lorraine Daston, and Fernando Vidal, pp. 361–383. Chicago, IL: University of Chicago Press, 2010.

Hájek z Hájku, Tadeáis. *Aphorismorum metoposcopicorum libellus unus*. Frankfurt am Main: Heirs of Andreas Wechel, 1584.

Hall, Bert S. "The Didactic and the Elegant: Some Thoughts on Scientific and Technological Illustrations in the Middle Ages and the Renaissance." In *Picturing Knowledge: Historical and Philosophical Problems Concerning the Use of Art in Science*, edited by Brian Scott Baigrie, pp. 3–39. Toronto: University of Toronto Press, 1996.

Henderson, Felicity. "Images." In *A Handbook of Editing Early Modern Texts*, edited by Claire Loffman and Harriet Philipps, pp. 70–73. New York: Routledge, 2018.

Hier Inn Begriffne Pis, Zaigen Clärlich an, Wie Ain Yedess Roß, Iüng Und Alt Soll Gezämbt, Dadurch Ime, Angenomene Pöse Gewonhait Ubelstandt, … Benomen … Werden Mag. VD 16 G 592. [Augsburg?], 1530.

Jordanova, Ludmilla. *The Look of the Past: Visual and Material Evidence in Historical Practice*. Cambridge, MA: Cambridge University Press, 2012.

Karr Schmidt, Suzanne. *Interactive and Sculptural Printmaking in the Renaissance*. Leiden: Brill, 2018.

Karr Schmidt, Suzanne, and Edward H. Wouk. *Prints in Translation, 1450–1750: Image, Materiality, Space*. Abingdon: Routledge, 2017.

Krohn, Deborah L. *Food and Knowledge in Renaissance Italy: Bartolomeo Scappi's Paper Kitchens*. New York: Routledge, Taylor & Francis Group, 2018.

Kruse, Britta-Juliane. "Fachliteratur–Fachprosa, X.1 Medizinische Sammelhandschriften/ Kräuterbücher, X.2 Wahrsagebücher." In *Aderlass und Seelentrost. Die*

Überlieferung deutscher Texte im Spiegel Berliner Handschriften und Inkunabeln, edited by Peter Jörg Becker and Eef Overgaauw, pp. 340–366. Mainz: P. von Zabern, 2003.

Kusukawa, Sachiko. *Picturing the Book of Nature: Image, Text, and Argument in Sixteenth–Century Human Anatomy and Medical Botany*. Chicago, IL: University of Chicago Press, 2012.

Kusukawa, Sachiko. "The Uses of Pictures in the Formation of Learned Knowledge: The Cases of Leonhard Fuchs and Andreas Vesalius." In *Transmitting Knowledge: Words, Images, and Instruments in Early Modern Europe*, edited by Sachiko Kusukawa and Ian Maclean, pp. 73–96. Oxford: Oxford University Press, 2006.

Leitch, Stephanie. "Getting to How-To: Chiromancy, Physiognomy, Metoscopy and Prints in Secrets' Service." In *Quid Est Secretum? Visual Representation of Secrets in Early Modern Europe, 1500–1700*, edited by Walter Melion, Ralph Dekoninck, and Agnès Guiderdoni, pp. 635–659. Leiden: Brill, 2020.

Leitch, Stephanie. "Visual Acuity and the Physiognomer's Art of Observation." *Oxford Art Journal* 38, no. 2 (2015): 187–206.

Marr, Alexander. "Ingenuity in Nuremberg: Dürer and Stabius's Instrument Prints." *The Art Bulletin* 100, no. 3 (2018): 48–79.

Marr, Alexander. "Knowing Images." *Renaissance Quarterly* 69, no. 3 (2016): 1000–1013.

Martin Mulsow, "History of Knowledge." In *Debating New Approaches to History*, edited by Marek Tamm and Peter Burke, pp. 159–188. London: Bloomsbury, 2019.

Ogilvie, Brian W. *The Science of Describing: Natural History in Renaissance Europe*. Chicago, IL: University of Chicago Press, 2006.

Panse, Melanie. *Hans von Gersdorffs Feldbuch der Wundarznei. Produktion, Präsentation und Rezeption von Wissen*. Wiesbaden: Reichert, 2012.

Pettegree, Andrew. "The Renaissance Library and the Challenge of Print." In *The Meaning of the Library: A Cultural History*, edited by Alice Crawford, pp. 72–102. Princeton, NJ: Princeton University Press, 2015.

Popplow, Marcus. "Why Draw Pictures of Machines?: The Social Contexts of Early Modern Machine Drawings." In *Picturing Machines, 1400–1700*, edited by *Edited by Wolfgang* Lefèvre, pp. 17–48. Cambridge, MA: MIT Press, 2004.

Raber, Karen, Treva J. Tucker, and Pia F. Cuneo, eds. *The Culture of the Horse: Status, Discipline, and Identity in the Early Modern World*. New York: Palgrave Macmillan, 2005.

Remmert, Volker R. *Picturing the Scientific Revolution: Title Engravings in Early Modern Scientific Publications*. Philadelphia, PA: Saint Joseph's University Press, 2011.

Rudolph, Pia. "Hausbücher. Bayerische Bild-Enzyklopädien." In *Katalog der deutschsprachigen illustrierten Handschriften des Mittelalters (KdiH)*, vol. 6., edited by Ulrike Bodemann, *et al.*, pp. 466–473. Munich: C.H. Beck, 2015.

Röttel, Karl. *Peter Apian: Astronomie, Kosmographie und Mathematik am Beginn der Neuzeit. Mit Ausstellungskatalog*. Buxheim: Polygon-Verlag, 1995.

Smith, Pamela H. *The Body of the Artisan: Art and Experience in the Scientific Revolution*. Chicago, IL: University of Chicago Press, 2004.

Sullivan, Erin, and Andrew Wear. “Materiality, Nature and the Body.” In *The Routledge Handbook of Material Culture in Early Modern Europe*, edited by Catherine Richardson, Tara Hamling and David Gaimster, pp. 141–157. New York: Routledge 2017.

About the Authors

Britta-Juliane Kruse, PD Dr. (Free University of Berlin), Medievalist, Literary and Art historian in the research department of the Herzog August Bibliothek Wolfenbüttel, analyzes the text-image tradition in the Liber Quodlibetarius, with emphasis on the history of the codex, chiromancy, and the inscriptions of the images.

Stephanie Leitch, associate professor of Art History at Florida State University, investigates how images in early printed books coached visual acuity and gave rise to new disciplinary practices by cuing observation and calibrating sightings.

Part 3

Imprint, Knowledge, and Affect

9. The Hydraulics of the Soul: Jacobus Meilingius's Allegorical Schemata

Anneke de Bont

Abstract

This case study considers a large single-sheet engraving by Jacobus Meilingius entitled *Figurata Meditatio Microcosmi* (1629). Its interlinking fountains and tiered flower vessel surrounded by allegorical figures suggest at first glance a moral theme, yet closer inspection proves this composition instead presents a faithful summary of the second-century physician Galen of Pergamum's writings on the organization and elaboration of the human soul via the digestive, vascular, arterial and nervous systems. Beyond parsing the pneumatic theme, this essay identifies this print's creator as a Dutch surgeon and argues his inspiration for this unique hydraulic metaphor for the soul was rooted in both medical and religious thought.

Keywords: print, Galen, soul, Fons Vitae, allegory

Enabled by the prosperity of printing technology and motivated by a "newly invigorated info-lust," early modern artists, scientists, and theologians produced numerous images that used diagrammatic and pictorial strategies to condense theories or practices into easily digestible graphic forms.[1] The resulting visual works were tools of transforming information into knowledge. They spatialized relationships, creating worlds in which physical and conceptual proximity were one and the same. They were instruments of pedagogy, championed by scholars such as Petrus Ramus and Johann Amos Comenius, who argued that comprehension could be better achieved and retained when the studied information was manifested spatially. Print designers, both renowned and unknown, found many ways of organizing

1 Blair, *Too Much to Know*, p. 6.

Noyes, R.S. (ed.), *Reassessing Epistemic Images in the Early Modern World.*
Taylor & Francis Group, 2023
DOI 10.5117/9789463723350_CH09

these spatialized relationships, modeling them after trees or wheels, and applying them to diverse genres from politics to theology.[2] Bold lines, text-heavy chains or trodden footpaths – the map-like skeletons of schematic prints – guide beholders through these complex images, inviting them to traverse imaginatively the pictorial space as if it were real, perhaps by tracing their chosen routes with their fingers, feeling the subtle topographic ridge of ink on the printed page.[3]

Despite their diversity and abundance, early modern single-page schemata, which exist somewhere between text and image, have long skirted the attentions of art historians. That they largely remain footnotes of visual culture can be attributed in part to their dry and convoluted nature, which has made them objects of study better suited to the histories of mathematics or science, who, however, focus primarily on the information represented rather than the manner of its representation.[4] Despite this imbalance in the scholarship, there is value in considering these images from an art historical perspective. As products of their own visual culture entwined with allusions and interdisciplinarity, there is much to be learned in their untangling. The case study that follows aims to provide a glimpse into the visual richness and complexity of schematic printmaking in early modern Europe by examining issues of influence and meaning in a previously unstudied Dutch print from the Rijksmuseum's collection: Jacobus Meilingius's *Figurata Meditatio Microcosmi* (The Hague, 1629) (Figure 9.1).

This is a curious image – a medical diagram rendered in an allegorical form that was influenced, as will be shown, by a popular religious motif. Because of this devotional gloss, the Rijksmuseum's online database at

2 This was, of course, no early modern invention. The notion of spatializing information as a tool of study traces back to antiquity, as can be observed in the mnemotechnics of Cicero, and the structural forms of trees and wheels have long histories as well.

3 Schematic forms were so closely associated with accurately relayed information in the early modern period that the structures themselves became sources of validation, lending authority to the content represented regardless of its accuracy or controversy. In her study of early modern English diagrams, Lori Anne Ferrell writes that "such graphics imparted a 'look of truth' to many controversial or complex new concepts, making them more acceptable, and believable, and thus rending them (as historical theologian Christine Helmer would have it) 'beautifully true.'" Ferrell, "Page *Techne*," p. 114.

4 Only recently have early modern art historians, bolstered by the foundations laid in the mid-1990s by scholars such as James Elkins, Horst Bredekamp, and Lorraine Daston and aided by the increased accessibility offered by museum digitization projects, begun to approach these types of images. Of particular note is Susanna Berger's book *The Art of Philosophy*, which examines the ways in which images were used for instruction and the generation of knowledge within early modern philosophical study. Much of this new scholarship exists under the umbrella term "epistemic image" and is reviewed in Marr, "Knowing Images."

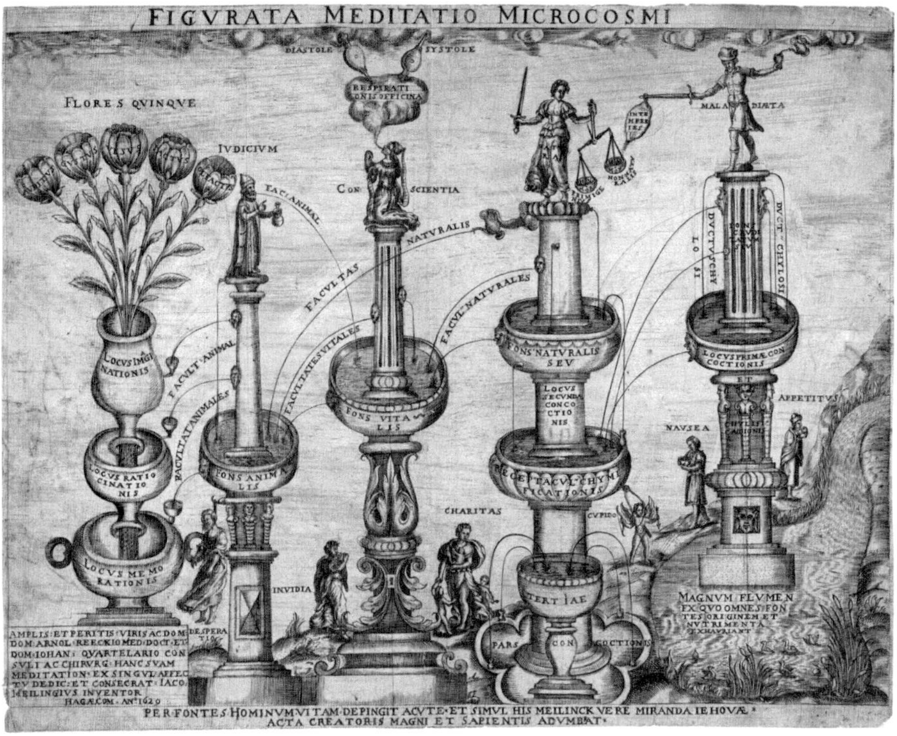

Figure 9.1 Jacobus Meilingius, *Figurata Meditatio Microcosmi*, 1629, engraving, 32.3 × 40.8 cm. Amsterdam, Rijksmuseum.

the time of this essay's completion classifies *Figurata*'s subject matter as "Christian doctrine," which is exactly what this image appears to be at first glance.[5] The interlinking Edenic fountains surrounded by Christian virtues and vices such as Charity and Envy are ostensibly more suggestive of Christian doctrine than any medical theory. The work's deliberately vague title offers no encouragement to the contrary. Only upon close examination of the Latin annotations does it become clear to the informed beholder that this image in fact allegorically relates the theory of second-century physician Claudius Galen on the tripartite organization and elaboration of the human soul via the vascular, arterial, and nervous systems.

This tension between devotional form and physiological content makes *Figurata* difficult to categorize in terms of genre or discipline. Its contextualization presents an equally challenging undertaking, as the engraving was designed by an amateur man of medicine with no other surviving prints attached to his name. Therefore, much of this study necessarily depends

5 http://hdl.handle.net/10934/RM0001.COLLECT.576072 (accessed 29 November 2021).

upon historically grounded speculation, framed in qualifying terminology such as "likely" and "perhaps" in order to explore how this amateur interacted with the visual arts. Yet, by examining such an elaborate image designed by a medical practitioner in terms of his biography and the broader visual culture with which he would have engaged – principally academic and religious images produced in the Low Countries – nuanced readings of this schematic image begin to emerge.

Meilingius's heavily annotated, single-sheet engraving uses the text-heavy water flowing between the basins of four tiered fountains and a flower vessel to plot the movements of the Galenic soul. In this way, the physiological, pneumatic sequence from digestion to sensory reception is imbued with a clear sense of sequence and directionality as it is allegorically visualized. That Meilingius conspicuously avoids naming Galen anywhere on his engraving suggests he anticipated beholders prepossessed of some understanding of Galenic theory. It is therefore unlikely that his intent with the image was fundamentally didactic with respect to the theories of Galen. Rather, the print's intended function was more likely to provide a condensed review and concise overview of the material to encourage beholders to assimilate and recall pre-existing knowledge, or to challenge beholders to link the depicted theory with its unnamed originator, all while demonstrating the designer's own mastery over the subject.[6]

To verify Meilingius's accuracy in summarizing the operational flow of Galen's linked digestive-pneumatic system, the original theory must first be sketched.[7] To begin, the Galenic soul was composed of three parts: the appetitive, spirited and rational souls, allocated respectively to the liver, heart and lungs, and brain. These pneumatic limbs each controlled a vaporous, regulatory substance (natural, vital, and psychic), which operated within the three duct systems of the body (vascular, arterial, and nervous). Galen referred to these substances as "spirits" or "faculties." Essential to his theory was the belief that the spirits' ability to function continually depended not on divine animation but on proper nutrition. Ingested food distilled

6 While Galen remained a staple of early modern medical study, it must be noted that the favor with which scholars looked upon Galenic physiology had greatly waned since the medieval period. On the precipitating debates over Galen in late medieval and Renaissance Italy, see Maclean, "Diagrams in the Defense of Galen." Challenges relevant to this study were also notably made by William Harvey, Robert Fludd, and Francis Bacon.

7 For more thorough analyses of the Galenic soul, see Hankinson, "Galen's Anatomy"; Schiefsky, "Galen and the Tripartite Soul"; Temkin, "On Galen's Pneumatology"; Vidal, *The Sciences of the Soul*, pp. 33–36. For a broader examination of the organic soul in early modern Europe: Park, "Psychology." On the contradictions within Galen's theories: Gill, *Naturalistic Psychology*.

into a milky white substance called "chyle," which, after passing through three stages of "concoction," became a crude form of blood. The liver then filtered out the impurities, making the blood a nourishing substrate for its "natural spirits," which moved through the body's veins and regulated growth. Once the blood reached the heart and mixed with air breathed in through the lungs, it became suitable to the "vital spirits," the substance operating in the arteries responsible for locomotion. Finally, when the blood reached the brain after passing through the *rete mirabile* (a network of small vessels), it was filled with "psychic spirits," which enlivened the body and held dominion over intellection and the five senses.[8]

The sequential motion of Galen's theory is clearly diagrammed from right to left in Meilingius's allegorical engraving, wherein each fountain embodies different parts of the digestive-pneumatic system and the organs that sustain them. Water spouting from the top of the first fountain on the right, the *fons cruditatum* (fountain of indigestion), trickles down through the *ductus chylosi* (chyle ducts) and the *locus primae concoctionis* (first stage of concoction) before flowing next into the *fons naturalis* (representing the liver-based appetitive soul), which contains the chyle reservoir and the second and third stages of concoction. The water, upon leaving this second fountain, is now heavy with the words *facultates naturales* (natural spirits). After passing through the third fountain, the *fons vitalis* (suggesting the spirited soul located in the heart and lungs), the water takes on the *facultates vitales* (vital spirits). Similarly, after the fourth fountain, the *fons animalis* (for the rational soul within the brain), the water is annotated with *facultates animales* (psychic spirits). A small portion of the liquid from this final fountain filters back through the personified mouth of *Iudicium* (Judgement) to replenish the *fons vitalis*, while the remainder flows forward to water the flowers of the five senses growing from the basins of imagination, reason and memory. To clarify each fountain's identity, personifications positioned around each represent the temperaments resulting from that part of the soul's proper or improper functioning; for example, the *fons animalis* is paired with the positive figure of *Iudicium* and the negative figure of suicidal *Desperatio*.[9] Meilingius's grasp on this Galenic theory, as evidenced within this image, is beyond dispute.

8 While it was not contentious in the time of Galen, the *rete mirabile* was a topic of noted debate in the early modern period. The existence of this arterial network in humans was disproved by Vesalius in the mid-sixteenth century, but discussions of the organ lingered in medical practice throughout the seventeenth century. Sugg, *The Smoke of the Soul*, pp. 66–68, 73–74.

9 Galen summarizes the ways in which imbalances within the various parts of the soul produce different emotions: Galen, *Claudii Galeni Pergameni*, pp. 103–104. These psychological theories

Beneath the flower vessel's lowest tier, which appropriately denotes the *locus memorationis* (place of memory), lies a heavily abbreviated commemorative plaque. In addition to identifying the date and place of printing, this plaque names *Figurata*'s two dedicatees, doctor of medicine "Arnol: Reeckio" (likely a Latinized and abbreviated form of Arnold Reek) and surgeon "Iohan: Quartelario" (perhaps Johannes Quartelar), as well as the print's inventor "Iaco: Meilingius" (Jacobus Meilingius), whose surname appears with the alternate spelling "Meilinck" in the image's caption.[10] Although the identities of the dedicatees remain elusive, the print's inventor was most likely Jacobus Meilingius, a surgeon and later medical doctor, who was the son of the wayward reform preacher and Remonstrant sympathizer Henricus Meilingius, and on occasion used the alternate spelling "Meilinck."

While brief biographies of Henricus exist, little is known of his son.[11] Jacobus Meilingius was born shortly after his parents' marriage in 1605 and presumably remained with them throughout his childhood in the southern Netherlands. The only concrete dates relating to his life are recorded in an unpublished, brief biography of Henricus based on since-decayed documents from the Central Bureau of Genealogy in The Hague. This document's compiler reports that Meilingius "became a surgeon, moved to Arnhem in 1636 and obtained a [medical] doctorate from [the University of] Franeker on 10 October 1637."[12] He married twice and with his second wife had a son, Henricus Jacobi Meilingius, who followed in his father's footsteps, dedicating his medical doctoral dissertation his father in 1660.[13] Meilingius likely died in 1665.[14]

Since he printed *Figurata* in 1629, eight years before earning his own doctorate, Meilingius's knowledge of Galenic physiology must have been acquired during his surgical training, perhaps under the tutelage of the elusive doctors Reek and Quartelar.[15] Where he trained is not recorded, but it is likely Meilingius completed that chapter of his education some time

are further developed in Galen and Guenther, "Secundus." For commentary and analyses, see Rocca, "Galen and Greek Neuroscience"; Gill, *Naturalistic Psychology*.

10 Plaque translation: "Jacobus Meilingius with particular affection commends and dedicates his meditation to the great and skilled men and masters, master Arnold Reek, doctor of medicine, and master Johannes Quartelar, consul [?] and surgeon." Translation assistance provided by Neil Wright.

11 On Henricus, see Nauta, *Biografisch Lexicon*, pp. 264–265; Van der Aa, *Biographisch Woordenboek*, pp. 543–544.

12 "Hij [Jacobus] werd chirurgijn, verhuisde in 1636 naar Arnhem en promoveerde te Franeker 10.10.1637 tot med.dr (zonder titel proefschrift)." Jaanus, *Een remonstrantse*, p. 6.

13 Dedication in Meilingius, *Disputatio Medica Inauguralis*, p. 2.

14 Jaanus, *Een remonstrantse*, p. 6.

15 The title of "surgeon" must, for the time being, remain a vague appellation since Jaanus's history of Henricus provides no further specification on Jacobus's education.

in or before 1629. This date can be deduced from a letter Henricus wrote to Hugo Grotius on the first of October that same year, in which he asks the Dutch humanist to look after his son, a "master in the medical arts," while he was in Paris.[16] Though perhaps exaggerated by a proud father, this title of "master" indicates a high level of medical accomplishment, and Meilingius's availability to travel suggests he had finished his studies. It is possible that Meilingius in fact designed *Figurata* to mark the completion of those studies, as a demonstration of his mastery over Galenic theory dedicated to his instructors in a manner similar to a thesis print.[17]

That Meilingius chose to represent this mastery (for whatever purpose) in this sequential schematic form is in itself unremarkable. As discussed above, numerous printmakers and publishers – especially those active in the Low Countries – used schematic compositional modes in the design of all types of images with which Meilingius would have been familiar, from political woodcuts to elaborate theological etchings.[18] He would have also encountered the simple linear diagrams in the course of his studies, as they often appeared in academic texts, functioning either as didactic tools to assist in comprehension or as spatially efficient mnemonic summarial schemata carefully arranged for easy memorization.[19]

Galen's writings were among those often glossed by diagrams, most commonly taking the format of unadorned branching or circular structures.[20] Theodor Zwinger's edition of Galen's *Ars medicae* is composed almost entirely of bracketed diagrams that connect large blocks of text. Their visual *ductus* – the leading of the mind through an argument using the eye – is rendered easily navigable through the application of bold perpendicular lines (Figure 9.2).[21]

16 "Magistro in arte medicinali." Henricus Meilingius, "Henricus Meilingius to Hugo de Groot," p. 108.

17 The theory that this print was produced for such a special occasion is supported by its large scale (32.3 × 40.8 cm). Printing at this scale would have incurred great cost and it was unlikely intended as an insert to a book, being instead nearer in scale to the thesis prints studied in Berger, *The Art of Philosophy*. On thesis prints, see also the essay by Gwendoline de Mûelenaere in this volume.

18 For example: Anonymous, *Arbor Haereseon*, 1550–1599 (Amsterdam, Rijksmuseum: RP-P-OB-78.838); Nicolaus van Aelst (publisher), *Arborem Divinae Doctrinae Summae Theologiae S. Thomae Aquinatis Utilissimam*, 1590 (Amsterdam, Rijksmuseum: RP-P-OB-38.769A, B, C and D).

19 On the use of diagrams in the early modern scientific study, see Franklin, "Diagrammatic Reasoning."

20 Late medieval and early modern examples (primarily Italian) cited in Maclean, "Diagrams in the Defense of Galen," esp. pp. 153–155.

21 Galen and Zwinger, *In Artem medicinalem Galeni*, 1. Zwinger's diagrams owe their heritage to branching schemata, the deep-seated form of European image making that is most thoroughly studied by medievalists. See, for example, Saxl, "A Spiritual Encyclopaedia"; Klapisch-Zuber,

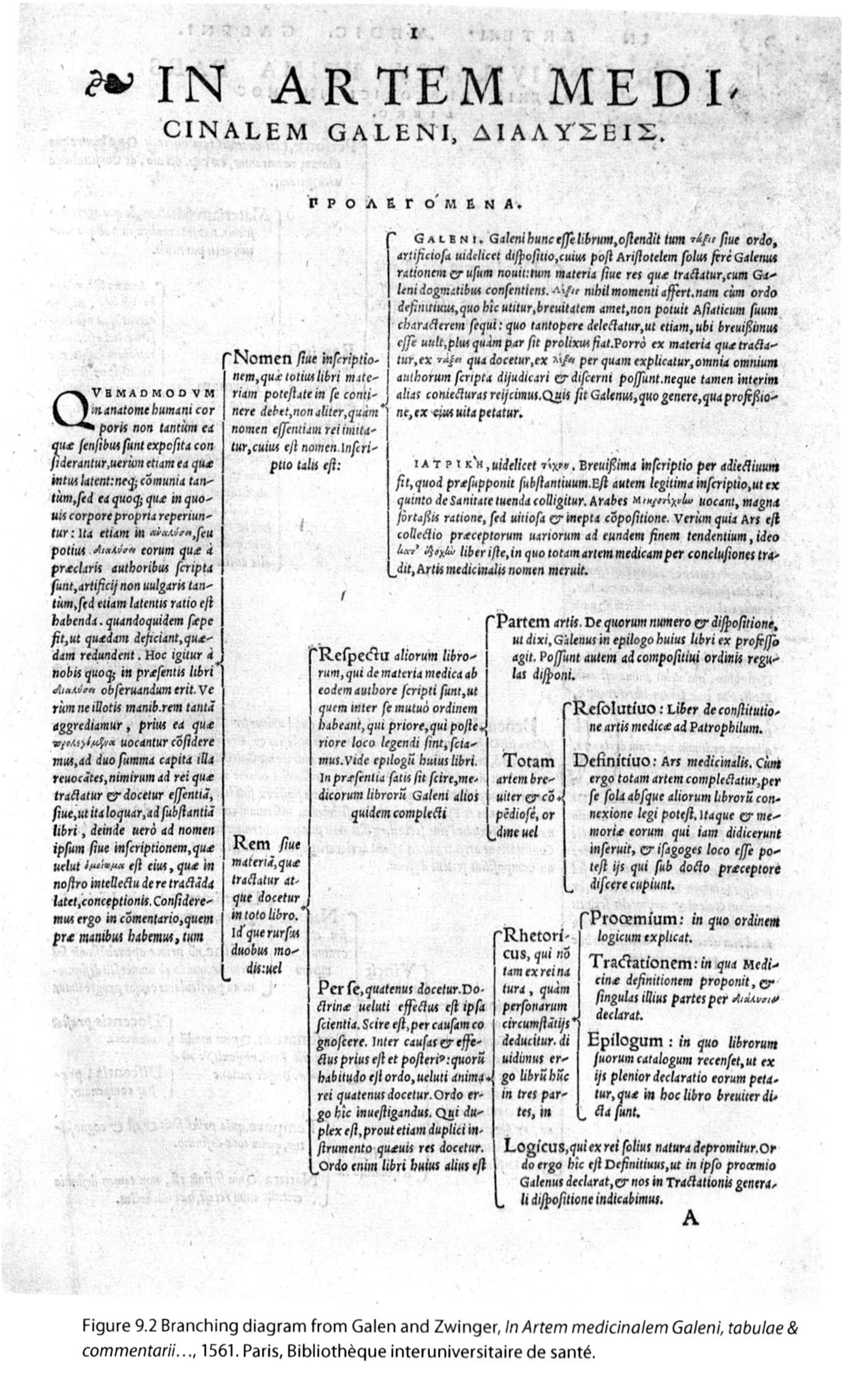

I

IN ARTEM MEDICINALEM GALENI, ΔΙΑΛΥΣΕΙΣ.

ΠΡΟΛΕΓΟΜΕΝΑ.

QUEMADMODUM in anatome humani corporis non tantùm ea quæ sensibus sunt exposita considerantur, uerùm etiam ea quæ intus latent: neq; cōmunia tantùm, sed ea quoq; quæ in quouis corpore propria reperiuntur: Ita etiam in ἀναλύσει, seu potius διαλύσει eorum quæ à præclaris authoribus scripta sunt, artificij non uulgaris tantùm, sed etiam latentis ratio est habenda. quandoquidem sæpe fit, ut quædam deficiant, quædam redundent. Hoc igitur à nobis quoq; in præsentis libri διαλύσει obseruandum erit. Verùm ne illotis manib. rem tantā aggrediamur, prius ea quæ προλεγόμενα uocantur cōsideremus, ad duo summa capita illa reuocātes, nimirum ad rei quæ tractatur & docetur essentiā, siue, ut ita loquar, ad substantiā libri, deinde uerò ad nomen ipsum siue inscriptionem, quæ uelut ὁμοίωμα est eius, quæ in nostro intellectu de re tractāda latet, conceptionis. Consideremus ergo in cōmentario, quem præ manibus habemus, tum

Nomen siue inscriptionem, quæ totius libri materiam potestate in se continere debet, non aliter, quàm nomen essentiam rei imitatur, cuius est nomen. Inscriptio talis est:

GALENI. Galeni hunc esse librum, ostendit tum τάξις siue ordo, artificiosa uidelicet dispositio, cuius post Aristotelem solus ferè Galenus rationem & usum nouit: tum materia siue res quæ tractatur, cum Galeni dogmatibus consentiens. λέξις nihil momenti affert. nam cùm ordo definitiuus, quo hìc utitur, breuitatem amet, non potuit Asiaticum suum characterem sequi: quo tantopere delectatur, ut etiam, ubi breuißimus esse uult, plus quàm par sit prolixus fiat. Porrò ex materia quæ tractatur, ex τάξει qua docetur, ex λέξει per quam explicatur, omnia omnium authorum scripta dijudicari & discerni possunt. neque tamen interim alias coniecturas reijcimus. Quis sit Galenus, quo genere, qua profeßione, ex eius uita petatur.

ΙΑΤΡΙΚΗ, uidelicet τέχνη. Breuißima inscriptio per adiectiuum fit, quod præsupponit substantiuum. Est autem legitima inscriptio, ut ex quinto de Sanitate tuenda colligitur. Arabes Μικροτέχνην uocant, magna fortaßis ratione, sed uitiosa & inepta cōpositione. Verùm quia Ars est collectio præceptorum uariorum ad eundem finem tendentium, ideo κατ' ἐξοχὴν liber iste, in quo totam artem medicam per conclusiones tradit, Artis medicinalis nomen meruit.

Rem siue materiā, quæ tractatur atque docetur in toto libro. Id'que rursus duobus modis: uel

Respectu aliorum librorum, qui de materia medica ab eodem authore scripti sunt, ut quem inter se mutuò ordinem habeant, qui priore, qui posteriore loco legendi sint, sciamus. Vide epilogū huius libri. In præsentia satis sit scire, medicorum librorū Galeni alios quidem complecti

Partem artis. De quorum numero & dispositione, ut dixi, Galenus in epilogo huius libri ex professo agit. Possunt autem ad compositiui ordinis regulas disponi.

Totam artem breuiter & cōpēdiosé, ordine uel

Resolutiuo: Liber de constitutione artis medicæ ad Patrophilum.

Definitiuo: Ars medicinalis. Cùm ergo totam artem complectatur, per se sola absque aliorum librorū connexione legi potest. Itaque & memoriæ eorum qui iam didicerunt inseruit, & isagoges loco esse potest ijs qui sub docto præceptore discere cupiunt.

Per se, quatenus docetur. Doctrinæ ueluti effectus est ipsa scientia. Scire est, per causam cognoscere. Inter causas & effectus prius est et posteri9: quorū habitudo est ordo, ueluti anima rei quatenus docetur. Ordo ergo hìc inuestigandus. Qui duplex est, prout etiam duplici instrumento quæuis res docetur. Ordo enim libri huius alius est

Rhetoricus, qui nō tam ex rei natura, quàm personarum circumstātijs deducitur. diuidimus ergo librū hūc in tres partes, in

Procemium: in quo ordinem logicum explicat.

Tractationem: in qua Medicinæ definitionem proponit, & singulas illius partes per διαλύσεις declarat.

Epilogum: in quo librorum suorum catalogum recenset, ut ex ijs plenior declaratio eorum petatur, quæ in hoc libro breuiter dicta sunt.

Logicus, qui ex rei solius natura depromitur. Ordo ergo hìc est Definitiuus, ut in ipso procemio Galenus declarat, & nos in Tractationis generali dispositione indicabimus.

A

Figure 9.2 Branching diagram from Galen and Zwinger, *In Artem medicinalem Galeni, tabulae & commentarii…*, 1561. Paris, Bibliothèque interuniversitaire de santé.

Although it does not follow such subdivisions, Meilingius's *Figurata* evinces an analogous sequential diagrammatic structure, similarly inviting beholders to trace its account along bold black lines. A degree of auto-reflexivity is introduced by the inscription of the word *ductus* seemingly dripping down from the upper left mask spouting water from the rightmost fountain's top tier. This compositional wordplay performed a dual referential function, denoting explicitly the chyle ducts while suggesting implicitly the rhetorical device structuring the overall directionality of the image.

Though Meilingius was a medical practitioner, it is possible that he himself not only designed *Figurata* but engraved the work as well.[22] The image was not engraved entirely without skill, however its technical execution reveals a certain naivety: visible traces of guidelines from the plate are transferred to the printed sheet, the illusionary perspective appears distorted, and the print's right-to-left sequence instead of the typical left-to-right suggests an engraver unfamiliar with period printmaking conventions and techniques. The many marks and errors indicating compositional changes made to the plate during the engraving process suggest the engraver was actively revising and finalizing the composition as he worked the plate. The majority of these marks appear around the far-right fountain: the uppermost pedestal was slightly lowered, and the bag of *intemperies* hung on the sword of *Mala Diaeta* (Bad Diet) is surrounded by pentimenti. The concentration of errata in this part of the image indicates that it was in this zone of the plate where Meilingius began engraving with uncertain hand. The most convincing evidence that the designer engraved the print himself can be seen between the first and second fountains (proceeding from right to left). Three faint but discernable streams of water arcing from right to left from the upper tier of the first fountain into the upper tier of the second were lightly incised into the plate, but appear to have been redirected into different basins, the subsequent incisions marked over by bolder lines. Such a change in content was unlikely made by a hired hand. However, lacking any other images attributed to Meilingius to compare to *Figurata*, this theory remains speculative.

Regardless of whether Meilingius engraved the print, that he designed it in a fashion similar to many printmakers before him can be gathered from

L'ombre des ancêtres; Katzenellenbogen, *Allegories of the Virtues and Vices*; and Schadt, *Die Darstellungen der Arbore Consanguinitatis*. On *ductus*, see Carruthers, "Rhetorical *Ductus*."

22 In the early modern period, scientists and medical practitioners did engage with image-making, most commonly etching and engraving. A notable example is the English physician Richard Haydocke. See Höltgen, "Richard Haydocke."

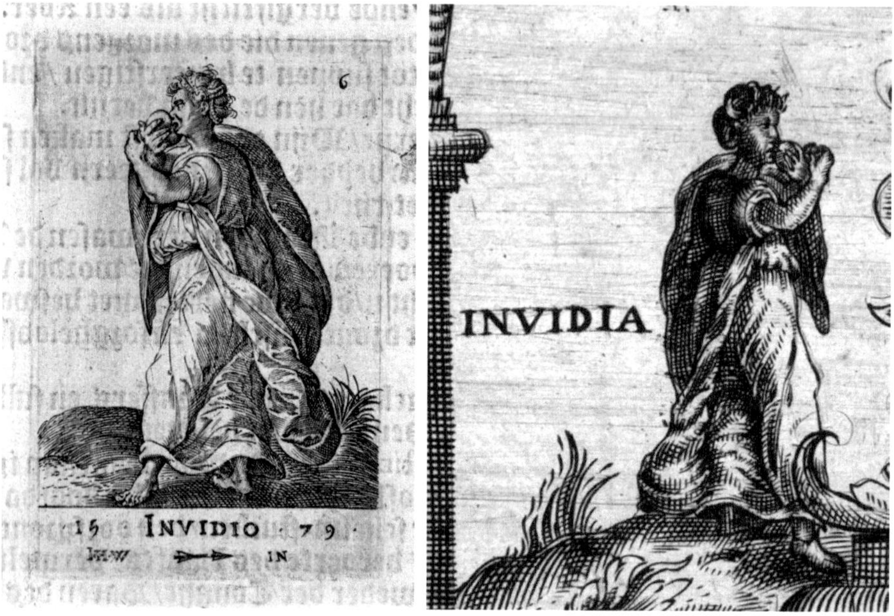

Figure 9.3 A–B A (Left): Johannes Wierix, *Invidio*, 1609, engraving, 5.9 × 3.9 cm. Amsterdam, Rijksmuseum. B (Right): detail from Meilingius, *Figurata Meditatio Microcosmi*.

his modeling of individual components, for which he turned to popular emblem books to source what might be called early modern stock images. His figure of heart-eating Envy, for example, clearly copied Johannes Wierix's "*Invidio*" (Figure 9.3 A–B), Charity bears striking similarities to Cesare Ripa's "*Carita*" emblem, and *Desperatio* shares source materials with a drawing by a follower of Jacque de Gheyn III.[23] Far from plagiarism, this act of repurposing was itself a tenet of the emblem book genre, wherein highly mobile iconographies and pictorial elements were often devised before the composition of the emblematic mottos and verse texts themselves, and occasionally republished later with new text when the old had grown stale. In the words of John Manning, emblematic images were "variously detachable."[24] While Meilingius does not change the identity of these personifications (for example, *Invidio* only becomes *Invidia*), his arrangement

23 "*Carita*" emblem in Ripa, *Della Novissima Iconologia*, 94. The follower of De Gheyn drawing is an allegory of the fate of painting and was sold at auction in Berlin at Grisebach on 25 October 2018.
24 Manning continues: "The same cut might at one time be emblematical and at another no more than an illustration of a fable's narrative. The implication of this is that the woodblock image was not emblematic in itself, but only when attached to emblematic verses. Further, the same cut can be read differently according to its context." Manning, "Towards an Emblematic Rhetoric," p. 85.

of them within the composition in relation to the fountains – much like the positioning of an emblematic image in relation to text – imbues them with Galenic meaning. *Invidia* here is not simply a sin but a product of disruption in the spirited soul.

Meilingius may have followed convention in appropriating elements from other works, but in designing *Figurata* he introduced an entirely novel way of visualizing Galenic pneumatology. Indeed, while early modern scientists, philosophers, and theologists produced many allegorical representations of the soul, most often depicting the substance as rays of light or small anthropomorphic bodies, to the best of my knowledge the motif of a fountain is not present among surviving designs.[25] With this image, Meilingius made something new, for which he appears to have drawn inspiration from two disparate sources, the first of which unsurprisingly lies in the writings of Galen.

When discussing the body's vascular and arterial networks, Galen often used aqueous analogies. In *De Naturalibus Facultatibus,* for example, he writes that the "[n]umerous conduits distributed through all the limbs bring [bodies] blood just as an irrigation system distributes water in gardens."[26] In *De Usu Partium Corporis Humani,* he describes the heart as the "fount" of innate heat.[27] When composing these analogies, Galen commonly used the Greek word πηγή, which was most often translated into the Latin *"fons"* meaning "source" or "fountain." For example, he writes in *Methodus Medendi*: "Concerning the psychic spirit, we have clearly shown that the brain is the fount [πηγή], so to speak, and that it is watered and nourished by inspiration and by the supply of the reticular plexus [*rete mirabile*]."[28] While it is unknown if Meilingius had this or a similar passage in mind when designing his image – which includes a fountain of the brain (*fons animalis*) watered by the fountain of the heart and lungs (*fons vitalis*) – the similarities are nonetheless striking. At the very least, a beholder intimately familiar with Galenic theory, perhaps Reek or Quartelar, would have recognized the selection of a hydraulic system to allegorize the flow of spirits through the

25 On early modern images of the soul, see de Brézé, "Picturing the Soul."

26 Translation from Brian, *Galen on Bloodletting*, p. 155. Without knowing where Meilingius undertook his surgical training, I cannot speculate on the editions of Galen to which he had access. For an example of an early modern printing of this passage, see Galen, "Galeni de Naturalibus Facultatibus," p. 308r. For a bibliography of Galen's texts printed in Renaissance Europe, see Durling, "A Chronological Census."

27 Galen, *De usu partium*, p. 52v.

28 Translation from Temkin, "On Galen's Pneumatology," p. 182. Passage from Galen, *Methodus Medendi*, p. 184r.

vascular, arterial and nervous systems as resonating with Galen's way of thinking.

The second and less expected source from which Meilingius likely drew inspiration is the predominantly Catholic textual and visual custom of representing the *Fons Vitae* (Fountain of Life) as a fountain of blood. Originating in late antiquity and rooted in the Gospel of John, the most common visual forms of this devotional motif depict a fountain or basin filled by the blood spilled from the five wounds of Christ within which the faithful are spiritually nourished.[29] Some images present Christ alone standing in a fountain of his own blood, while others include the bodies of the faithful bathing in the savior's blood, receiving the sacrament of baptism, or filling up bowls and pressing them to their lips.[30] These types of *Fons Vitae* existed in many artistic media, including painting and poetry; however, in the sixteenth century, the Low Countries saw a surge in demand for *Fons Vitae* prints, which were quickly exported across Europe, serving as models for paintings and other printed images.[31]

As well as representing fountains of blood, Meilingius uses the inscription captioning *Figurata* to introduce another subtle reference to the *Fons Vitae* tradition. These three lines of hexameter verse running along the bottom of the page begin with the words "*Per fontes hominum vitam.*"[32]

29 Scholarship on the *Fons Vitae* exists mostly within analyses of specific works of art: for example, Krebber and Kotting, "Jean Bellegambe"; Jurkowlaniec, "The Crucified Christ." The only extensive transhistorical investigation can be found in a published PhD dissertation from Gothenburg University in which these works are referred to as *Fons Pietatis*: Wadell, *Fons Pietatis*. For another subtle reframing, see Knipping's distinction between the 'Fountain of Life' and the 'Well of Life': Knipping, *Iconography*, pp. 466–473. While the fountains of blood were the most common visual manifestations of the *Fons Vitae* in medieval and early modern Europe, they did not all appear in this manner, nor did they invite the same interpretation. Regional, temporal, interpretive, and pictorial variations are documented throughout Wadell's text. On variation in examples from late antique and early medieval Europe, see Underwood, "The Fountain of Life," p. 49.

30 The images of imbibition tend to fall into two categories: those which reference the sacrament of the Lord's Supper, being linked closely to the motif of Christ in the wine press, and those which recall the story of Christ's encounter with the Samaritan woman at the well (John 4:14). The latter is of particular note here, as there are two Galenic commentaries from the mid-sixteenth century which adorn their frontispieces with emblems from this episode: Galen and Fuchs, *Claudii Galeni Pergameni* and Tomai, *De febribus commentarius*. Similar passages from John often appeared on *Fons Vitae* prints with Christ on his own, such as Marten de Keyser's woodcut frontispiece to *Fons Vitae* on which is inscribed directly below Christ's blood-filled fountain: "Si quis sitit, ueniat ad me & bibat. Ioan. 7."

31 Wadell, *Fons Pietatis*, p. 61.

32 The caption in its entirety translates: "Through the fountains, Meilinck accurately depicts the life of men at the same time in them represents the truly wonderful deeds of Jehovah, the great and wise creator." Translation assistance provided by Wright.

The grammar of this sentence does not allow these words to be translated *per se*, since *fontes* belongs to the prepositional phrase following from *per*, while *hominum* and *vitam* are part of the main clause. However, Meilingius's deliberate arrangement of the distinct order of these terms verbally links the religious motif with his Galenic image, while also distinguishing *Figurata* as containing *fontes humanae vitae*, fountains of distinctly human life.[33]

The parallels between Meilingius's *Figurata* and the *Fons Vitae* extend beyond the realms of form and text into that of depicted function, as they both document a similar phenomenon, the blood-based purification of the soul. A particularly clear example of this process in *Fons Vitae* is illustrated in an engraving by Dick Volckertsz. Coornhert after Maarten van Heemskerck (Figure 9.4).[34] The statuesque muscular female figure of *Fides* wields a cruciform broom, stirring human hearts in a large basin filled with blood poured forth from Christ's side wound.[35] The hearts nearer to the woman's intervention appear cleansed, while those farther away retain features of vice, grotesque animal heads, and symbols of greed. The meaning is clear: when embraced by faith, these synecdochical souls are purified with the blood of Christ. The purification depicted by Meilingius, on the other hand, occurs without divine intervention. This mechanical purification is hydraulic rather than spiritual.[36] Instead of the blood of Christ purifying the soul of the sinner, it is the organs (in fountain form) purifying and spiriting the blood of man.[37]

33 Similarly, Meilingius positioned the *"lis"* of *"fons vitalis"* (the third fountain) on a separate line, thus allowing the top line to read simply *"fons vita"*.

34 The British Museum possesses a copy of this image printed by Adriaen Collaert after Ambrosius Francken, published by Philips Galle (London, British Museum: 1901,0611.87).

35 This image is placed in a larger theological and artistic context in Veldman, "Maarten van Heemskercks," pp. 197–199.

36 While the most prominent religious influence in Meilingius's print seems to originate in the fountain of blood *Fons Vitae* images, he may have drawn inspiration as well from the *Fons Vitae* concerned with the fountain in Eden. This is suggested in *Figurata*'s first fountain on the right which rises from the "Great River from which all fountains originate and draw nourishment" (*Magnum flumen ex quo omnes fontes originem et nutrimenta exhavriant*), thus recalling the River in Eden (Genesis 2:10). On Edenic *Fons Vitae*, see Bakker, "The Amsterdam 'Paradise'" and Underwood, "The Fountain of Life," p. 47. The narrow, tiered forms of Meilingius's fountains also recall images of the Edenic fountain, examples of which are reproduced in Bakker's text.

37 While the principal meanings of blood in this image relate to specific facets of medicine and religion, Rebecca Zorach's scholarship on Renaissance France shows that the symbolic features of blood at this time extended much further and into often contradicting spheres of meaning, such as "purity and impurity, disproportion and dissolution [...] [and] continuity and discontinuity." Zorach, *Blood, Milk, Ink, Gold*, p. 33.

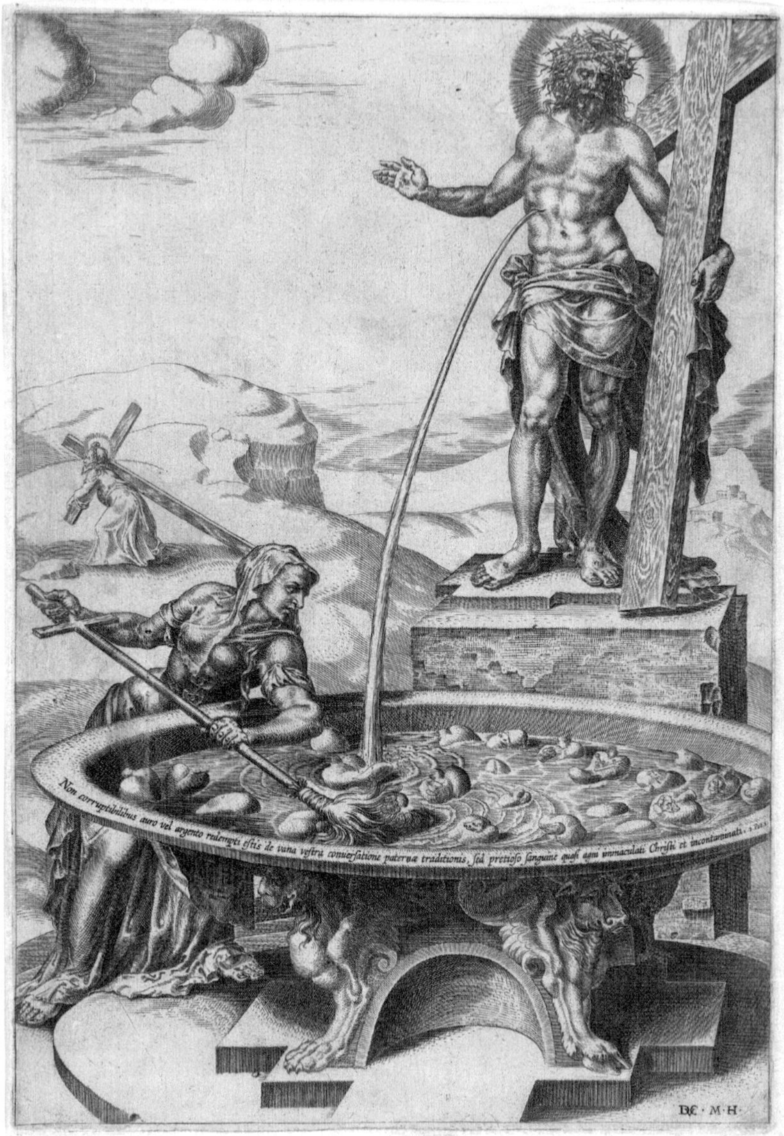

Figure 9.4 Dirck Volckertsz. Coornhert, after Maarten van Heemskerck, *Faith purifying the hearts of men with the blood of Christ*, 1557–1561, engraving, 25.7 × 36.6 cm. Amsterdam, Rijksmuseum.

It is possible that *Figurata*'s purification occurs not only without but also despite divine intervention. At the top right corner of Meilingius's engraving, the vernacular hand of God is shown pouring wine into the unbalanced personification of *Mala Diaeta*, who spills the seeds of intemperance into the *non naturalis* bowl of *Natura*'s humoral scales

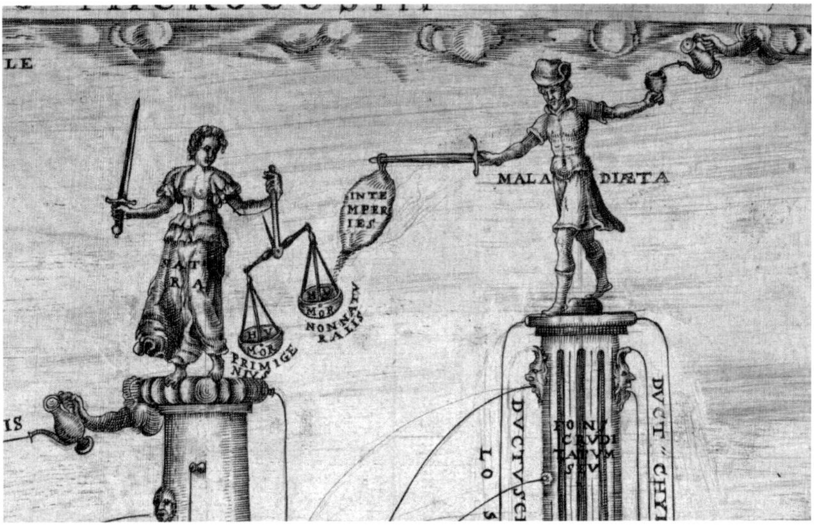

Figure 9.5 Detail from Meilingius, *Figurata Meditatio Microcosmi.*

(Figure 9.5). Could this be a subtle reference to the sacrament of the Lord's Supper? If he followed the beliefs of his father, Meilingius observed one of the Reformed faiths that rebuked the Catholic tenet of transubstantiation, only observing the sacrament when paired with and confirmed by the preachings of the Word of God.[38] This can be tentatively read in Meilingius's figure of *Mala Diaeta*, who accepts the sacramental wine from on high without the Word of God, thus finding not a balm but an intoxicant that widens his eyes and tips his balance.[39] Nevertheless, spiritual and dogmatic failures aside, this man's soul would be continually purified by the hydraulic power of his interlinked network of digestive, vascular, arterial and nervous systems.[40]

38 It cannot be determined with certainty that Meilingius followed his father's confession, but, with little doubt, he retained a Reformed faith, and, in general terms, Reformed faiths accepted that when practiced "apart from the preaching of the gospel, the sacraments would be empty and meaningless signs." Venema, "The Doctrine of the Lord's Supper," p. 137.

39 In early modern European thought, an interesting relationship exists between medical spirits and alcoholic spirits, both in their effect and their process of distillation, which would be an interesting topic of further study in relation to this image. On this relationship, see Kodera, "The Art of the Distillation."

40 Or, to put an even more subversive, anti-Catholic twist to this reading, *Mala Diaeta* can be understood as an intoxicated Catholic, mocked for his folly in believing that the wine from God was transubstantiated into the blood of Christ. Although intent in this case cannot be proven, there are precedents of anti-Papist prints that subtly attack this sacrament. For example, Hendrik Hondius's c. 1599 engraving *Piramide Papistique* (Amsterdam, Rijksmuseum: RP-P-OB-78.843),

While such a reading must remain in the realm of speculation, inflecting Meilingius's image through a religious gloss that in turn places the work in conversation with *Fons Vitae* prints invites such a nuanced interpretation. This very blending of medicine and religion allows this image to "mean more freely," while still performing its utilitarian function of spatially condensing and schematizing Galenic pneumatology.[41] To this end, the *Fons Vitae* motif presented the perfect iconographic corollary for Meilingius, as it allowed him to expand the breadth of interpretation available within his medical diagram, inviting beholders to draw connections between an ancient physiological pneumatic theory and contemporary religious iconography, without betraying Galen's original texts analogizing the body as a network of fountains. It also facilitated Meilingius's introduction of the topic of divine creation to this Galenic theory. *Figurata* contends that even though God may have designed the body, the daily health of the soul is not entirely within his purview.[42] The Galenic soul requires proper nutrition, not divine intervention, to function. Instead of God or Christ, the soul in *Figurata* is watched over by the personification of *Iudicium*, who looks out over the fountains holding a half-full matula in one hand and raising the other in a locutionary gesture, much like a physician announcing a uroscopic diagnosis.[43] This figure, the only personification engaged in the hydraulics of this image, has replaced Christ on the *Fons Vitae*; the rejuvenating substance he here provides is not his blood but his words, written by Meilingius's hand. This distinctive figure appears to be loosely based on Wierix's emblem of John the Evangelist from the same series as his above-cited *Invidio* (Figure 9.6 A–B). Could it perhaps be a self-portrait or surrogate for the image's inventor? Instead of Johannes the Evangelist, Meilingius presents Jacobus the Physician – watching, as it were, over the health of the souls of humanity.

presents a papist snake balancing the Eucharist on his tongue, which is about to be struck by a lightning bolt of *Verbum Dei*.

41 Elkins uses this phrase in his well-known essay in which he challenges the art historical discipline's treatment of "nonart" images as "half-pictures, or hobbled versions of full pictures, bound by the necessity of performing some utilitarian function and therefore unable to mean more freely." Elkins, "Art History," p. 443. Much of the recent scholarship on epistemic and schematic images is indebted to this publication.

42 For those of Remonstrant sympathies, like Meilingius's father and perhaps Meilingius himself, this reading would be made clearer by the doctrine of free will.

43 Compare this detail with, for example, the physician at the center of the uroscopy wheel in Ulrich Pinder's *Epiphanie medicorum*.

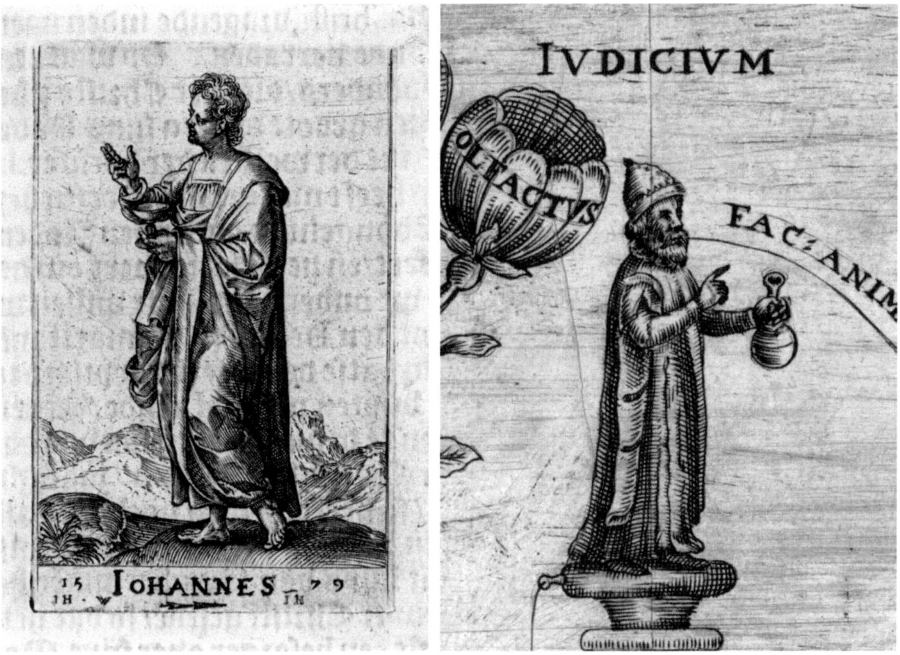

Figure 9.6 A–B A (Left): Johannes Wierix, *John the Evangelist*, 1609, engraving, 5.7 × 3.7 cm. Amsterdam, Rijksmuseum. B (Right): detail from Meilingius, *Figurata Meditatio Microcosmi*.

Bibliography

Bakker, Boudewijn. "The Amsterdam 'Paradise' by Herri met de Bles and the Fountain of Life." *The Rijksmuseum Bulletin* 63, no. 2 (2015): 114–133.

Berger, Susanna. *The Art of Philosophy: Visual Thinking in Europe from the Renaissance to the Early Enlightenment.* Princeton, NJ: Princeton University Press, 2017.

Blair, Ann. *Too Much to Know: Managing Scholarly Information Before the Modern Age.* New Haven, CT: Yale University Press, 2010.

Brian, Peter. *Galen on Bloodletting: A Study of the Origins, Development and Validity of his Opinions, with a Translation of the Three Works.* Cambridge: Cambridge University Press, 1986.

Carruthers, Mary. "Rhetorical *Ductus*, or, Moving through a Composition." In *Acting on the Past: Historical Performances Across the Disciplines*, edited by Mark Franko and Annette Richards, pp. 99–125. Hanover, NH: University Press of New England, Wesleyan University, 2000.

de Brézé, Nathalie. "Picturing the Soul, Living and Departed." In *The Anthropomorphic Lens: Anthropomorphism, Microcosmism, and Analogy in Early Modern Thought and Visual Arts*, edited by Walter Melion, Bret Rothstein and Michel Weemans, pp. 119–144. Leiden: Brill, 2015.

de Keyser, Marten. *Fons Vitae, ex quo Scaturiunt Suauissimae Co[n]solationes, Adflictis Me[n]tibus in Primis Necessariae, ne in Aduersitate & Dolore Protinus Animum Despondeant.* Antwerp: apud Martinum Caesarem, 1533.

Durling, Richard J. "A Chronological Census of Renaissance Editions and Translations of Galen." *Journal of the Warburg and Courtauld Institutes* 24, no. 3/4 (1961): 230–305.

Elkins, James. "Art History and Images That Are Not Art." *The Art Bulletin* 77, No. 4 (1995): 553–571.

Ferrell, Lori Anne. "Page *Techne*: Interpreting Diagrams in Early Modern English 'How-To' Books." In *Printed Images in Early Modern Britain: Essays in Interpretation*, edited by Michael Hunter, pp. 113–126. Farnham: Ashgate, 2010.

Franklin, James. "Diagrammatic Reasoning and Modeling in the Imagination: The Secret Weapons of the Scientific Revolution." In *1543 and All That: Image and Word, Change and Continuity in the Proto-Scientific Revolution*, edited by Guy Freeland and Anthony Corones, pp. 53–115. Dordrecht: Kluwer Academic Publishers, 2010.

Galen, Claudius. *Methodus Medendi.* Commentary by T. L. Croanglo. Paris, 1526.

Galen, Claudius. *De Usu Partium Corporis Humani Libri 17.* Basel: Andreas Cratander, 1533.

Galen, Claudius. *Claudii Galeni Pergameni de Hippocratis et Platonis Placitis Opus Eruditum, et Philosophis & Medicis Utilissimum....* Paris: Apud Simonem Colinaeum, 1534.

Galen, Claudius. "Galeni de Naturalibus Facultatibus Libri Tres, Thoma Linacro Anglo interprete." In *Galeni Omnia qual Extant Opera in Latinum Sermonem Conuersa*, 290v–308r. Venice: apud haeredes Lucaeantonij Iuntae, 1565.

Galen, Claudius, and Johann Guenther. "Secundus". In *Libri tres.* Paris: Apud Simonem Colinaeum, 1528.

Galen, Claudius, and Leonhart Fuchs. *Claudii Galeni Pergameni.* Paris: Apud Iacobum Dupuys, 1549.

Galen, Claudius, and Theodor Zwinger. *In Artem Medicinalem Galeni, Tabulae & Commentarii....* Basil: apud Iacobum Parcum, expensis Ionnis Oporini, 1561.

Gill, Christopher. *Naturalistic Psychology in Galen and Stoicism.* Oxford: Oxford University Press, 2010.

Hankinson, R. J. "Galen's Anatomy of the Soul." *Phronesis* 36, no. 2 (1991): 197–233.

Höltgen, Karl Josef. "Richard Haydocke: Translator, Engraver, Physician." *The Library* 33 (1978): 15–32.

Jaanus, H. J. *Een Remonstrantse Voorzaat: Henricus Meilingius.* The Hague: Central Bureau of Genealogy, s.d. (mid-twentieth century).

Jurkowlaniec, Grażyna. "The Crucified Christ as the Source of the Seven Sacraments. Patters of Reception of a Sixteenth-Century Image on Both Sides of the Alps and on Both Sides of the Atlantic Ocean." Transl. by Zuzanna Sarnecka. In *Artistic Translations Between Fourteenth and Sixteenth Centuries: International Seminar for Young Researchers. Proceedings*, edited by Zuzanna Sarnecka and Aleksandra Fedorowicz-Jackowska, pp. 187–209. Warsaw: University of Warsaw, 2013.

Katzenellenbogen, Adolf. *Allegories of the Virtues and Vices in Mediaeval Art from Early Christian Times to the Thirteenth Century*. London: The Warburg Institute, 1939.

Knipping, John B. *Iconography of the Counter Reformation in the Netherlands: Heaven on Earth, Vol. II.* Nieuwkoop: B. de Graaf; Leiden: A. W. Sijthoff, 1974.

Klapisch-Zuber, Christiane. *L'Ombre des Ancêtres. Essai sur l'Imaginaire Médiéval de la Parenté.* Paris: Fayard, 2000.

Kodera, Sergius. "The Art of the Distillation of 'Spirits' as a Technological Model for Human Physiology: The Cases of Marsilio Ficino, Joseph Duchesne and Francis Bacon." In *Blood, Sweat and Tears: The Changing Concepts of Physiology from Antiquity into Early Modern Europe*, edited by Manfred Horstmanshoff, Helen King, and Claus Zittel, pp. 139–170. Leiden: Brill, 2012.

Krebber, G. B. and Kotting, K. "Jean Bellegambe en zijn 'Mystiek Bad' voor Anchin." *Oud Holland* 104, no. 3/4 (1990): 123–139.

Maclean, Ian. "Diagrams in the Defense of Galen: Medical Uses of Tables, Squares, Dichotomies, Wheels, and Latitudes, 1480–1574." In *Transmitting Knowledge: Words, Images, and Instruments in Early Modern Europe*, edited by Sachiko Kusukawa and Ian Maclean, pp. 135–164. Oxford: Oxford University Press, 2006.

Manning, John. "Towards an Emblematic Rhetoric." In *The Emblem*, pp. 90–101. London: Reaktion, 2002.

Marr, Alexander. "Knowing Images." *Renaissance Quarterly* 69, no. 3 (2016): 1000–1013.

Meilingius, Henricus Jacobi. *Disputatio Medica Inauguralis.* Trajecti ad Rhenum, 1660.

Meilingius, Henricus Jacobi. "Henricus Meilingius (Borculo, NL) to Hugo de Groot (Paris, France), 1 October 1629." In *Briefwisseling van Hugo Grotius. Deel 4*, edited by B.L. Meulenbroek, pp. 107–108. The Hague: Martinus Nijhoff, 1964.

Nauta, D., *et al.*, eds. *Biografisch Lexicon voor de Geschiedenis van het Nederlandse Protestantisme, Deel 3.* Kampen: Uitgeversmaatschappij J.H. Kok, 1988.

Park, Katharine. "Psychology: The organic soul." In *The Cambridge History of Renaissance Philosophy*, edited by Charles B. Schmitt, Quentin Skinner, Eckhard Kessler and Jill Kraye, pp. 464–484. Cambridge: Cambridge University Press, 1988.

Pinder, Ulrich. *Epiphanie Medicorum.* Nuremburg: [Friedrich Peypus], 1506.

Ripa, Cesare. *Della Novissima Iconologia di Cesare Ripa Perugino Cavalier de SS. Mauritio & Lazzaro...* Padua: Per Pietro Paolo Tozzi, 1625.

Rocca, Julius. "Galen and Greek Neuroscience (Notes Towards a Preliminary Survey)." *Early Science and Medicine* 3, no. 3 (1998): 216–240.

Saxl, F. "A Spiritual Encyclopaedia of the Later Middle Ages." *Journal of the Warburg and Courtauld Institutes* 5 (1942): 82–142.

Schadt, Hermann. *Die Darstellungen der Arbore Consanguinitatis und der Arbore Affinitatis Bildschemata in Juristischen Handschriften.* Tübingen: Wasmuth, 1982.

Schiefsky, Mark. "Galen and the Tripartite Soul." In *Plato and the Divided Self*, edited by Rachel Barney, Tad Brennan, and Charles Brittain, pp. 331–349. Cambridge: Cambridge University Press, 2012.

Sugg, Richard. *The Smoke of the Soul: Medicine, Physiology and Religion in Early Modern England.* Basingstoke: Palgrave Macmillan, 2013.

Temkin, Owsei. "On Galen's Pneumatology." *Gesnerus: Swiss Journal of the History of Medicine and Sciences* 8, no. 1–2 (1951): 180–189.

Tomai, Camillo. *De Febribus Commentarius.* Venice: Andrea Arrivabene, 1555.

Underwood, Paul A. "The Fountain of Life in Manuscripts of the Gospels." *Dumbarton Oaks Papers* 5 (1950): 41–148.

van der Aa, A. J. *Biographisch Woordenboek der Nederlanden.* Haarlem: J. J. van Brederode, 1869.

Veldman, Ijla M. "Maarten van Heemskercks visie op het geloof." *Bulletin van het Rijksmuseum* 35, no. 3 (1987): 193–210.

Venema, Cornelis P. "The Doctrine of the Lord's Supper in the Reformed Confessions." *Mid-America Journal of Theology* 12 (2001): 135–199.

Vidal, Fernando. *The Sciences of the Soul: The Early Modern Origins of Psychology.* Transl. by Saskia Brown. Chicago, IL: University of Chicago Press, 2011.

Wadell, Maj-Brit. *Fons Pietatis: Eine Ikonographische Studie.* Transl. by Dieter Rosenthal and Anne-Marie Thiberg. Göteborg: Elanders Boktryckeri Aktiebolag, 1969.

Zorach, Rebecca. *Blood, Milk, Ink, Gold: Abundance and Excess in the French Renaissance.* Chicago, IL: The University of Chicago Press, 2005.

About the Author

Anneke de Bont is a PhD candidate in the History of Art at the University of Cambridge where she specializes in early modern European print culture. Her dissertation explores interdisciplinarity in Christian printed visualizations of order from early modern Northern Europe through the framework of the Christian epistemic image.

10. Images of the Eye from Vesalius to Fabricius ab Aquapendente

The Rise of Metrical Representation in Anatomical Diagrams and the Cross-Fertilization of Visual Traditions

Tawrin Baker

Abstract
Illustrations of the eye were rare in printed medical and anatomical works prior to Vesalius's 1543 *Fabrica*. Subsequent epistemic images of the eye, seen in figures such as Juan Valverde, Felix Platter, Johannes Kepler, and Fabricius ab Aquapendente, had a complex relationship with the evolving, and increasingly intertwined, disciplines of anatomy, natural philosophy, and optics. Images of the eye reflected, and to some degree helped to generate, new practices of reading, approaches to creating knowledge, and the formation of communities. This analysis challenges a widely-held distinction in early modern historiography between the role of pictures in the anatomical/medical sciences and that of diagrams in the mathematical sciences.

Keywords: philosophy, optics, anatomy, Vesalius, Kepler

This chapter provides an abbreviated survey of sixteenth- and early seventeenth-century printed images of the interior of the eye, tracing the impact of illustrations of the eye in Vesalius's 1543 *De humani corporis fabrica libri septem* through Johannes Kepler's 1604 *Ad Vitellionem paralipomena, quibus astronomiae pars optica traditur* and Hieronymus Fabricius ab Aquapendente's 1600 *De visione*. This is read against a persuasive narrative in recent historiography that posits clear differences between the impact of epistemic images in, on the one hand, the domains of medicine, anatomy, and natural history generally, and in the mathematical sciences,

Noyes, R.S. (ed.), *Reassessing Epistemic Images in the Early Modern World.*
Taylor & Francis Group, 2023
DOI 10.5117/9789463723350_CH10

particularly astronomy, on the other. Vesalius looms large in accounts of the former, Copernicus and his wake (Tycho, Kepler, Galileo, etc.) in the latter.[1] These studies in part responded to earlier, rather sweeping accounts of the importance of new kinds of graphic techniques, technology (print, foremost), and ways of seeing to the so-called Scientific Revolution. Such works, by, among others, William Ivins, Elizabeth Eisenstein, and Bruno Latour, made no clear distinctions between the visual traditions of medicine and mathematics.[2]

Depictions of the interior of the eye, however, challenge aspects of both accounts.[3] This chapter considers how entanglements of visual depiction, textual content, and experimental activities through developments in anatomy, medicine, and mathematical optics might revise broader narratives about the role of visualization in the development of modern science.

I. Metrication of Eye Illustrations in Vesalius

Though less famous than other now-canonical woodcuts in the *Fabrica*, within their domain Vesalius's images of the eye exercised enormous influence.[4] There were pre-existing image traditions depicting the interior of the eye in the Renaissance, but no established models in learned anatomy until Vesalius.[5] Thus, Vesalius's *Fabrica* drew primarily from models found in mathematical optics (or *perspectiva*), natural philosophy textbooks, and, to some extent, ophthalmology, all of which featured schematic depictions of the ocular interior.[6] This corpus of ocular imagery reinforced the notion, found in Galenic medicine and perspectivist optics, of a transparent passage from the outside world through (in successive order) the cornea,

1 See especially Kemp, "Temples of the Body," pp. 40–85; Pantin, "Analogy and Difference," pp. 9–44.

2 A short list might consist of the following: Panofsky, "Die Perspektive Als 'Symbolische Form'"; *Idem*, *Perspective as Symbolic Form;* Ivins, *Rationalization of Sight;* Ivins, *Prints and Visual Communication;* Eisenstein, *Printing Press as an Agent of Change;* Edgerton, "The Renaissance Artist as Quantifier"; Latour, "Visualisation and Cognition."

3 This challenge is, however, implicit in Bertoloni Meli, *Mechanism.*

4 There is a huge secondary literature on Vesalius's illustrations, but, in addition to the above, see the following, especially their bibliographies: Kusukawa, *Picturing the Book of Nature*; Margócsy, Somos, and Joffe, *The Fabrica of Andreas Vesalius.*

5 Though illustrations appeared in late fifteenth- and early sixteenth-century printed books on learned anatomy and medicine, the eye was nowhere depicted. E.g., Ketham, *Fascicolo di medicina*; Berengario da Carpi, *Isagoge breves.*

6 Raynaud, *Eye Representation.*

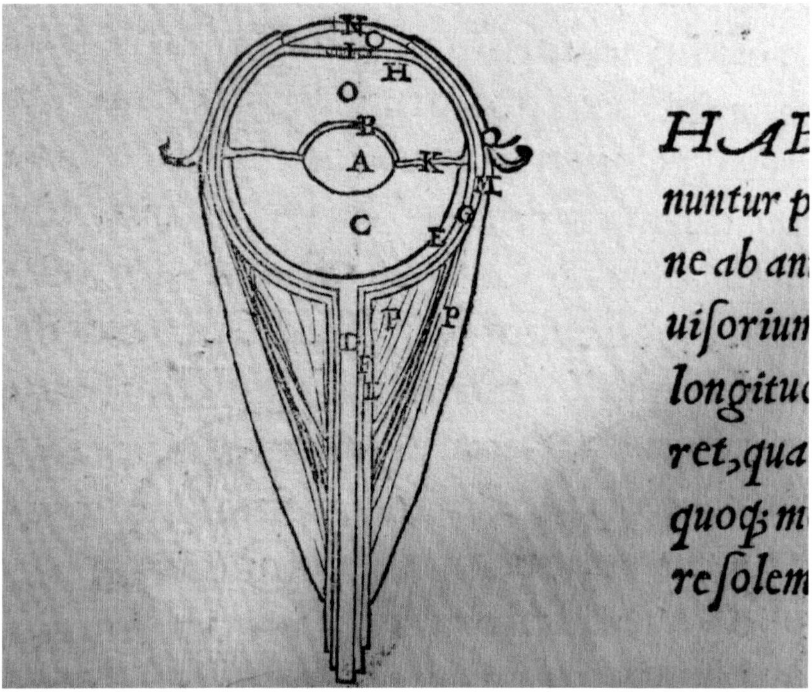

Figure 10.1 Vesalius's "cosmic eye." Vesalius, *Fabrica*, 643. Courtesy of the Newberry Library, Chicago. Special Collections, Case 6A 156.

aqueous, crystalline, and vitreous humors, then the spirits filling the supposedly hollow optic nerves, and, finally, to the ventricles of the brain, i.e. the seat of the cognitive faculties. Vesalius drew from this visual corpus but transformed it, initiating a new ocular image tradition that dominated anatomical illustrations for perhaps a century, while also spreading well beyond anatomical and medical books.[7]

Book 7, Chapter 14, *On the Eye, the Instrument of Vision*, opens with an inset woodcut of the eye (Figure 10.1).[8] His laborious clarification of this image, "a single section made from the front through the back [...] just as one would illustrate half an onion divided lengthwise to show the surface where it had been attached and continuous with the other half," bespeaks the novelty of this true cross-sectional representation.[9] Accompanying text links this illustration to geocentric cosmological diagrams with which readers would have been familiar, making this a "cosmic" figuration of the

7 *Ibid.*, esp. pp. 81–134, 459.
8 Vesalius, *Fabrica*.
9 *Ibid.*, p. 643; *Idem*, *The Fabric of the Human Body*, p. 1303.

eye; he writes, "this is also how we are accustomed to depict the heavens and the four elements in a plane."[10]

His first image of the eye is furthermore linked to his professorship in surgery and anatomy, hinting at the visual culture of learned surgeons and physicians in university classrooms and public anatomies.[11] Namely, he used illustration as an orderly framework to organize a messy dissection:

> I have described the eye using the account I give in the schools, first explaining the construction of the eye and drawing (*delineans*) a large picture on a sheet of paper as I go, like the one I have tried to show in the first figure placed at the beginning of this chapter. After my explanation, I perform my dissection, as I shall add at the appropriate place in the final chapter of this book.[12]

On the page just after the first "cosmic" eye are eighteen labeled illustrations (Figure 10.2), employing perspectival techniques, multiple orientations, and naturalistic shading to indicate dimensionality and texture. Note that both depictions are of an ideal, *living*, human eye. In an actual dissection the flexible tunics, delicate connecting structures, and fluid humors presented to sight via the anatomist's knife would appear differently. Moreover, Vesalius followed Galen's pneumatic physiology whereby subtle visual spirits inflated the eye, giving the parts their animate conformation; upon death these spirits were supposed to escaped immediately, removing a key structuring element of the eye and rendering it impossible to grasp the living shapes and locations of the parts without theoretical reconstruction.

Read in sequence, Figure 10.2 offers a "demiurgic plan" – assembly instructions in the mind of a divine craftsman, perhaps. It begins with the crystalline humor (or lens), held to be the principal seat of vision in the eye, proceeding outward layer by layer. Read in reverse order, however, the sequence illustrates Vesalius's instructions for the dissection of the eye given in a later chapter.[13] The single longitudinal cut in his Figure XVI also corresponds to his actual method of eye dissection used in his 1540 public anatomy in Bologna, as recorded by the medical student Baldasar

10 Vesalius, *Fabrica*, p. 643.
11 See also Shotwell, "Animals, Pictures, and Skeletons."
12 Vesalius, *The Fabric of the Human Body*, p. 1311. Original at *Idem*, *Fabrica*, p. 649.
13 Vesalius, *Fabrica*, p. 655.

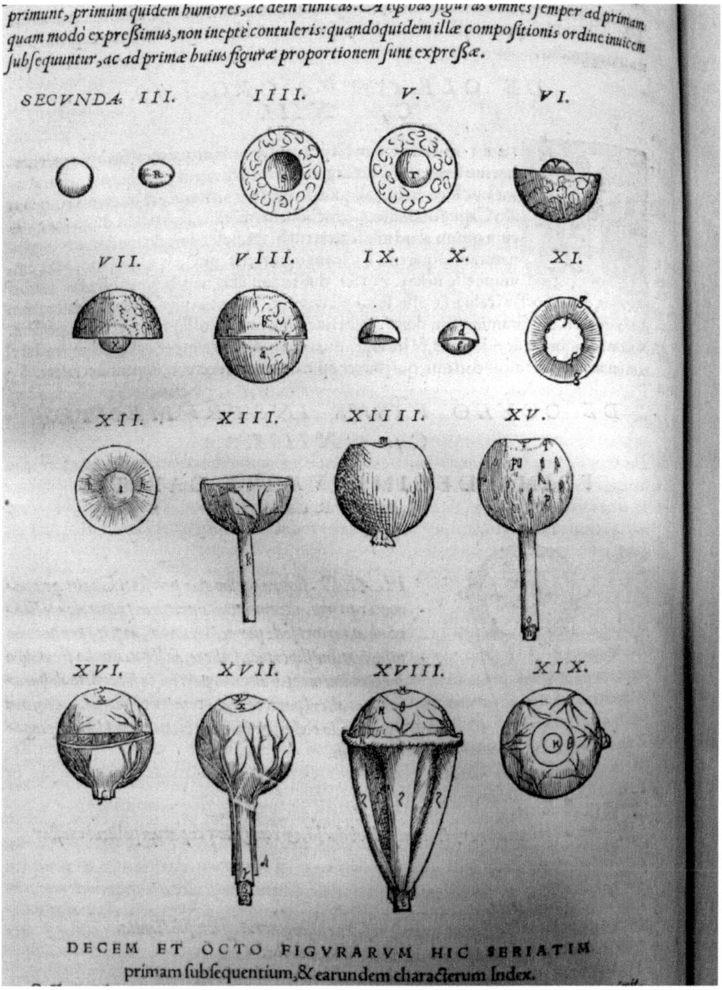

Figure 10.2. Andreas Vesalius, *De humani corporis fabrica libri septem* (Basel: Oporinus, 1543), 644. Courtesy of the Newberry Library, Chicago. Special Collections, Case 6A 156.

Heseler.[14] But this reverse reading is secondary to the account of how the eye is *fabricated*, as the title of the book itself emphasizes.

Perhaps the most noteworthy innovation in Vesalius's cross-sectional images is that they were the first printed depictions, based on extensive first-hand experience with dissected eyes, that claimed to accurately represent the sizes, shapes, and measurable quantities of the parts of the eye in a living human being. These woodcuts clearly influenced

14 Heseler and Eriksson, *Andreas Vesalius' First Public Anatomy*. See also Baker, "Dissection, Instruction, and Debate."

depictions in other genres over ensuing centuries, but whether his near-contemporaries recognized this new epistemic power in Vesalius's images is difficult to determine. Texts continued to carry more authority than images among the learned throughout the sixteenth century. However, among anatomists, surgeons, and physicians prior to Vesalius first-hand experience with bodies and dissection was not wholly subservient to textual knowledge. Generally speaking, for a physician or learned surgeon just prior to Vesalius, knowledge of the body involved coordinating authoritative texts, first-hand anatomical experiences, verbal instruction, discussion, and debate in medical school, and practical and professional experiences with bodies. Theory permeated all of this. Some clearly identifiable uses for images in medical contexts prior to Vesalius included learning the names of parts, reinforcing and remembering textual or verbal information, and helping to organize one's direct sensory experiences with bodies while simultaneously aiding recall of these experiences.[15] After Vesalius all of this continued, as it continues today, to be an important function of medical illustrations, but after Vesalius an additional element was added.

Put differently, while in the sixteenth century reading an anatomical text required learned physicians and surgeons to bring a number of extra-textual factors to bear, any new text nevertheless was capable of prompting a change in opinions about some aspect of anatomy and physiology. The newly-revived, edited, and translated works of Galen, especially *On Anatomical Procedures* and *Usefulness of the Parts,* clearly did so. Anatomical images, however, lacked this specific power prior to Vesalius. After Vesalius an added layer of precise metrical, spatial, and mechanical knowledge could potentially be inscribed and read in illustrations. Anatomists, physicians, oculists (lower-status surgical specialists), mathematicians, and others were indeed affected by this change in image culture at some point. While physicians appear to have been slow to embrace this new capacity generally,[16] we have clearer evidence that it was indeed employed in eye illustrations.

Notable aspects of the *Fabrica* archetype were altered when others – especially anatomists – used them as basis for their own illustrations. The depiction of the *conjunctiva,* for example, was unstable. More salient here, the location and shape of the crystalline lens was soon redrawn in anatomies critical of the *Fabrica*. Over several decades, it was moved

15 Nutton, "Representation and Memory."
16 Margócsy, "From Vesalius through Ivins to Latour."

toward the front of the eye and was described as asymmetric, with a flattened front and gibbous posterior.[17] The *Fabrica* model for ocular illustrations was used, first, as a mirror of anatomical claims and criticism in the text, then as a more independent source for anatomical corrections, and, finally, as a superior form of information compared to the textual description.

In his anatomy, first published in Spanish 1556 and later in Latin and Italian, Valverde, a physician and once student of Realdo Colombo, corrected the Vesalian eye in his text by placing the crystalline humor significantly forward in the eye.[18] His engraved copperplate images were largely copied from those of the *Fabrica*, but altered to reflect Valverde's textual corrections to Vesalius concerning the shapes, quantities, and locations of the parts. In the text, Valverde says that he borrowed Vesalius images in part because it would allow readers to easily notice the numerous differences between his and Vesalius's anatomy. Yet, in the dedication to the 1559 Italian edition, Valverde complains that readers ignorant of Spanish assumed that, because the illustrations were similar, the work was merely a translation of Vesalius's *Fabrica*.[19] This gives some evidence that, at that time it was published, images themselves were not read as containing anatomical arguments or criticism, particularly regarding the spatial relationships between parts. In contrast, in his 1583 work, the physician and anatomist Felix Platte presented his text in sparse dichotomies or trees, separated from the copper engravings at the end of the volume. He introduced an asymmetrical convexity to the crystalline humor, but this information was only given via illustration – it was not mirrored in the text. About 25 years after Valverde, Platter's appropriation of the Veasalian illustration of the eye indeed contained independent criticism of, and corrections to, past anatomical *historiae* (Figure 10.3).

Johannes Kepler had Felix Platter's plates copied for his 1604 *Ad Vitellionem Paralipomena.* (Figure 10.4). These images show that he was a part of a newly expanded community of readers and authors that had absorbed the new capacity for illustrations to accurately convey metrical information; it also shows that he took for granted that his readers would interpret the images as such. In an endnote referencing a page describing

17 Note that criticism of Vesalius's depiction of the shape and location of the lens was, in some ways, a reversion to the pre-Vesalian consensus in anatomical texts. See Baker, "Dissection, Instruction, and Debate," p. 132.

18 Juan Valverde, *Historia de la composicion del cuerpo humano*, p. 82v.

19 Valverde, Juan, *Anatomia del corpo humano*, n.p. See also Skaarup, "Unexpected Success."

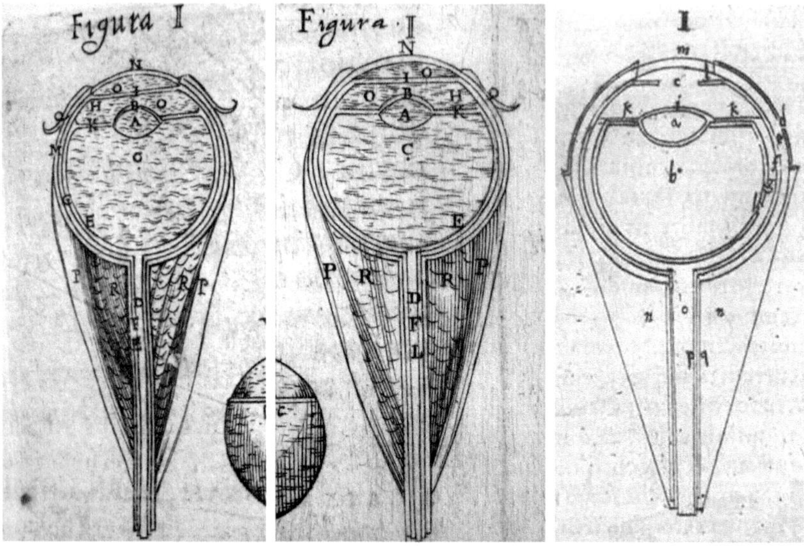

Figure 10.3 Left: Valverde, *Anatomia,* 116r, RB 498768. Center: Valverde, *Anatome Corporis Humani Auctore Joanne Valverdo Nunc Primum a Michaele Columbo Latine Reddita et Additis Novis Aliquot Tabulis Exornata* (Venice: Giunta, 1607), 261, RB 624900. Right, Platter, *De corporis,* 1583, Plate 49. Left and center courtesy of The Huntington Library, San Marino, California.

the eye using the Vesalian commonplace of an "image of an onion," he writes:

> Forgive the crude analogy, reader. For when I wrote this, I had not yet hit upon the idea of copying Platter's drawing, which depicts the parts of the eye true to life, in print [...]. And so I wanted to accomplish here with words and comparisons, what I did not then intend to include in a picture.
>
> You should therefore disregard this comparison and repeatedly look at the drawing referred to, and carefully compare what has been said in this entire argument with that drawing.[20]

The dotted lines above his Figures I, II, IIII, and V were added at Kepler's direction, as was the letter "m" above his Figure I; this represents the corneal bulge, which had only recently been described by anatomists.[21]

20 Kepler, *Optics*, p. 175. The original note appears in Kepler, *Paralipomena* (1604), pp. 438–439, referencing p. 163.
21 Kepler, *Paralipomena*, p. 189.

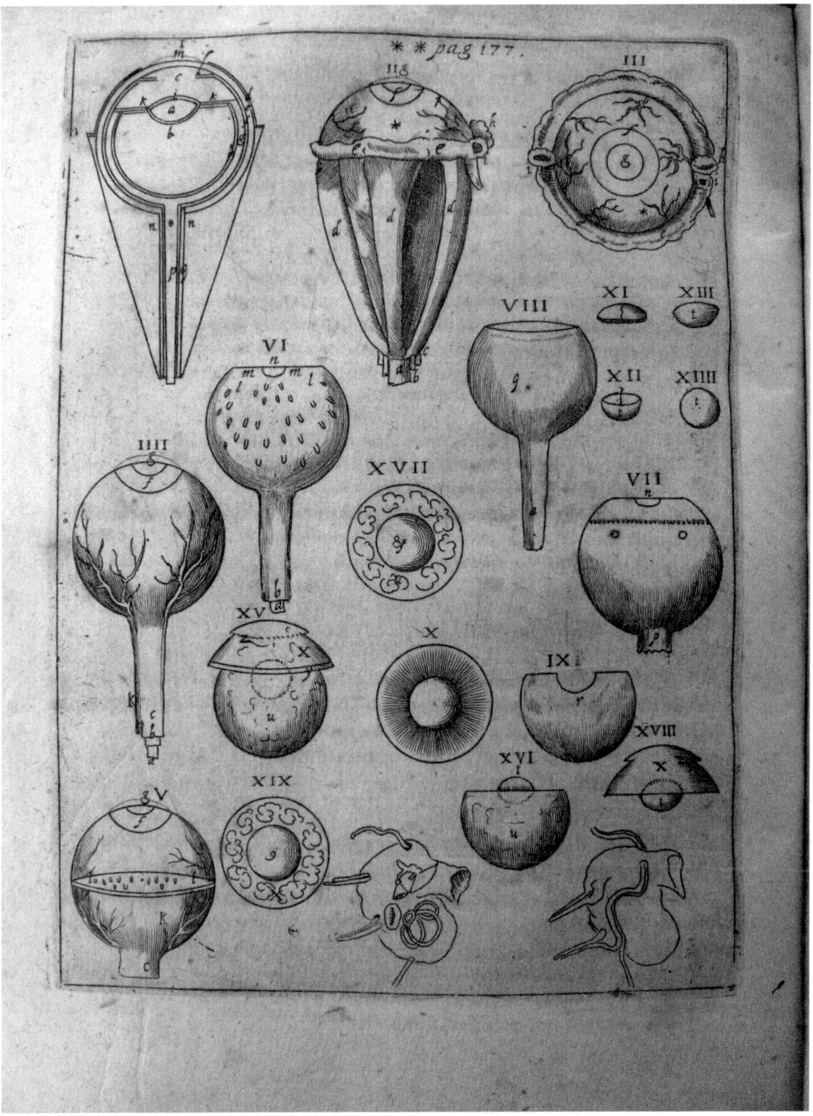

Figure 10.4 Johannes Kepler, *Ad Vitellionem paralipomena, quibus astronomiae pars optica traditur* (Frankfurt: Marnius and heirs of Aubrius, 1604), 177. Courtesy of the Huntington Library, San Marino, California, RB 708372.

Nevertheless, we see Kepler operating within the limits of the Vesalian visual framework. His groundbreaking mathematical analyses were accompanied by diagrams that stayed squarely within a conservative mathematical visual tradition, and Kepler's claims that his geometrical results account for how *picturae* are actually projected upon the retina – a revolution visual theory – involved verbal analogies and arguments, not visual ones; the mathematics of the eye and the anatomy of the eye were not connected by visual style. Kepler's engagement with the new epistemic possibilities of anatomical images was, as with the eye illustrations of Vesalius, Valverde, and Platter, restricted to anatomical *historiae* – to facts and not causes, be they mathematical/formal or natural/teleological.

II. Bifurcation of Ocular Illustrations in Fabricius ab Aquapendente

The Vesalian eye and its progeny reigned in the second half of the sixteenth century. While it continued to be copied in the seventeenth it began to function schematically, particularly when the "cosmic" eye diagram of 1543 was used, e.g. in Thomas Bartholin's 1641 edition of his father's *Institutiones anatomicae*.[22] This shift in the use of the Vesalian model is due, in part, to developments in the use of anatomical images of the eye around 1600, the most significant of which can be seen in Fabricius ab Aquapendente's (1533–1619) first officially published work. It consists of three treatises, sold both as separate pamphlets for his students (to be bound as they saw fit) and together as a single volume: *De visione*, *De voce*, and *De auditu* (*On Vision, On Voice, and On Hearing*).[23] Fabricius called his overall project, incomplete at the time of his death, the *Theater of the Craftsmanship of the Whole Animal* (*Totius animalis fabricae theatrum*).[24]

Fabricius framed his anatomical project as an improvement upon Vesalius's. This included surpassing Fabrica's illustrations, but in general by systematically relaying not just the fabric or *historia* of the body, but also the activities (*actiones*) and uses or teleological ends (*utilitates*) of the parts. Fabricius also aimed to broaden the scope of anatomy to the investigation of animals, and the animal soul, as a whole.[25] Fabricius wrote that anatomy, properly considered, has four aspects: *dissectio*, *historia*, *actio*, and *usus* (or *utilitas*).[26] Fabricius's Galeno-Aristotelian investigation of the whole animal linked Galen's *On Anatomical Procedures* and Aristotle's *Dissections*

22 Bartholin, *Institutiones anatomicae*, p. 293. See also Raynaud, *Eye Representation*, p. 116.

23 Fabricius, *De visione*.

24 On Fabricius's life, see Muccilo, "Fabrici d'Acquapendente, Girolamo." In *Dizionario Biographico Degli Italiani*, vol. 43, 1993, https://www.treccani.it/enciclopedia/girolamo-fabrici-d-acquapendente_%28Dizionario-Biografico%29/ (accessed 18 February 2021).

25 Cunningham, "Fabricius and the 'Aristotle Project.'"

26 Fabricius, *De voce*, p. *ii.

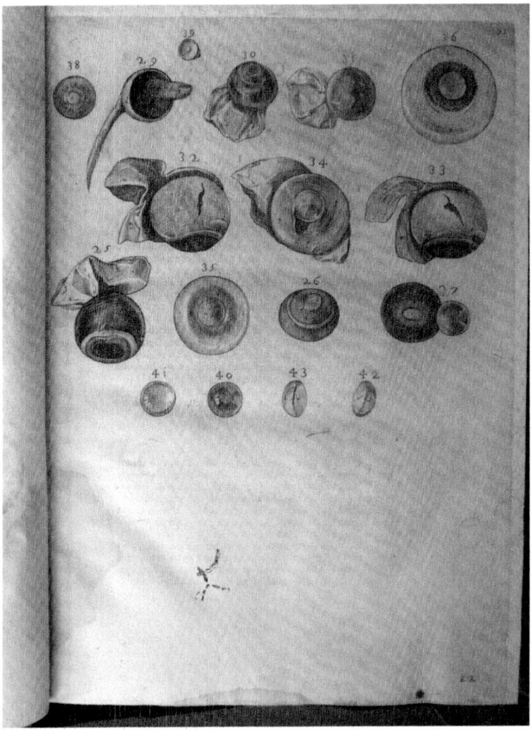

Figure 10.5 Hieronymus Fabricius ab Aquapendente, *De visione, voce, auditu* (Venice: Franciscus Bolzetta, 1600), 35. Courtesy of the Newberry Library, Chicago. Stanton A. Friedberg Collection.

to comprise the *dissectio* section of his anatomical project; often lumped into *historia* by anatomists, he says, *dissectio* primarily concerns *how* one ought to dissect as the basis for a reliable natural history. In this case, however, Galen's section on the dissection of the eye in his *On Anatomical Procedures* was not available to Fabricius (premodern copies were extant only in Arabic), and the entirety of Aristotle's *Dissections* is lost after antiquity.

Fabricius's illustrations were tailored to support and complement this precise scientific method in a variety of ways. Just two sets of images are examined here. The first is contained on a single copperplate engraving, the last of four tables located at the end of the *historia* of the eye. The first three tables (not reproduced here) show the dissection of the orbit, fat, and muscles of the eye in humans and several other animals, while the final table (Figure 10.5) shows the dissection of human, sheep, and ox eyeballs. The second image (Figure 10.6) is a single woodcut accompanying the text at the end of Part 3, Chapter 8 on the usefulness of the *aranea* or cobweb-like tunic (today the lens capsule).

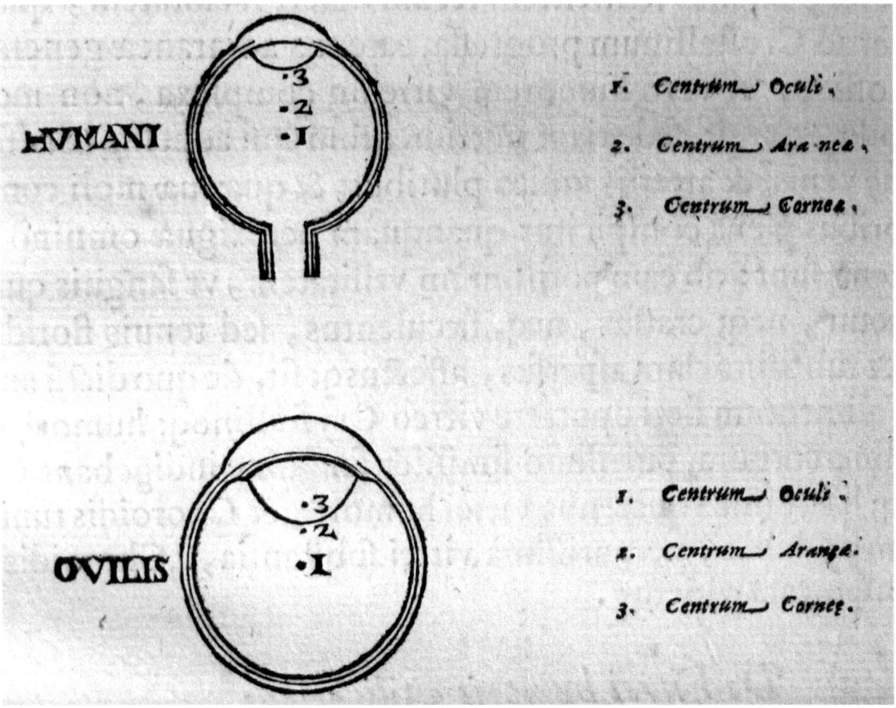

Figure 10.6 Fabricius, *De visione*, 105. Courtesy of The Newberry Library, Chicago, Stanton A. Friedberg Collection.

Fabricius was intimately involved with the illustrations in his works; he and his anatomical rival in Padua, Julius Casserius, both had artists living in their residences at various points to sketch the private anatomies they frequently performed. Fabricius's *dissectio* table (Figure 10.5) evinces a heightened naturalism compared to Vesalius's images, partly as function of the copperplate medium versus the earlier woodcut, with more detailed naturalistic renderings from delicate lines and shading techniques. These illustrations serve largely as dissection guides, preliminaries to the textual *historia*. As the text makes clear, they are intended to convey the eye as one might actually see them in a dissection; for example, in the key to Figures labeled 32 and 33 we read:

> 32. & 33. Both of these are figures of an ox [eye], presenting to sight the retina separated from the choroid [i.e., posterior *uvea*] as much as possible.
>
> m. m. the internal surface of the choroid.
>
> n. the cornea.

o. o. o. the retina torn somewhere.

r. r. the vitreous humor coming into view via a hole in the torn retina.[27]

The use of naturalistic techniques does not guarantee that they are faithfully depicted from life.[28] Yet, why are we given *two* figures here? We know that Fabricius had artists in his home to sketch the results of his private dissections, and we might surmise that several such drawings were made from life, two of which were engraved so that the reader might see what was common in several particular experiences.

This visual *dissectio* stage in his anatomical method indeed depicts particulars (whether idealized or invented), and not universalized matters of fact presented in the textual *historia*. The choroid (or posterior *uvea*) is a dark, thin, but somewhat tough membrane covering the retina. In my experience with eye dissection, removing this without damaging the underlying retina is challenging. These illustrations highlight this: in both images we see that the delicate retina has been ruptured differently, gesturing to two unique courses that the anatomist's knife's took through tough and slippery membranes (first the *sclerea,* then the *uvea* or *choroid*) that surround a fragile one (the retina); the former are pulled away and the latter only given shape by the delicate gel of the vitreous body, itself deformed by gravity. When Fabricius says that he is "presenting to sight the retina separated from the choroid as much as possible," he telegraphs that this image represents the choroid removed from the retina as much as possible in an actual, and indeed difficult, dissection. Compare this to Vesalius's "demiurgic plan" woodcut, where an ideal retina and choroid are reconstituted into their supposed living configurations around the vitreous body.

The text of *De visione* contains a chapter on how to dissect the eye, but it is short and confusing apart from the images. His four copperplates thus visually guide the reader toward extra-textual experiences. The images of the eye in the *Fabrica* certainly point beyond the text as well, but they are not tied to an underlying, systematic research program, and they assume, *a priori*, a Galenic pneumatic physiology of the eye. While Vesalius's images

27 "32. & 33. Bovis est utraque figura, separata choroide retinam potissimum visui obiiciens.

m. m. interna choroidis facies.

n. cornea.

o. o. o. retina alicubi lacera.

r. r. Vitreus humor per lacerae retinae foramina in conspectum veniens."

Fabricius, *De visione*, p. 34.

28 Kemp, "Temples of the Body," pp. 44–51; Kusukawa, *Picturing the Book of Nature*, pp. 4–8.

of the eye are dissection guides of a sort if read in reverse, they are not intended to represent particular experiences rooted to singular events in space and time, and they gloss over the difficulty of investigating the recalcitrant matter of the eye.

His account of vision unites mathematics and natural philosophy with anatomy, simultaneously moving beyond previous accounts given by each. The woodcut that is shown here (Figure 10.6) does this as well, introducing several epistemic possibilities new to both anatomy and optics. One is to challenge fundamental aspects of *perspectiva* itself. The woodcut is offered as a resource for mathematicians to generate better accounts of the rays passing through the eye compared to that given by the medieval *perspectivae:*

> But so that those who produce works of optical science can accurately observe the diverse progression of rays, which are called visual, while they cross over from one humor into another; and [so that] they can accurately measure off the angles of refraction, and thence grasp the innumerable *utilitates* of the parts: we provide, with the most exact care, human and sheep eyes divided through the middle. And the whole magnitude and that of the individual parts, including their situations and figures, are described, and the place that each of their centers occupy is revealed, and everything is sketched out in the tables below.[29]

The vertical dots in these diagrams are labeled: "1. Center of the eye / 2. Center of the [anterior surface of the] *aranea* / 3. Center of the cornea." These points, determined by experience, follow graphic features similar to sixteenth-century editions of John Peckham's *Perspectiva communis,* the standard introduction to perspectivist optics. Crucially, they contradict Peckham's account. Mathematical education, including optics, was becoming more common toward the end of the sixteenth century, and Fabricius used the visual language of optics via seventeen woodcuts interspersed with the text in the previous seven chapters, diagrams that were copied or otherwise derived from editions of Peckham's *Perspectiva.*[30] The viewer is primed

29 "Ut autem qui Opticae scientiae operam dant, accuratè obervare possint, progressum varium radiorum, quos visuales appellant, dum ab uno in alium humorem transeunt; atque angulos refractionis dimetiri, & inde innumeras utilitates partium excepere: curavimus exactissima diligentia, oculum humanum & ovilem per medium secari, & magnitudinem totius, ac singularum partium, nec non earundem situs, & figuras describi, & loca qua eorum centra obtinent inveniri, & omnia in subiecta tabella delineari." Fabricius, *De visione*, p. 105.

30 The images most closely resemble those found in either the 1504 Venice edition, or else the 1580/1592 Cologne editions. For a list of the published editions, see Pecham, *John Pecham and*

to read our Figure 10.6 using the visual language in this mathematical tradition, but Fabricius leverages this language to launch an attack on the geometry of the eye held by the *perspectivae,* in the process refuting key aspects of perspectivist visual theory. Modern scholars have overlooked this refutation, perhaps because the only textual (as opposed to visual) indication of it is Fabricius's cryptic phrase just below the diagram.[31] Fabricius's attack is not obvious to those who have not learned the visual language of sixteenth-century optics.

The context of Figure 10.6 is the determination of the *utilitates* of the parts of the eye, that is, the reason why Nature constructed human and animal eyes with specific temperaments, shapes, sizes, situations, degrees of refraction, etc. The explicit *utilitates* that Fabricius highlights with this diagram are ones that he was, as he acknowledged, only partly competent to determine, and thus he invited mathematicians to deduce these purposes more precisely, using his diagram to trace the "diverse progression of rays" in the eye. Prior to the seventeenth century, *perspectivist* optics structured the eye quite apart from anatomical experience; derived from Ibn al-Haytham (Alhacen/Alhazen in Latin), they argued that the cornea and the anterior crystalline lens surface are portions of concentric spheres by *necessity,* and thus *must* have identical centers of curvature. Such a geometry was needed to "save" the mathematical cone of vision described by Euclid, Galen, and especially Ptolemy while uniting it with a roughly Aristotelian intromission visual theory. The eye's geometry was thus established *a priori*.[32] Fabricius reversed this, inviting investigators to use his empirically derived diagram of the eye to put mathematical optics on a new basis. The locations, relative sizes, and shapes of the parts of the eye in *De visione* were gathered and abstracted from anatomical experience, but these mathematical details were presented only visually.

Fabricius's images appear to be novel in the histories of anatomy, natural philosophy, and mathematical optics. Earlier cases of mathematical and philosophical *aporia* encoded diagrammatically may exist, but this seems to be the first applied to the living word.[33] Here, a complex, natural body has been mathematized and depicted, a ray-analysis problem whose solution was not easily forthcoming – indeed, one that would not be solved either geometrically

the Science of Optics, pp. 56–57. See also Raynaud, *Eye Representation.*

31 "Habebunt enim curiosi indagatores operum naturae, ubi multa contemplari possint." Fabricius, *De visione,* p. 105.

32 Smith, *Sight to Light,* pp. 181–227.

33 Something similar regarding products of art does seem to have appeared earlier. See Dupré, "Ausonio's Mirrors and Galileo's Lenses; Borrelli, "Optical Diagrams as 'Paper Tools'."

or analytically without subsequent developments in the mathematics of refraction. Figure 10.6 can also be seen as what historians of science have termed a *paper tool*,[34] i.e. "visible and maneuverable substitutes for scientific objects under investigation."[35] Importantly, the epistemic status of these woodcut diagrams builds upon the foundation established via the former copperplate engravings – instructions for performing dissections that would enable the reader, as Fabricius did, to understand the mathematical structure of the eye via an Aristotelian process of abstraction, experience, and intuition. The copperplate engravings were also testimony that Fabricius had the skill and experience to arrive at the properties depicted in his later cross-sectional diagrams. Together, these images implicitly attack earlier models of the eye by embedding improved (in the case of the copperplate) or new (in the case of the woodblock) epistemic functions within the images. Moreover, the images are grounded in a rigorous and well-respected (at the time) Galeno-Aristotelian scientific methodology, they are practical in the sense of guiding the reader toward in performing their own anatomies, aid in the discovery of both material and formal causes, and they call for those proficient in mathematical optics to wrestle with a problem about the final causes of nature, a problem embedded in the diagram and only hinted at textually.

III. Conclusion

Fabricius simultaneously bifurcated and extended the possibilities present in the Vesalian tradition of ocular illustrations. These new uses are striking when compared to the more gradual developments of the *Fabrica* visual model we saw in Valverde, Platter, and Kepler.

These changes to the kind of work that images of the eye could perform rested upon the assimilation of new visual conventions and expectations by an expanding community of authors, artists, and readers. From the sixteenth through the early seventeenth century images of the eye moved through the boundaries separating natural philosophy, mathematics, medicine, and ophthalmology. While the history of ocular anatomy and illustration is not generalizable in all its aspects, it is nevertheless a counter-example to recent historiography that makes a sharp distinction between mathematical diagrams and anatomical pictures in this period. The notion that illustrations in anatomy, medicine, and natural history played a decisive

34 Klein, *Experiments, Models, Paper Tools*.

35 Hentschel, *Visual Cultures in Science and Technology*, p. 75.

role in the development of those sciences in the sixteenth century, while the mathematical sciences were hardly affected by print technology or new visual conventions, must at least be amended.[36]

Let us return to the broad narratives that this distinction between anatomical illustrations and mathematical diagrams was supposed to problematize: that the exact reproduction of images, made possible by print, was foundational for the development of modern science. William Ivins posited that two changes were crucial to the Scientific Revolution. Firstly, the "rationalization of sight" began in the Renaissance, "a reciprocal, or two-way, metrical correspondence between the pictorial representations and the shapes of those objects located in space." Secondly, print introduced the possibility of "the exact duplication of pictorial symbols for visual awareness."[37] Vesalius's woodblocks in particular allowed, he says, for "the first grammar of the human figure," defined by exact reference to invariant, repeatable pictures; this "power of invariant pictorial symbolization," according to Ivins, lies at the heart of modern science and technology.[38]

Bruno Latour famously built upon this dual thesis, offering his own about so-called *immutable mobiles.* Addressing the question of how to "define what is specific to our modern scientific culture,"[39] he proposed that the widespread use of inscriptions with certain capacities was fundamental: "you have to invent objects which have the properties of being *mobile* but also *immutable, presentable, readable* and *combinable* with one another."[40] Expanding also upon Elizabeth Eisenstein,[41] he argued that the printing press allowed for multiple pictorial representations to be reproduced without corruption, and to be collected together in one place, "no matter how wrong, strange, or wild," and consequently, "no matter how inaccurate these traces may be at first, they will all become accurate just *as a consequence* of more mobilization and more immutability."[42]

More recently, Dániel Margócsy analyzed the history and historiography of Vesalius's woodblocks, annotations to the *Fabrica*, and copies (or plagiarisms) of the *Fabrica's* images. He concluded that what characterizes this period is not the exact repeatability or immutability of Vesalius's images; instead,

36 Notably, Kemp, "Temples of the Body" and Pantin, "Analogy and Difference."

37 Ivins, *Rationalization of Sight*, p. 10. See also Ivins, *Prints and Visual Communication*, pp. 160–164.

38 Ivins, *Rationalization of Sight*, p. 13; Ivins, *Prints and Visual Communication*, pp. 179–180.

39 Latour, "Visualisation and Cognition," p. 1.

40 *Ibid.*, "Visualisation and Cognition," p. 7. Emphasis Latour's.

41 Eisenstein, *Printing Press as an Agent of Change.*

42 Latour, "Visualisation and Cognition," p. 12. Emphasis Latour's.

"innovation in Early Modern anatomy happened when authors decided to emulate and improve upon the *Fabrica*" – at least in the sixteenth and seventeenth centuries.[43] While Vesalius's *text* was readily reworked, corrected, or refuted, engagement with the illustrations was more ambivalent. To the extent that they *were* reproduced without alteration, he says, this was because authors and publishers were unsure of what to do with this new visual approach to medicine, not because they were "exactly repeatable" or "immutable" in the senses of Ivins and Latour. According to Margócsy, in the sixteenth and seventeenth centuries medical innovation connected to Vesalius's images took place via *inexact* reproduction and *mutability*, such as when the *Fabrica's* illustrations were poorly emulated, pirated, or reinscribed as a critique.

The present study does not refute Margócsy's account, but he underestimated the role *Fabrica's* images played within the narrower domain examined here. Preexisting visual practices within mathematical optics (and to a lesser extent ophthalmology) meant that the *Fabrica's* images of the eye were read less ambivalently than others in the book. Vesalius's ocular illustrations initiated cascading changes to eye images, sparking the expectation within a growing community that certain kinds of depictions would carry new representational capacities (metrical ones, especially). Reinforcing Margócsy's position, this resulted in some mathematical publications reverting to schematic depictions to avoid potential conflict with anatomical authorities, even while medical authors altered the *Fabrica*'s ocular illustrations as a form of commentary and critique; that is the Fabrica's ocular illustrations seemed to precipitate *less exact* representations in some cases, and to contribute to a visual tradition characterized by *mutability* from author to author. This visual response occurred just as these medicine and mathematical optics, and the visual cultures associated with them, began to intertwine.[44]

Nevertheless, this is not a refutation of Ivins and Latour. If "exact repeatability" and "immutability" merely characterizes publication runs their theses are perhaps useful. Only, it seems, when hundreds or thousands of impressions could be made from a single woodblock or copperplate does the intensive intellectual and artistic labor required to produce metrically accurate representations pay off; only then can the reader trust that, just like the text, the illustration one views is, in some key respects, exactly like those found in other copies; only then could printed illustrations be authoritative in a manner similar to a text. For these new capacities to be

43 Dániel Margócsy, "From Vesalius through Ivins to Latour," p. 321.

44 This is seen especially in the works of Christoph Scheiner and Descartes.

activated, wherever the book itself was transported, there needed to be readers or communities with shared ways of reading, seeing, or decoding the images. With respect to the images of the eye, this does indeed seem to have occurred in a generation or two after the *Fabrica's* publication, as evidenced by the minute alterations to the Vesalian ocular illustration which served to critique Vesalius's, and anyone else's, ocular anatomy. Thus, some degree of "repeatability" or "immutability" among the copies of (at a minimum) a single edition of a text laid the foundation for new capacities of images of the eye. These include that the measurable sizes, shapes, etc. of the marks on the page were understood to refer to measurable properties in a living eye; the ability to visually comment on, correct, or overturn the factual claims – and even the theoretical systems – of one's predecessors, alone or in conjunction with text; to visually integrate disparate stages of a scientific methodology; and to make visual arguments about the formal and final causes of living bodies.

Bibliography

Early modern published sources

Bartholin, Caspar. *Institutiones Anatomicae.* Leiden: Franciscus Hackius, 1641.

Fabricius ab Aquapendente, Hieronymus. *De Visione, Voce, Auditu.* Venice: Franciscus Bolzetta, 1600.

Kepler, Johannes. *Ad Vitellionem Paralipomena, quibus Astronomiae Pars Optica Traditur.* Frankfurt: Marnius and heirs of Aubrius, 1604.

Peckham, John. *Perspectiva Communis.* [Milan]: Petrus de Corneno [Bonus Accursius?], 1482.

Peckham, John. *John Pecham and the Science of Optics: Perspectiva Communis.* Edited by David C. Lindberg. Madison, WI: University of Wisconsin Press, 1970.

Valverde, Juan. *Historia de la Composición del Cuerpo Humano.* Rome: Impressa por Antonio Salamanca, y Antonio Lafrerij, 1556.

Valverde, Juan. *Anatomia del Corpo Humano.* Rome: Salamanca et Lafreri, 1559.

Valverde, Juan. *Anatome Corporis Humani Auctore Joanne Valverdo Nunc Primum a Michaele Columbo Latine Reddita et Additis Novis Aliquot Taulis Exornata.* Venice: Giunta, 1607.

Vesalius, Andreas. *De Humani Corporis Fabrica Libri Septem.* Basel: Ex Officina Ioannis Oporini, 1543.

Vesalius, Andreas. *The Fabric of the Human Body: An Annotated Translation of the 1543 and 1555 Editions of "De Humani Corporis Fabrica Libri Septem."* Transl. by Daniel H. Garrison and Malcolm H. Hast. 2 vols. Basel: Karger, 2014.

Secondary sources

Baker, Tawrin. "Dissection, Instruction, and Debate: Visual Theory at the Anatomy Theatre in the Sixteenth Century." In *Perspective as Practice: Renaissance Cultures of Optics*, edited by Sven Dupré, pp. 123–147. Turnhout, Belgium: Brepols, 2019.

Bertoloni Meli, Domenico. *Mechanism: A Visual, Lexical, and Conceptual History*. Pittsburgh, PA: Pittsburgh University Press, 2019.

Borrelli, Arianna. "Optical Diagrams as 'Paper Tools': Della Porta's Analysis of Biconvex Lenses from De Refractione to De Telescopio." In *The Optics of Giambattista Della Porta (ca. 1535–1615): A Reassessment*, edited by Arianna Borrelli, Giora Hon, and Yaakov Zik, pp. 57–96. Cham: Springer, 2017.

Dupré, Sven. "Ausonio's Mirrors and Galileo's Lenses: The Telescope and Sixteenth-Century Practical Optical Knowledge." *Galilaeana* 2 (2005): 145–180.

Edgerton, Samuel Y. "The Renaissance Artist as Quantifier." In *The Perception of Pictures*, edited by Margaret A. Hagen, 2 vols., I: 179–211. New York: Academic Press, 1980.

Eisenstein, Elizabeth L. *The Printing Press as an Agent of Change: Communications and Cultural Transformations in Early-Modern Europe*, 2 vols. Cambridge: Cambridge University Press, 1979.

Hentschel, Klaus. *Visual Cultures in Science and Technology: A Comparative History*. Oxford: Oxford University Press, 2014.

Heseler, Baldasar, and Ruben Eriksson. *Andreas Vesalius' First Public Anatomy at Bologna: 1540*. Uppsala: Almqvist & Wiksell, 1959.

Ivins, William M. *On the Rationalization of Sight, with an Examination of Three Renaissance Texts on Perspective*. New York: Metropolitan Museum of Art, 1938.

Ivins, William M. *Prints and Visual Communication*. Cambridge, MA: MIT Press, 1969.

Kemp, Martin. "Temples of the Body and Temples of the Cosmos: Vision and Visualization in the Vesalian and Copernican Revolutions." In *Picturing Knowledge: Historical and Philosophical Problems Concerning the Use of Art in Science*, pp. 40–85. Toronto: University of Toronto Press, 1996.

Klein, Ursula. *Experiments, Models, Paper Tools: Cultures of Organic Chemistry in the Nineteenth Century*. Stanford, CA: Stanford University Press, 2003.

Kusukawa, Sachiko. *Picturing the Book of Nature: Image, Text, and Argument in Sixteenth-Century Human Anatomy and Medical Botany*. Chicago, IL and London: The University of Chicago Press, 2012.

Latour, Bruno. "Visualisation and Cognition: Drawing Things Together." Edited by H. Kuklick. *Knowledge and Society Studies in the Sociology of Culture Past and Present* 6 (1986): 1–40.

Margócsy, Dániel. "From Vesalius through Ivins to Latour: Imitation, Emulation and Exactly Repeatable Pictorial Statements in the *Fabrica*." *Word & Image* 35, no. 3 (2019): 315–333.

Margócsy, Dániel, Mark Somos, and Stephen N. Joffe. *The Fabrica of Andreas Vesalius: A Worldwide Descriptive Census, Ownership, and Annotations of the 1543 and 1555 Editions.* Leiden: Brill, 2018.

Muccilo, Maria. "Fabrici d'Acquapendente, Girolamo." In *Dizionario Biographico Degli Italiani*, Vol. 43, 1993. https://www.treccani.it/enciclopedia/girolamo-fabrici-d-acquapendente_%28Dizionario-Biografico%29/ (accessed 18 February 2021).

Nutton, Vivian. "Representation and Memory in Renaissance Anatomical Illustration." In *Immagini per Conoscere: Dal Rinascimento alla Rivoluzione Scientifica*, edited by Fabrizio Meroi and Claudio Pogliano, pp. 61–80. Florence: Olschki, 2001.

Panofsky, Erwin. "Die Perspektive Als 'Symbolische Form.'" In *Vorträge Der Bibliothek Warburg 1924/25*, pp. 258–330. Liepzig/Berlin: Teubner, 1927.

Panofsky, Erwin. *Perspective as Symbolic Form.* Transl. by Christopher S. Wood. New York: Zone Books, 1997.

Pantin, Isabelle. "Analogy and Difference: A Comparative Study of Medical and Astronomical Images in Books, 1470–1550." *Early Science and Medicine* 18, no. 1–2 (2013): 9–44.

Raynaud, Dominique. *Eye Representation and Ocular Terminology from Antiquity to Helmholtz.* Amsterdam: Wayenborgh Publications, 2020.

Shotwell, R. Allen. "Animals, Pictures, and Skeletons: Andreas Vesalius's Reinvention of the Public Anatomy Lesson." *Journal of the History of Medicine and Allied Sciences* 71, no. 1 (2016): 1–18.

Skaarup, Bjørn Okholm. "The Unexpected Success of a Spanish Anatomy Book: Juan Valverde de Amusco's Historia de la Composicion del Cuerpo Humano (Rome, 1556), and its Many Later Editions." In *Specialist Markets in the Early Modern Book World*, edited by Richard Kirwan and Sophie Mullins, pp. 123–141. Leiden: Brill, 2015.

About the Author

Tawrin Baker is a historian of early modern science, natural philosophy, and mathematics. He is a visiting assistant professor in the History and Philosophy of Science at the University of Notre Dame.

11. Illustrating the Vernacular Body: Juan Valverde de Amusco and the Art of Embodied Anatomy

Emily Monty

Abstract
Andreas Vesalius's *Fabrica* (Basel, 1543) presented knowledge of the human body in woodcut pictures and Latin text. This chapter analyzes a Spanish commentary on and translation of the *Fabrica* written by Spanish physician Juan Valverde de Amusco and published in 1556 in Rome. Valverde's important contributions to the history of anatomical illustration are understood by identifying a stylistic shift from Vesalius's classicizing idealism towards clarity in communicating practical applications and sensory cognition. New engravings strengthen the epistemic potential of Vesalius's images for audiences encountering up-to-date anatomical information in a vernacular register. Considering related publications issued in London and Antwerp, Valverde's treatise is positioned within a trans-alpine epistemology of the body formed at the intersection of craft and humanist knowledge.

Keywords: anatomy, intaglio, vernacular, book illustration

I. Introduction

The treatise *De humani corporis fabrica* by Andreas Vesalius (1514–1564) was a landmark atlas of human anatomy that provided readers in early modern Europe with a new standard in medical illustration and

Noyes, R.S. (ed.), *Reassessing Epistemic Images in the Early Modern World.*
Taylor & Francis Group, 2023
DOI 10.5117/9789463723350_CH11

instruction from the moment that it was published in Basel in 1543.[1] The book represented an epistemic shift from a tradition that privileged the literary canon of medical texts, centered on the writings of the second-century Greek physician Galen, to a practice that depended on the author's purportedly direct observation of dissected cadavers.[2] From an early moment in the book's history, contemporary readers saw Vesalius's anatomy as a mutable font of knowledge ready to be folded into an ever-growing mass of information in text and image about the human body.[3]

In 1556, the Spanish physician Juan Valverde de Amusco (1525–1587) published in Rome *Historia de la composición del cuerpo humano*, a Spanish commentary on and translation of the Latin *Fabrica*.[4] With the support of his patron, Spanish cardinal Juan Álvarez de Toledo (1488–1557), who was called to Rome in 1553 to serve as inquisitor general, Valverde contracted with the reputable publishers Antoine Lafréry (1512–1577) and Antonio Salamanca (1478–1562) to print the engravings, and papal printer Antonio Blado (1490–1567) to print the text. This combination of patron and publishers implied an element of orthodoxy for the publication and opened access to international networks of potential readers beyond Rome. Working with Salamanca and Lafréry, who specialized in the production of single sheet engravings, also ensured that the illustrations would be of the highest possible quality.[5] The composition

1 Jan Steven van Calcar is accepted as the artist of Vesalius's illustrations. For more on their attribution, see Saunders and O'Malley, *Illustrations from Andreas Vesalius*, pp. 25–29. For complimentary studies on aspects of Vesalius, see the chapters by Baker and by Joffe, Margócsy, and Somos in this volume. I am grateful to Evelyn Lincoln for her comments on this chapter and Ruth Noyes for her insightful editorial eye. I also wish to thank Sietske Fransen and the members of the Bibliotheca Hertziana who provided feedback as part of a session of the Visualizing Science in Media Revolutions research group. All translations, unless otherwise noted, are my own.

2 Carlino, *Books of the Body*, pp. 187–225 describes this shift as a long process. López Piñero, "Disección en España," pp. 67–68 stresses similar themes.

3 Houtzager, "Editions, Copies and Replicas," esp. pp. 196–198 documents the dizzying array of editions, compilations, and emulations of the *Fabrica* and *Epitome* that preceded Valverde. Margócsy, "Vesalius through Ivins," shows that it was not until the eighteenth century that publishers became interested in reproducing Vesalius's illustrations and text exactly. He helpfully describes early modern imitations of Vesalius in terms of the rhetorical concept of *emulation*, as opposed to "slavish imitation."

4 For medical culture in sixteenth-century Rome, see Andretta, *Roma medica*.

5 Salamanca was of Spanish birth. Carlino, "Tre piste" analyzes how the *Historia* was produced at the nexus of Spanish and Italian cultural networks. Sánchez-Molero, "Roma y España," argues that Salamanca's print shop was a hub of Spanish culture in Rome serving the large community of Spaniards in the city.

of this publishing team suggests the central role that the illustrations were expected to play in the overall project of translating Vesalius for a new vernacular audience.

Spanish artist Gaspar Becerra (c. 1520–1570) supervised production of the drawings for the plates and French printmaker Nicolas Beatrizet (1507–1565) led a workshop to produce 42 full-page engravings.[6] Printed with intaglio illustrations and moveable type, pages thus passed between the two publishing houses, each with specialized skills, machinery, and materials necessary for printing copperplates and letter press type. To streamline the process, the publishers diverged from Vesalius, who used woodcut images that could be printed at the same time as the moveable type. Instead, they extracted the images from the text and assembled them in plates at the end of each chapter. In doing so, they followed the pictorial groupings created by Flemish engraver Thomas Geminus (1510–1562), whose publication of 1545 was the first to translate Vesalius's woodcuts into copper engravings.[7] Valverde's artists further reimagined the pictorial space on the printed pages by adding surprising details that invite the viewer to linger in new ways.

Valverde's volume is often described as a case of plagiarism and its illustrations dismissed as derivative.[8] In contrast, I approach Valverde as a humanist author, commentator, and editor, following Cynthia Jennifer

6 Throughout this chapter I refer to Valverde as a representative for the collective authorship of the writer, artists, printers, and publishers. Beatrizet's monogram appears in Plate IV (sig. M6r) and V (sig. 2Mr) of Book II as well as in a portrait of Valverde appearing in later editions, but a number of unidentified hands are evident throughout. For stylistic and circumstantial evidence supporting the accepted attribution to Gaspar Becerra, see Marías, *El largo siglo XVI*, pp. 594–596 and Redín Michaus, *Pintores españoles en Roma*, pp. 159–251, esp. 211, summarizing ideas about their authorship and giving compelling reasons to attribute the drawings to Becerra.

7 Geminus, *Compendiosa totius anatomie delineatio*. Nicholas Udall translated the Latin text into English in 1553. See Larkey, "Geminus and Nicholas Udall's Translation" for more on how the edition relates to Vesalius.

8 For a bibliographic assessment of the relationship between the *Fabrica* and *Historia* and a discussion of the sources of Valverde's illustrations, see Meyer and Wirt, "The Amuscan Illustrations." Guerra, "Juan de Valverde de Amusco," pp. 353–354, describes how Valverde's illustrations differ from Vesalius and summarizes the ways that these changes have been discussed by scholars. On so-called plagiarisms of Vesalius see O'Malley, *Andreas Vesalius*, pp. 88–92, where he alludes to the fact that authorship and originality were conceived much differently in the early modern period than today. It should nevertheless be noted that Vesalius tried to prevent pirated copies by obtaining printing privileges and was upset when so-called plagiarisms did appear. Dillon, *Michelangelo and the English Martyrs*, pp. 215–216, discusses Vesalius's response to Valverde and the history of Valverde's reputation as a plagiarist.

Klestinec, who characterizes him as such in her textual analysis to account for the fact that the *Historia* offers a simplified and reformatted version of the *Fabrica* rather than a direct translation.[9] The resulting publication is more concise in its teachings and, being less than half the size of Vesalius's book, is more manageable as a material product.[10] Valverde's relationship to Vesalius is therefore best understood in terms of the freedom that Renaissance translators felt to add to, subtract from, and reimagine their textual and visual sources using words and images.[11] The originality of Valverde's illustrations lies in the juxtapositions they create with their sources, and in the visual solutions they contribute to the broader project of translating Vesalian knowledge from Latin to various vernaculars. This shift is consistent with Vesalius's interest in making anatomical knowledge accessible to a growing intellectual elite who collected luxury publications for their libraries.[12]

As cultural historians have emphasized, early modern authors confronted a political choice when choosing to publish in either Latin or the vernacular. The latter, though more limited than Latin in its geographic reach, allowed them to disseminate knowledge of the natural world widely among non-university educated readers.[13] Valverde wrote in a vernacular that would reach an international audience of Spanish readers unbound from place; that is, he produced his Spanish text at the international crossroads of papal Rome so that it could be consumed by readers throughout an empire on which the proverbial sun never set. Just as Valverde adapted, amended, and translated Vesalius's text, the artists working on his project reimagined the illustrations for their new environment through an economical program of changes. In this translation project he and his publishing team gave

9 Klestinec, "Valverde and Print Culture," pp. 78–94.

10 These facts often lead to speculation that the book was less expensive, such as in Roberts and Tomlinson, *Fabric of the Body*, p. 211, although I am unaware of any extant sale records. Margócsy, Somos, and Joffe in *Fabrica of Andreas Vesalius* document prices paid for Vesalius's *Fabrica* in the sixteenth century.

11 For more on the role of the Renaissance translator, see Burke, "Renaissance Translator," p. 26. Among the rich literature dealing with the question of so-called reproductive printmaking in the early modern period, Michael Bury's essay on Beatrizet is particularly important for how I think about Valverde's reuse of images: Bury, "Beatrizet and Antique Relief Sculpture."

12 Carlino, "Vesalio e la cultura visiva," addresses this growing audience.

13 Fransen, "Latin in a Time of Change." Her discussions of Galileo Galilei (1564–1642) and René Descartes (1596–1650) show that these writers thought about their communication in the vernacular as a way of reaching people who operated outside of university spaces. Historian of medicine Mary Fissell has analyzed how the circulation of cheap, accessible, vernacular texts about human reproduction formed "vernacular beliefs" and "vernacular ideas," about women's bodies: Fissell, *Vernacular Bodies*.

the illustrations equal weight relative to the text, using the former to shift Vesalius's important eye-witness images into a visual register more consistent with Valverde's vernacular writing.

The value of this translation of the Vesalian visual world into a vernacular register can be elucidated within a framework of art historical studies about the linguistic structures that shape the production and reception of art. For example, Michael Baxandall highlighted how humanists approached Giotto through the rhetoric of Ciceronian Latin, and Elizabeth Cropper described Parmigianino's paintings in terms of a vernacular style exhibiting ideas of female beauty established in Tuscan poetry.[14] As recent work has emphasized, the vernacular in the visual arts can deliver a region-specific or popular style while still maintaining an ability to communicate a classical or cosmopolitan aesthetic.[15] That is to say that classical and vernacular modes of visual expression could easily co-exist in the early modern period, and they are in fact both at play in the aesthetic of Valverde's treatise.

Valverde's ideal audience comprised an emerging category of learned readers lacking both proficiency in Latin and in-person guided experience of dissections. The author expresses interest in reaching readers who struggled because, "[...] Vesalius wrote so obscurely that he can only be understood with difficulty except by those who have first had a body in front of their eyes a few times and a very good teacher to explain it to them [...]."[16] Though such readers found themselves in a different position than medical students enrolled at university, they nevertheless operated in learned spaces with varying degrees of ease and familiarity. Valverde's Italian edition of 1559–1560, for example, soon found an audience among artists in Italy who were participating in formalized dissections and intellectual activities at the newly-founded academies of art.[17]

14 Baxandall, *Giotto and the Orators*; Cropper, "On Beautiful Women."

15 Keizer and Richardson, *Vernacular Expression in Early Modern Arts*. The essays by Alexandra Onuf and James Bloom are particularly useful for reconsidering the relationship between vernacular and classical registers.

16 "[...] auer escrito el Vesalio tan escuramente, q*ue* con difficultad puede ser entendido, sino de aquellos que primero algunas vezes an tenido el cuerpo delante de sus ojos, y muy buen maestro que se le declare [...]." Valverde de Amusco, *Historia,* sig. 2r.

17 Okholm Skaarup, *Anatomy in Early Modern Spain*, p. 242 describes how a number of artists, many associated with the Accademia del Disegno in Florence referred to Valverde in their own writings and art. He also notes that Valverde's treatise appears in the 1608 inventory of the Accademia di San Luca in Rome after being acquired from the estate of Ottaviano Mascherino. ASR, TNC, uff. 11, 1608, pt. 1, vol. 76, fol. 818r, accessed through digital platform of The History of the Accademia di San Luca, c. 1590–1635: Documents from the Archivio di Stato di Roma: https://www.

In the preface to his book, Valverde writes to his patron that he is particularly interested in reaching surgeons in Spain in urgent need of anatomical knowledge in the vernacular:

> Considering, most illustrious Señor, the great lack that our nation has of men who understand anatomy [...] and seeing the damage that follows from this to the whole Spanish nation, in part because surgeons (who most need to understand anatomy) know little Latin [...] it seemed to me a very advisable thing to write this history in our language [...].[18]

In prioritizing these readers, Valverde departed from Vesalius's original goal of incorporating the duties of surgeons into the role of the physician/anatomist. Tradition held that barber-surgeons performed surgeries and bloodletting at the direction of a physician. They also assisted in university dissections, being called upon to handle the decaying cadaver while the professor recited the writings of ancient authors from a respectable distance at the cathedra.[19] Vesalius aimed to reclaim the lowly manual work of dissection for the educated elite to integrate empirical and textual knowledge of the body in the single figure of the professor. Whereas Vesalius expanded the intellectual and practical domains of the physician, Valverde ensured that the knowledge communicated in the professor's Latin text was available to medical practitioners tasked with caring for human bodies under different circumstances.[20] Imaginative adjustments to the illustrations helped to achieve this shift in purpose.

nga.gov/accademia/en/intro.html. For more on artists' academies and anatomical study in Italy see Laurenza, *Art and Anatomy*, p. 33. For information on the types of readers – overwhelmingly physicians – who owned Vesalius's *Fabrica* in the sixteenth century see Margócsy, Somos, and Joffe, *Fabrica of Andreas Vesalius*, pp. 30–37.

18 "Considerando Illustrissimo Señor la gran falta, que la nacion nuestra tiene de hombres, que entiendan la Anatomia [...] y visto el daño, que dessto se sigue a toda la nacion Española, parte por los Cirujanos (a quien mas falta haze no entenderla) saber poco latín [...] pareciame cosa muy conueniente, escriuir esta historia en nuestra lengua [...]." Valverde de Amusco, *Historia*, sig. F2r.

19 Vesalius describes this division of labor, and his interest in overcoming it through a more holistic approach to anatomy and dissection, in his prefatory letter to Charles V, pp. ii–iii. Carlino, "The Book, the Body, the Scalpel," pp. 33–36 analyzes the visual representations of this tradition that confined the physician-professor to the cathedra while another man performed the dissection.

20 Lincoln, *Brilliant Discourse*, pp. 115–161 analyzes the relationship between the physician and barber-surgeon as it was negotiated through printed images in the sixteenth century. See, especially, pp. 127–128 and n. 33, where she describes the complicated social position of surgeons – even those trained at the university – in Italy.

Valverde's decision to use the more expensive and labor-intensive medium of engraving – increasingly preferred for the precision it could provide in visual aids like maps and architectural designs – can be understood as a choice to strengthen the epistemic potential of Vesalius's images for an audience without access to either a cadaver or a teacher. The focus on increased clarity and dynamism in the illustrations finds parallels in contemporary ideas about the vernacular as both a more accessible and descriptive register.[21] In fact, Valverde directed his readers' attention to his work's increased legibility and vividness relative to the *Fabrica* in his description of the first figure showing the muscles: "it should be known that this figure is different than those in Vesalius because the shadows show the fibre of the flesh running in a specific direction in each muscle."[22]

While many of Valverde's illustrations hew closely to the Vesalian model, they exhibit a new attention to naturalism, liveliness, and sensory perception effected through minor changes. For example, to illustrate the articulation of the bones, Vesalius presented a full skeleton with its right forearm supported by the handle of a spade, its head tilted back as if looking up from digging its own grave (Figure 11.1 A–B). The rigid spinal column suggests that the image was made from a specimen prepared and mounted with an iron rod according to the practices for preparing bones, to which Vesalius devoted a subchapter that Valverde subsequently omitted from his translation.[23] In contrast, Valverde's corresponding figure grips a staff and meets the viewer's gaze with an upright skull as a traveller rather than an undertaker. Carefully placed shadows bring out the natural curve of the lumbar spine, and an increased tilt of the pelvis allows the sacrum and coccyx bones to protrude forward for the inspection of the viewer. The skeleton seems to stand of its own volition in a livelier and more naturalistic iteration of contrapposto. Weight shifted into the right hip produces a relaxed curve in the vertebrae, illustrating the interlocking relationships of the spine, pelvis, and shoulders. Later in the book, newly vivified specimens probe their own bodies, holding organs aside to facilitate viewing or guiding attention by peering at their own innards.

21 Serjeantson, "Proof and Persuasion," pp. 158–159 links the rise of "the fact" with the increasing tendency to write about natural history in the vernacular, allowing authors to "thereby escape the expectations about philosophical terminology and argument generated by the Latin of the schools." I am thankful to Pia Bornus for bringing this source to my attention.

22 "[...] y es de saber, que esta figura es differente delas del Vesalio, en que en esta las sombras muestra*n* el andar del hilo dela carne, segun que en cada morzillo particularmente caminan." Valverde de Amusco, *Historia*, sig. M1v.

23 Vesalius, *Fabrica*, Book 1, Chapter XXXIX.

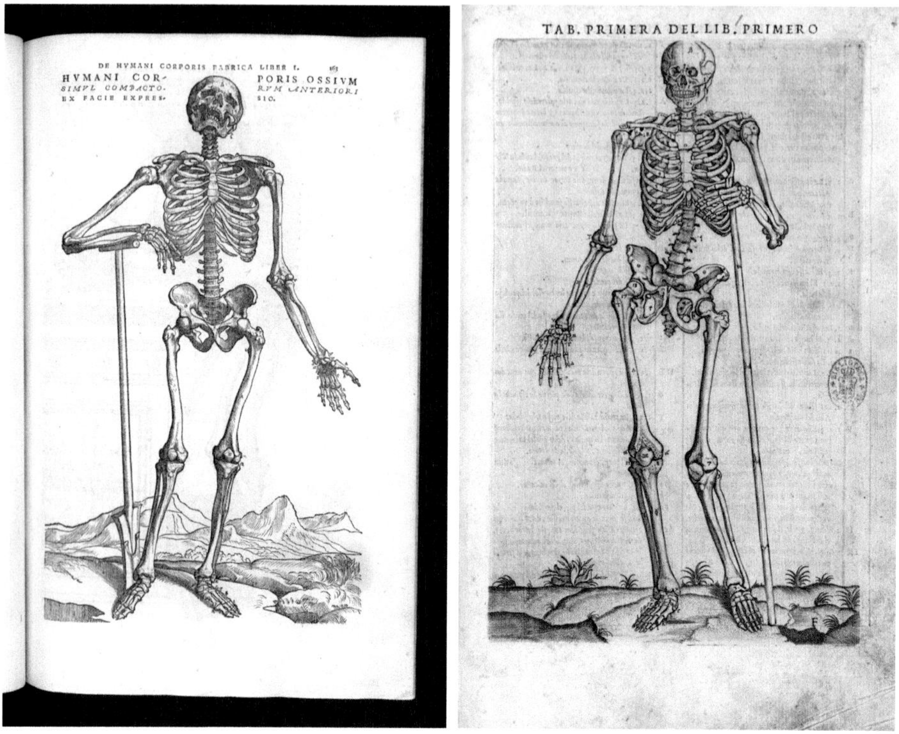

Figure 11.1 A–B A (left): Jan Steven van Calcar?, *Skeleton*, from Andreas Vesalius, *De humani corporis fabrica* (Basel: Johannes Oporinus, 1543), p. 163, woodcut, 41.5 × 27 cm, image property of the Biblioteca Nacional de España. B: (right) Nicholas Beatrizet after Gaspar Becerra, *Skeleton*, from Juan Valverde de Amusco, *Historia de la composición del cuerpo humano* (Rome: Antonio Salamanca and Antoine Lafréry, 1556), engraving, sig. F1r, 29 × 19 cm, image property of the Biblioteca Nacional de España.

II. From Stone to Flesh

One of the most legible ways that Valverde's artists translate Vesalius's images into a vernacular register is by breaking down the antique sculptures that served as pictorial containers for Vesalius's anatomized bodies. As Glenn Harcourt has described, these classicizing shells mask the organic body, allowing Vesalius to encode his specimens as idealized empirical fonts of knowledge rather than variable human subjects.[24] Such stylized representation was also consistent with humanist ideology guiding Vesalius's critique of Galen and underwriting his authority as university professor.

24 Harcourt, "Anatomy of Antique Sculpture." Kusukawa, *Picturing the Book of Nature*, pp. 199–227 addresses Vesalius's ideas about the canonical body.

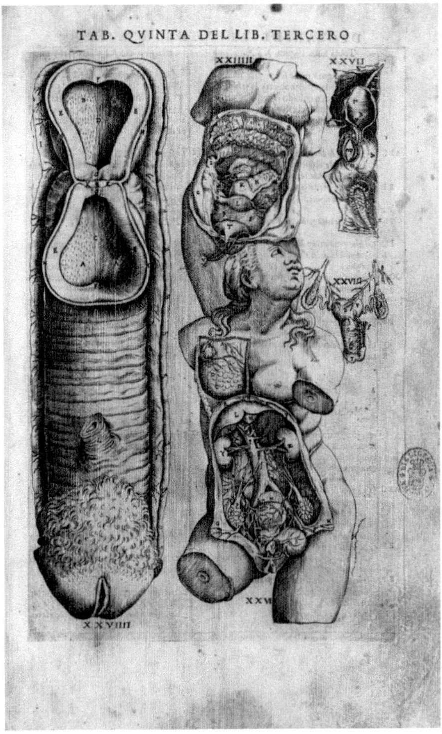

Figure 11.2 Nicholas Beatrizet after Gaspar Becerra, *Views of Female Reproductive Organs*, from Juan Valverde de Amusco, *Historia de la composición del cuerpo humano* (Rome: Antonio Salamanca and Antoine Lafréry, 1556), engraving, sig. 2P3r, 29 × 19 cm, image property of the Biblioteca Nacional de España.

Valverde transformed the specimens from antique sculpture fragments into images of living bodies.[25] Though many of his torsos retain the flat, hard surfaces of broken stone, the female body provided the appropriate intellectual space for a more organic approach (Figure 11.2). Consecutive circles of bone and marrow materialize the bisected arm and leg and delicate rings appear for the first time to trace the dermis and epidermis. Soft ripples denote a fatty layer along the inner bend of the torso, and an

25 Rose Marie San Juan also writes about how Valverde breaks the illusions of Vesalius's sculpted bodies and returns them to a state of flesh. She interprets this adjustment as part of a larger attempt by Valverde to restore a sense of wholeness and embodiment to the dissected figures while commenting on the fraught place of dissection in society: San Juan, "Restoration and Translation," esp. p. 60.

individual likeness emerges from the crowded picture. The fleshy aspect of these female torsos encourages meditation on biological generation. These stylistic modifications emphasize the humanity inherent in the images of female reproductive organs also appearing on the plate, including a striking representation of the vaginal canal and uterus excised from the body to visualize the Galenic argument that female genitals are the inverse of male reproductive organs.[26]

At times, Valverde presents his bodies as alert and evidently uncomfortable, conditions undoubtedly familiar to surgeon-readers. Such expressions were foreign to Vesalius, who animates bodies to practical rather than emotional or psychological ends: when demonstrating the function of muscles, for example, or to emphasize an allegorical message like a *memento mori*.[27] In place of one of Vesalius's noble stony torsos, readers of Valverde's *Historia* encounter a bald man who appears very much alive, even though his intestines tumble out of his body (Figure 11.3 A–B). He has the most unpleasant task of restraining his own viscera with his left hand, seemingly attempting to contain his innards while displaying them for inspection. Holding an organ removed from his body in his right hand, he allows a better view of the digestive system. His twisting posture is not only stylistically appropriate for illustrations created in mid-sixteenth-century Rome, but also calls to mind the contorted positions that patients may have been forced to assume during surgery.[28]

Many of Valverde's bodies join the bald man in their willingness to aid the viewer. In the chapter on nutritional organs, limbless fragments found in Vesalius become three men peeling back their own abdominal skins to display their dissected interiors (Figure 11.4).[29] The figure on the bottom left grips the cut flap of skin in his teeth to reveal the mesentery, a fatty sheet of tissue and vessels covering the abdominal cavity, depicted in isolation in the fourth quadrant. He pulls down on the loose skin around his waist with both hands to offer a better view of his lower intestines. Whereas Vesalius used a passive tense and narrative voice to convey the distanced objectivity of the anatomist forcibly rearranging organs to accommodate the viewer ("the upper membrane of the omentum [...] is pulled up during dissection"), Valverde emphasized the cadaver's active participation in its

26 For more on these ideas about the female genitals see Park, *Secrets of Women*, p. 186.

27 For example, the second assembled skeleton, on page 164 rests on a plinth inscribed VIVITVR INGENIO, CAETERA MORTIS ERVNT: "talent lives on, the rest will die."

28 For more on Becerra's Roman *maniera* and approach to the human body see Redín Michaus, *Pintores españoles en Roma*, p. 248.

29 The corresponding illustrations appear in Vesalius, *Fabrica*, pp. 355–357.

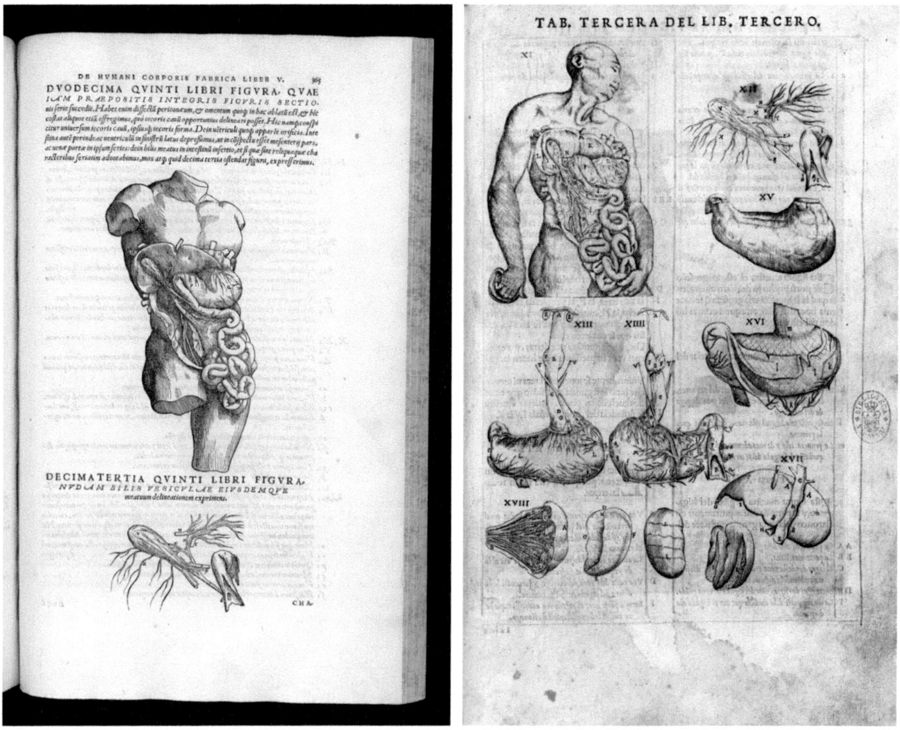

Figure 11.3 A–B A (left): Jan Steven van Calcar?, *Dissected Torso,* from Andreas Vesalius, *De humani corporis fabrica* (Basel: Johannes Oporinus, 1543), p. 365 (detail), woodcut, image property of the Biblioteca Nacional de España. B (right): Nicholas Beatrizet after Gaspar Becerra, *Dissected Torso,* from Juan Valverde de Amusco, *Historia de la composición del cuerpo humano* (Rome: Antonio Salamanca and Antoine Lafréry, 1556), sig. P5r (detail), engraving, image property of the Biblioteca Nacional de España.

own didactic dismantling: "This figure shows the location of the intestines, with the mesentery opened out and turned upwards, pulled up by the teeth, making the stomach a bit higher than it should be."[30]

The imagined misfortunes endured by these figures emphasize the role of touch both in dissection and surgery, an idea visualized most clearly in the arrangement of two torsos that create a macabre vignette in which the anatomized body becomes the anatomist, and vice versa (Figure 11.5).[31]

30 The translation of Vesalius is given in Saunders and O'Malley, *Illustrations from Andreas Vesalius,* Plate 54, p. 158. "Esta figura muestra el sitio delas tripas, y el redaño desdoblado y buelto hazia arriba, y tirado conlos dientes y por esso el estomago esta vn poco mas alto que no devria." Valverde de Amusco, *Historia,* sig. P2r. The omentum and the mesentery are closely related sets of abdominal membranes covering the visceral organs.

31 Whereas Vesalius mainly emphasized sight as a tool for empirical validation, earlier Renaissance writers such as Niccolò Massa and Jacopo Berengario da Carpi also privileged touch.

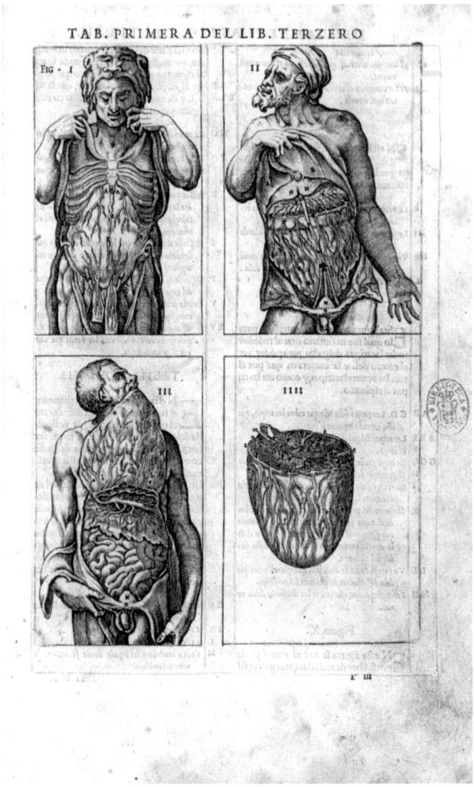

Figure 11.4 Nicholas Beatrizet after Gaspar Becerra, *Self-Demonstrating Torsos,* from Juan Valverde de Amusco, *Historia de la composición del cuerpo humano* (Rome: Antonio Salamanca and Antoine Lafréry, 1556), sig. P3r, engraving, 29 × 19 cm, image property of the Biblioteca Nacional de España.

The vertical figure has escaped both the noose around his neck and the rope that we can imagine binding his wrists behind his back in Vesalius's rendering. He uses his newly unrestrained limbs to probe his companion's thoracic cavity, turning away from the sight and evoking imagined sensations of warm flesh, odour, and cracking bones. This appeal to the senses is yet another way in which Valverde's illustrations speak to readers who learned anatomy outside of the lecture hall. Whereas university-trained Latinate physicians tended to prescribe treatments from a distance, barber-surgeons, who learned through practical experience and apprenticeship, gained embodied knowledge when laying hands on ill or broken bodies.

Carlino, "Vesalio e la cultura visiva," esp. pp. 78–79.

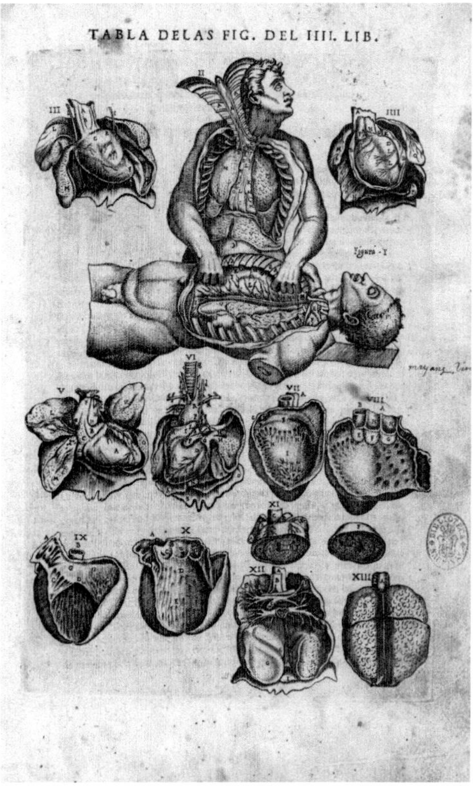

Figure 11.5 Nicholas Beatrizet after Gaspar Becerra, *Views of the Dissected Ribcage,* from Juan Valverde de Amusco, *Historia de la composición del cuerpo humano* (Rome: Antonio Salamanca and Antoine Lafréry, 1556), sig. R2r (detail), engraving, 29 × 19 cm, image property of the Biblioteca Nacional de España.

III. Re-siting Anatomical Knowledge

Such visual transformations suggest that Valverde was thinking of the many contexts in which anatomical knowledge was produced and consumed in the early modern period. Whereas Vesalius's *Fabrica* opens with an image of an anatomy theater, the *Historia* draws on printing conventions that use architectural portals as thresholds for entering the space of the book (Figure 11.6 A–B).[32] Muscular *ignudi* appear to fly out of the picture plane and away from a classicizing portico or tabernacle ornamented with grotesque

32 See Sherman, "On the Threshold" for architectural metaphors, both visual and written, used to welcome readers into the space of the Renaissance book.

masks and bucranium, associated in period architecture with death and sacrifice. Trailed by shrouds that billow around them like a parted curtain, they seemingly deliver their bodies to the ensuing pages for dissection. On the other hand, Valverde's patron is promised fame beyond his death, as the *ignudi* support an escutcheon bearing Álvarez de Toledo's chequered heraldic device, while putti above carefully hold the cardinal's wide-brimmed hat surmounting a central cross. The articulated midsections and powerful limbs of the *ignudi* recall Michelangelo's athletic souls and angels swirling above the Sistine chapel altar in the *Last Judgement* while the blind window at their back reminds us of Michelangelo's Medici tombs in the Basilica of San Lorenzo in Florence. Their apparently active participatory function and voluntary address to the reader contrast startlingly with the central body of Vesalius's frontispiece that lies helplessly in death awaiting transformation into an object of knowledge.[33]

Vesalius greets readers in the frontispiece while performing a dissection on a female cadaver at the center of a crowded rotunda. Visual cues reinforce his authority over the anatomical specimen: raising one hand in a rhetorical gesture while pointing to the opened torso with the other, he is the only one among a group of eager spectators permitted to touch the corpse. A nearby scalpel, pen, and paper illustrate his conviction that the human body should not be viewed through the lens of the text but instead be a means of generating written knowledge.[34] Vesalius's authority to speak on behalf of the human body is asserted again in a separate author portrait appearing between the proem and the text that uses iconography similar to the frontispiece, showing him gripping the flayed forearm of a cadaver wherein the tendons of the hand have been separated from the muscle. The lifeless anatomized limb contrasts with Vesalius's own, reemphasizing his ability to use both scalpel and pen to produce knowledge. Valverde, in contrast, is visually absent from his frontispiece. In fact, his first author portrait only appears three decades later in a 1586 edition of the *Historia* published in Venice.[35] Rather than the author, human bodies take center stage as animated participants that can speak for themselves.

Lacking an image of teacher or writer, Valverde's specimens gain an unexpected agency, recruited as collaborators in the anatomical project.

33 Park, *Secrets of Women*, pp. 207–260 gives a brilliant reading of the gendered subject in this frontispiece.

34 Andrea Carlino analyzes the iconography and its epistemological innovation in centring the body and the anatomist (rather than the text) in Vesalius's frontispiece in "The Book, the Body, the Scalpel."

35 Valverde de Amusco, *La anatomia del corpo umano*.

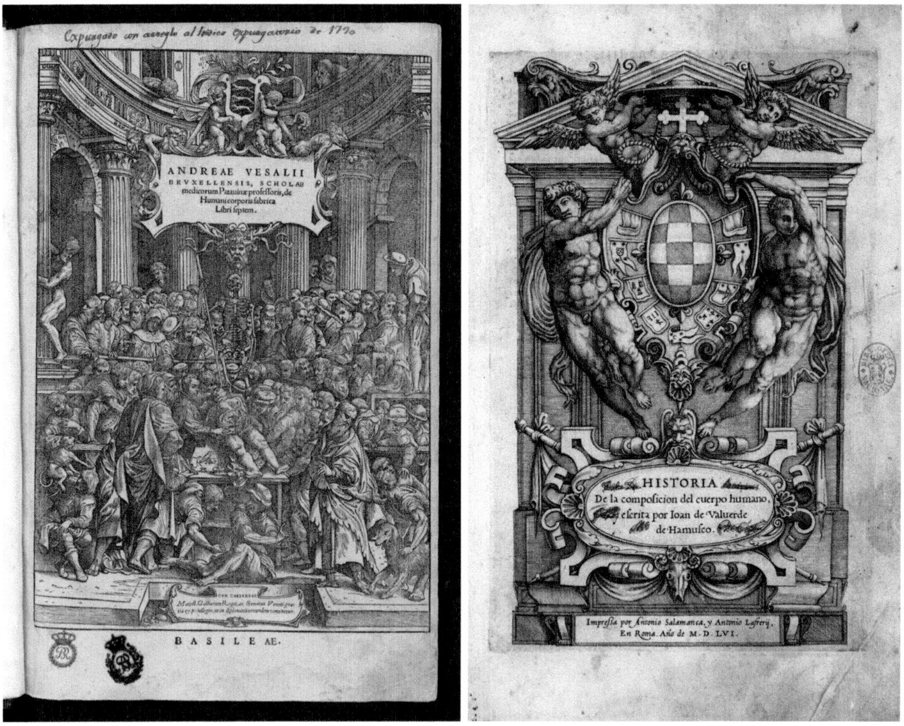

Figure 11.6 A–B A (left): Jan Steven van Calcar?, *Frontispiece,* from Andreas Vesalius, *De humani corporis fabrica* (Basel: Johannes Oporinus, 1543), woodcut, 41.5 × 27 cm, image property of the Biblioteca Nacional de España. B (right): Nicholas Beatrizet after Gaspar Becerra, *Frontispiece,* Juan Valverde de Amusco, *Historia de la composición del cuerpo humano* (Rome: Antonio Salamanca and Antoine Lafréry, 1556), engraving, 29 × 19 cm, image property of the Biblioteca Nacional de España.

They pick up a knife to flay their own skin or peel back a thigh muscle to better show its contours as distinct from the muscle that lies below.[36] Such self-demonstrating images can be found earlier – in Jacopo Berengario da Carpi's anatomical treatises of the 1520s published in Bologna – and later animate the imaginative and emotive bodies enlivening subsequent Baroque anatomical treatises by Giulio Cesare Casseri and John Browne.[37] Valverde's

36 Valverde describes how this adjustment aids our comprehension, compared to the figure shown by Vesalius, in his description of the previous image (Plate 2 of Book 2), where the muscle first sits on top of the leg: "Esta figura […] diffiere dela del Vesalio, en que […] la tela del sesto morzillo que mueue la pierna, esta alçada, paraque se vean mejor los morzillos." Valverde de Amusco, *Historia,* sig. M2v. "This figure […] differs from Vesalius's in that […] the fabric of the sixth muscle that moves the leg is lifted to better show the muscles." The self-flaying man appears in Plate 1 of Book 2.

37 Baroque examples include: Casserio, *Tabulae Anatomicae*; Browne, *A Compleat treatise.*

undead join a tradition of participatory anatomical imagery, revealing their secrets and speaking back to the more restrained classicizing models developed by Vesalius. The *Historia* thus creates a more engaging dialog between the reader and the bodies pictured in the text.

IV. Conclusion

Valverde recasts dissected corpses that rely on the author to make them speak as primary sources, shared informants, and co-authors. His newly awakened bodies gain consciousness of their liminal states and begin to produce knowledge. As Valverde's image of one dissected figure peering into another reminds us, anatomical study is a corporeal practice of uncovering natural knowledge within human bodies. The newfound agency of Valverde's specimens works to transfer intellectual authority from the humanist text back to the body, where it is ultimately produced, activating the practical and sensorial aspects of anatomical learning that were in fact stressed by Vesalius and leading anatomists of the day.[38]

Valverde is therefore successful in extending Vesalius's anatomy to a new audience, furthering, while remaining consistent with, the original project of the *Fabrica*. Though Vesalius developed his work in the context of his role as a university professor at a time when academic training was carried out in Latin, he acknowledged a vernacular audience for this material by publishing in German in 1543 his *Epitome*, a condensed and less expensive volume meant primarily for medical students. Likewise, later publishers saw a market for Valverde's volume among Latinate readers, resulting in a Latin translation (Venice, 1589).[39]

Valverde's alterations to the Vesalian imagery emphasize the importance of manual skill and sensory cognition in anatomical study. As Jacqueline Vons has shown, Vesalius imagined the hand as a "an organ of verification, of proof," even inviting readers to experience the articulation of their own bodies through a series of practical exercises.[40] This aspect of the Vesalian project is lost in his illustrations, which aimed above all at clearly representing the author's anatomical knowledge with a degree of elevating decorum to

38 Carlino, "Vesalio e la cultura visiva," p. 79.

39 Valverde de Amusco, *Anatome corporis humani.*

40 Vons, "Books for Learning," pp. 66–70, esp. p. 68. Pamela Smith interprets Vesalius's emphasis on direct engagement with nature and its material properties in relation to the alchemical tradition: Smith, *The Body of the Artisan*, pp. 156–157.

recuperate the low and unsavoury act of dissection. In contrast, Valverde's bald man gripping his intestines and dissected man reaching into his companion's ribcage resonate with Pamela Smith's understanding of artisanal epistemology as an embodied practice: "knowledge can be gained by observing and experiencing – often by bodily struggle – the particularity of nature."[41] In espousing these values, Vesalius took on work formerly allocated to barber-surgeons, but his illustrations give little hint of this. By filling Vesalius's classicizing contours with the flesh of the human body, and by making these newly embodied specimens participate in their own dissections, Valverde's images perform the dialog between artisans and humanists that Smith describes as a crucial aspect in the production of natural knowledge in the early modern period. This exchange of knowledge within the different kinds of anatomical spaces would be especially legible to readers like barber-surgeons, whom Sandra Cavallo has described as "artisans of the body."[42]

Valverde and his artists emphasized this continuity between the visual and intellectual worlds of professors and barber-surgeons in their illustrations. The coexisting and compatible registers of anatomical knowledge, communicated in their respectively preferred Latin and vernacular registers, are most visible in a pair of antique cuirasses illustrated in Plate 2 of Book 3 (sig. P4r) in *Historia*, where the finely worked armour houses a fleshy, organic interior that reveals views of the intestines at varying stages of dissection.[43] These images assert the overlapping realms of sculpture and flesh notionally underlying contemporary artistic *paragone* of art and nature, proving that embodied vernacular expression can exist in the frame of a hard classicizing form.[44]

Publishers continued to perceive these typically classicizing and vernacular visual registers as compatible in later anatomy books. In Antwerp, Christophe Plantin commissioned an expensive set of copperplates copied from Valverde's Italian edition of 1559–1560 to illustrate an anatomical text with its own complicated relationship to Vesalius's writings.[45] Plantin's

41 Smith, *The Body of the Artisan*, p. 6.

42 Cavallo, *Artisans of the Body*. While Cavallo is concerned with seventeenth- and early eighteenth-century Italy, she points to scholars who have shown similar circumstances in the sixteenth century.

43 San Juan, "Restoration and Translation," p. 58 interprets the cuirasses as part of Valverde's larger project to restore Vesalius's fragmented bodies, stressing "the antique body as shaped by the violent physical struggles of war and athletics."

44 For example, Cole, "Cellini's Blood" discusses the spiritual and organic elements of life conveyed in the casting of Medusa's blood as it is represented draining from her severed head.

45 Plantin published editions in Latin (1566) and Dutch (1568), with a Spanish translation planned but never issued. The text is taken from a Vesalian anatomy by Jacques Grévin illustrated

substantial investment in these new images suggests the favorable reception of Valverde's pictorial adjustments, a conclusion particularly important in light of Dániel Margócsy's recent discussion of the surprisingly tepid reception of Vesalius's images among early modern readers.[46] Other publishers adopted Valverde's images in a more piecemeal fashion. For example, Valverde's image of one dissected body inspecting another appears on the title page of Jan Wouters's 1569 Dutch translation of Vesalius's *Epitome*.[47] Helkiah Crooke's *Mikrokosmographia* used Valverde's popular image of a self-flaying man on the title page of the second edition, published in 1631.[48] The book also contained a prefatory letter to the Company of Barber Surgeons in London. Finally, a 1604 Venice edition of the *Fabrica* is remarkably faithful to Vesalius in the presentation of the text and images, displaying a visible effort to follow the layout and typographical design of the original (Figure 11.7).[49] However, the frontispiece includes Valverde among a group of men discussing a dissection conducted by Vesalius, in a frame ornamented with a selection of animated torsos taken from the *Historia*. The appearance of Valverde's self–demonstrating torsos and anatomized women in the architectural title page speaks to what was perhaps the Spanish author's most important contribution to Vesalian anatomy: reorienting the viewer to the role of the body, both in the materiality of the specimen and the labor of the anatomist in producing anatomical knowledge.

Looking beyond Valverde's treatise as plagiarism and attending instead to the qualities of his work that continued to attract later readers and viewers opens our eyes to how visual knowledge of the human body converged and multiplied as anatomical texts were appropriated and transformed to meet the expectations of new audiences. It also explains why Valverde's images, reimagined for a vernacular audience, quickly became a natural part of anatomical knowledge. In this context, we can more fully understand the strangely familiar visual world that Valverde left as his legacy.

with the same copperplates used by Geminus. Plantin included Valverde's explanatory notes for the images. For more information on Plantin's publications see Bowen and Imhof, *Christopher Plantin*, pp. 67–84. *Ibid.*, pp. 78–79, outlines the expenses of producing these new copperplates and the profits Plantin made from the sale of the first edition.

46 Margócsy, "Vesalius through Ivins."

47 Wouters, *Dat epitome*. The frontispiece is illustrated in Houtzager, "Editions, Copies and Replicas," fig. 152. An impression of this rare volume is held at the Universiteitsbibliotheek Gent BHSL.RES.1080. The only other image in the volume is a portrait of the writer/translator.

48 Crooke, *Mikrokosmographia*. For more on the image, see footnote 36.

49 Vesalius, *Andreae Vesalii anatomia*.

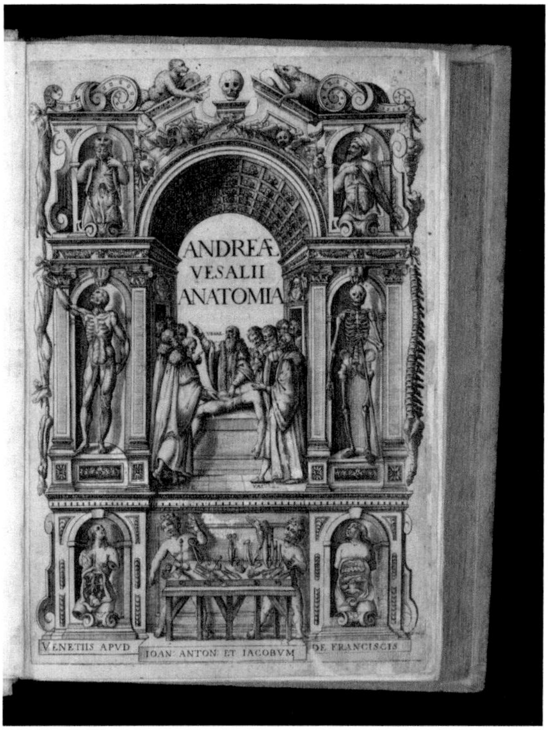

Figure 11.7 Francesco Valeggio, Frontispiece, from Andreas Vesalius, *Andreae Vesalii Anatomia* (Venice: Giovanni Antonio and Giacomo de Franceschi, 1604), engraving, 21 × 31 cm, image property of the Bibliotheca Hertziana.

Bibliography

Early modern published sources

Crooke, Helkiah. *Mikrokosmographia: A Description of the Body of Man*. London: Thomas and Richard Cotes, 1631.

Geminus, Thomas. *Compendiosa Totius Anatomie Delineatio*. London: John Herford, 1545.

Valverde de Amusco, Juan. *Historia de la Composición del Cuerpo Humano*. Rome: Impressa por Antonio Salamanca, y Antonio Lafrerij, 1556.

Valverde de Amusco, Juan. *La Anatomia del Corpo Umano Composta da m. Giouanni Valuerde*. Venice: Giunti, 1586.

Vesalius, Andreas. *De humani corporis fabrica*. Basel: Johannes Oporinus, 1543 and 1555.

Vesalius, Andreas. *Andreae Vesalii Anatomia*. Venice: Apvd J. A. et J. de Franciscis, 1604.

Wouters, Jan. *Dat Epitome ofte Cort Begriip der Anatomien*. Bruges: P. de Clerck, 1569.

Secondary Sources

Andretta, Elisa. *Roma Medica: Anatomie d'un Système Médical au XVIe Siècle.* Rome: École française de Rome, 2011.

Baxandall, Michael. *Giotto and the Orators: Humanist Observers of Painting in Italy and the Discovery of Pictorial Composition.* Oxford: Clarendon, 1971.

Bowen, Karen Lee, and Dirk Imhof. *Christopher Plantin and Engraved Book Illustrations in Sixteenth-Century Europe.* Cambridge: Cambridge University Press, 2008.

Burke, Peter. "The Renaissance Translator as Go-Between." In *Renaissance Go-Betweens: Cultural Exchange in Early Modern Europe*, edited by Andreas Höfele and Werner Von Koppenfels, pp. 17–31. Berlin: De Gruyter, Inc., 2005.

Bury, Michael. "Beatrizet and the 'Reproduction' of Antique Relief Sculpture." *Print Quarterly* 13, no. 2 (1996): 111–126.

Carlino, Andrea. "The Book, the Body, the Scalpel: Six Engraved Title Pages for Anatomical Treatises of the First Half of the Sixteenth Century." *RES: Anthropology and Aesthetics* 16 (1988): 33–50.

Carlino, Andrea. *Books of the Body: Anatomical Ritual and Renaissance Learning.* Transl. by John Tedeschi and Anne Tedeschi. Chicago, IL: University of Chicago Press, 1999.

Carlino, Andrea. "Tre piste per l'Anatomia di Juan de Valverde: Logiche d'Edizione, Solidarietà Nazionali e Cultura Artistica a Roma nel Rinascimento." *Mélanges de l'Ecole française de Rome. Italie et Méditerranée* 114, no. 2 (2002): 513–541.

Carlino, Andrea. "Vesalio e la Cultura Visiva delle Anatomie a Stampa del Rinascimento." In *Rappresentare il Corpo: Arte e Anatomia da Leonardo all'Illuminismo*, edited by Giuseppe Olmi, pp. 75–91. Bologna: Bononia University Press, 2004.

Cavallo, Sandra. *Artisans of the Body in Early Modern Italy: Identities, Families and Masculinities.* Manchester: Manchester University Press, 2010.

Cole, Michael. "Cellini's Blood." *Art Bulletin* 81, no. 2 (1999): 215–235.

Cropper, Elizabeth. "On Beautiful Women, Parmigianino, Petrarchismo, and the Vernacular Style." *The Art Bulletin* 58, no. 3 (1976): 374–394.

Dillon, Anne. *Michelangelo and the English Martyrs.* Burlington, VT: Ashgate, 2012.

Fissell, Mary Elizabeth. *Vernacular Bodies: The Politics of Reproduction in Early Modern England.* Oxford: Oxford University Press, 2004.

Fransen, Sietske. "Latin in a Time of Change: The Choice of Language as Signifier of a New Science?" *Isis* 108, no. 3 (2017): 629–635.

Guerra, Francisco. "Juan de Valverde de Amusco." *Clio Medica* 2 (1967): 339–362.

Harcourt, Glenn. "Andreas Vesalius and the Anatomy of Antique Sculpture." *Representations* 17 (1987): 28–61.

Houtzager, Hans. "Editions, Copies and Replicas of Fabrica and Epitome, Published Between 1543 and 1725." In *Art of Vesalius*, edited by Robrecht van Hee, pp. 191–222. Antwerp: Garant, 2014.

Keizer, Joost M. and Todd M. Richardson, eds. *The Transformation of Vernacular Expression in Early Modern Arts.* Leiden: Brill, 2012.

Klestinec, Cynthia Jennifer. "Juan Valverde de (H)Amusco and Print Culture: the Editorial Apparatus in Vernacular Anatomy Texts." *Zeitsprünge: Forschungen zur Frühen Neuzeit* 9 (2005): 78–94.

Kusukawa, Sachiko. *Picturing the Book of Nature: Image, Text, and Argument in Sixteenth-Century Human Anatomy and Medical Botany.* Chicago, IL: University of Chicago Press, 2012.

Larkey, Sanford V. "The Vesalian Compendium of Geminus and Nicholas Udall's Translation: Their Relation to Vesalius, Caius, Vicary and de Mondeville." *The Library* 13, no. 4 (1933): 367–394.

Laurenza, Domenico. *Art and Anatomy in Renaissance Italy: Images from a Scientific Revolution.* New York: Metropolitan Museum of Art, 2012.

Lincoln, Evelyn. *Brilliant Discourse: Pictures and Readers in Early Modern Rome.* New Haven, CT: Yale University Press, 2014.

López Piñero, José María. "La Disección y el Saber Anatómico en la España de la Primera Mitad del Siglo XVI." *Cuadernos de Historia de la Medicina Española* 13 (1974): 51–110.

Margócsy, Dániel. "From Vesalius through Ivins to Latour: Imitation, Emulation and Exactly Repeatable Pictorial Statements in the Fabrica." *Word & Image* 35, no. 3 (2019): 315–333.

Margócsy, Dániel, Mark Somos and Stephen N. Joffe. *The Fabrica of Andreas Vesalius: A Worldwide Descriptive Census, Ownership, and Annotations of the 1543 and 1555 Editions.* Leiden: Brill, 2018.

Marías, Fernando. *El Largo Siglo XVI. Los Usos Artísticos del Renacimiento Español.* Madrid: Taurus, 1989.

Meyer, A. W. and Sheldon K. Wirt. "The Amuscan Illustrations." *Bulletin of the History of Medicine* 14, no. 5 (1943): 667–687.

O'Malley, Charles Donald. *Andreas Vesalius of Brussels, 1514–1564.* Berkeley, CA: University of California Press, 1964.

Park, Katharine. *Secrets of Women: Gender, Generation, and the Origins of Human Dissection.* New York: Zone Books, 2006.

Redín Michaus, Gonzalo. *Pedro Rubiales, Gaspar Becerra y los Pintores Españoles en Roma, 1527–1600.* Madrid: Departamento de Historia del Arte, Instituto de Historia, Consejo Superior de Investigaciones Científicas, 2007.

Roberts, K. B. and J. D. W. Tomlinson. *The Fabric of the Body: European Traditions of Anatomical Illustrations.* Oxford: Clarendon Press, 1992.

San Juan, Rose Marie. "Restoration and Translation in Juan de Valverde's *Historia de la composision del cuerpo humano.*" In *The Virtual Tourist in Renaissance Rome: Printing and Collecting the Speculum Romanae Magnificentiae,* edited by Rebecca Zorach, pp. 53–61. Chicago, IL: University of Chicago Library, 2008.

Sánchez-Molero, José Luis Gonzalo. "Antonio de Salamanca y los Libros Españoles en la Roma del Siglo XVI." In *Roma y España un Crisol de la Cultura Europea en la Edad Moderna*, edited by Carlos José Hernando Sánchez, pp. 335–365, vol. I. Madrid: Sociedad Estatal para la Acción Cultural Exterior, 2007.

Saunders, J. B. de C M. and Charles Donald O'Malley. *The Illustrations from the Works of Andreas Vesalius of Brussels with Annotations and Translations, a Discussion of the Plates and Their Background, Authorship and Influence, and a Biographical Sketch of Vesalius.* Cleveland, OH: World Publishing Company, 1950.

Serjeantson, R. W. "Proof and Persuasion." In *The Cambridge History of Science*, edited by Katharine Park and Lorraine Daston, pp. 132–176. Cambridge: Cambridge University Press, 2006.

Sherman, William H. "On the Threshold: Architecture, Paratext, and Early Print Culture." In *Agent of Change: Print Culture Studies after Elizabeth L. Eisenstein*, edited by Sabrina A. Baron, Eric N. Lindquist, and Eleanor F. Shevlin, pp. 67–81. Amherst, MA: University of Massachusetts Press, 2007.

Skaarup, Bjørn Okholm. *Anatomy and Anatomists in Early Modern Spain.* Farnham: Ashgate, 2015.

Smith, Pamela H. *The Body of the Artisan: Art and Experience in the Scientific Revolution.* Chicago, IL: University of Chicago Press, 2004.

Vons, Jacqueline. "Books for Learning about the Body." In *Art of Vesalius*, edited by R. van Hee, pp. 51–70. Antwerp: Garant, 2014.

About the Author

Emily Monty received her Ph.D. in the History of Art and Architecture at Brown University (2021). Currently, she is the Fagel Collection Visiting Research Fellow at Trinity College Dublin. Her research has been supported by the Samuel H. Kress Foundation and the Fulbright Program among others.

12. Epilogue

Forgetting How to See

Stephanie Porras

Abstract
This chapter considers what happens to the "knowing imprint" when it moves outside of its original epistemological habitat. Wherever early modern Europeans traveled they brought printed images with them, to explain political and religious systems, to introduce customs, stories and artistic styles. Reverse engineering early modern prints' epistemic opacity in colonial contexts may not be fully achievable, but as the discipline of art history embraces computer vision, certain methodological parallels emerge. Computers have to be taught to see, and digitization ventures like the Getty Research Institute's PhotoTech project seek to automate metadata generation, to recognize important information and visual similarities. This chapter explores print's ontological operations and epistemic limits, drawing on the operative challenges facing the ontology of computer vision.

Keywords: ontology, engraving, globalization, colonial, vision, computer vision

I. Introduction

What does it mean to forget, learn, or relearn how to see? Reflecting on these questions, this essay emerges from the intersection of digital methodologies and my own art historical interest in the ontological operations of early modern print. These two apparently unrelated fields of inquiry come together as I consider the role of new image technologies and the concomitant processes of replication, circulation, and standardization, today and at the beginnings of globalization around 1600. Here, I briefly sketch

Noyes, R.S. (ed.), *Reassessing Epistemic Images in the Early Modern World.*
Taylor & Francis Group, 2023
DOI 10.5117/9789463723350_CH12

some parallels between the infrastructural role of early modern print and twenty-first century attempts to automate and instrumentalize vision, both as efforts to deconstruct and reconstruct ways of seeing.

II. Indulgent Images, Imperial Infrastructures

Wherever Europeans landed in the sixteenth and seventeenth centuries, they brought printed images with them. Stacked and bundled, these woodcuts and engravings were used to introduce foreign rulers, customs, stories and artistic styles. Here is a king, an animal, a city you will never see. The Christian God looks like this; this is his mother (Figure 12.1). Accounting for the diverse roles of prints in the colonization of the Americas, Tom Cummins has described prints as the "indulgent medium par excellence," as "simultaneously the most democratic and oppressive of all forms of images."[1] Cummins identifies the ubiquity of prints in the colonial world as symptomatic both of colonial power and its resistance.

In this vein, I would suggest that prints acted as what John Durham Peters has called "infrastructural media" managing time and space, organizing and orienting people, places and property into communities, markets, subjects; put differently, prints operated epistemologically but also ontologically to create the very idea of Spain, of Christianity, of Europe.[2] Prints were not just aimed at picturing the New World for European consumption, but also served as models for painting, sculpture and architecture across the globe, generating new artworks based on the compositions of artists like Raphael and Peter Paul Rubens,[3] replications of Serlian architectural forms,[4] royal likenesses,[5] and holy sites.[6] Wherever prints moved, they rendered visible the abstract concepts of sovereignty and commerce that underpinned these images' mobility. In recognition of this fact, when commenting on the administrative burden of his vast geopolitical empire, Phillip II supposedly quoted the Latin dictum: *Quod non est in actis, non est in mundo* (What is not in the documents is not in the world).[7] The movement of people and goods across the Spanish empire, as Bernhard

1 Cummins, "The Indulgent Image," p. 225.
2 Durham Peters, *The Marvelous Clouds*, pp. 37, 176.
3 Hyman, *Rubens in Repeat*; Pon, *Raphael*.
4 Heuer, *The City Rehearsed*, pp. 117–122 and Verdi Webster, "Secret Life of Buildings."
5 Osorio, "The King in Lima."
6 Giffin, "Replicas of the Santa Casa."
7 As quoted in Durham Peters, *The Marvelous Clouds*, p. 20.

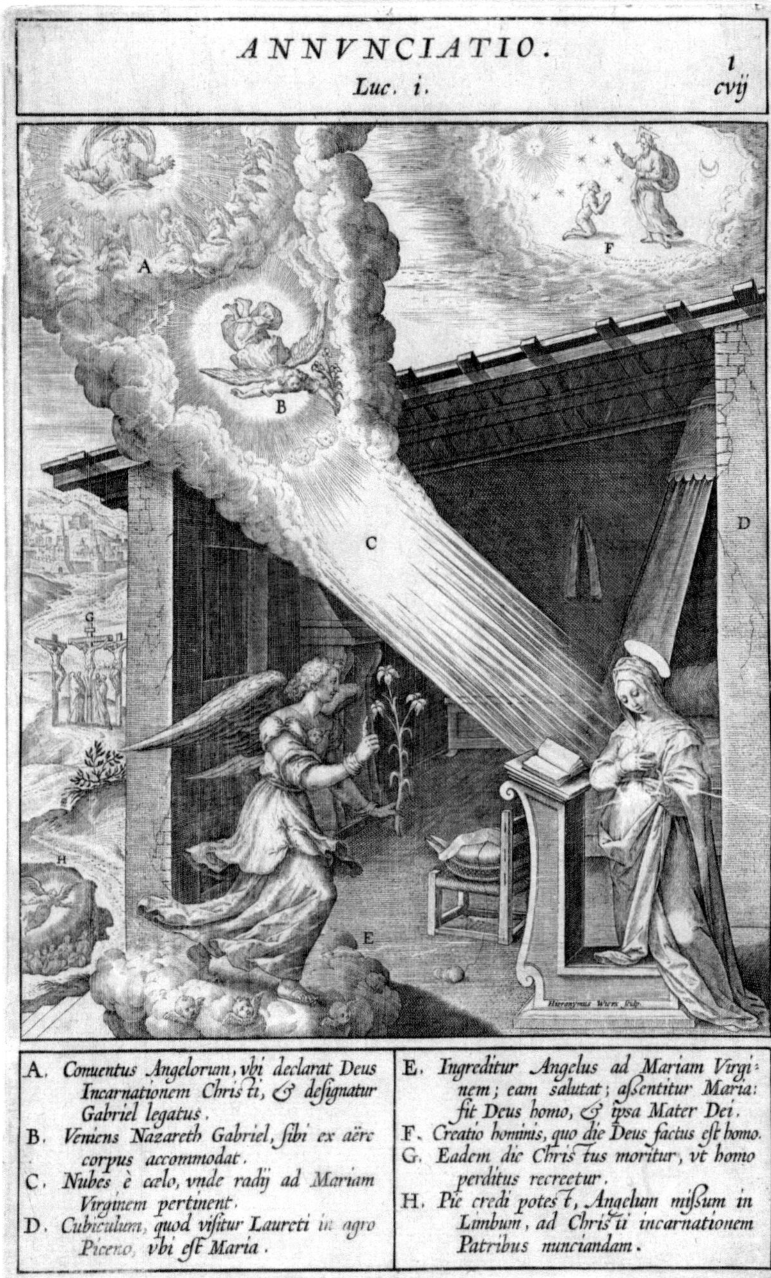

Figure 12.1 Hieronymus Wierix, *Annunciation*, 1593, engraving from Jerónimo *Nadal*, S.J., *Evangelicae historiae imagines (Antwerp:* Societas Iesu, 1593). Brussels, Bibliothèque royale de Belgique.

Siegert has elegantly described, necessitated the development of a host of paper mechanisms: inventories, manifests, identity papers, bills of lading.[8] Printed images, too, served the global imperial reach of both crown and Church, as pedagogical tools for conversion and conquest and objects used to model allegiance.

Printed images on paper were familiar in both Europe and China before the sixteenth- and seventeenth-century expansions of European military, missionary, colonial, and commercial power. During the invasion of New Spain, the conquistador Hernán Cortés placed images of the Virgin and Child at former sites of Aztec veneration, testifying to how European actors invested prints with both political and ecclesiastical authority.[9] The earliest Franciscan missions in New Spain used prints to train indigenous artists in Christian iconography;[10] the seminary and art school established by the Jesuit mission in Japan also employed prints as models for local artists.[11] The duplication, movement, and copying of printed images meant that the monochrome linearity of engraved images became synonymous with colonial and missionary authority. William MacGregor has described how the metaphoric language of printmaking (impression, incision, engraving) came to dominate early modern descriptions of memory and cognition,[12] but outside of Europe these metaphors took on additional valence.

That colonial authority mobilized print in specific ways to make present the absent ruler (and the Catholic God), and to enforce conformity is made explicit in the work of Diego Valadés, a former pupil at San José de los Naturales, the school for indigenous elites founded by the Franciscans in the capital of New Spain. At San José de los Naturales, printed and painted images were used as *loci* for memory to explain and memorize the catechism and key concepts of the Christian worldview. When he later traveled to Italy on assignment for the Order, Valadés wrote the *Rhetorica Christiana*, and he most likely designed and/or etched the 27 engravings and etchings included in the text. In the *Rhetorica*, Valadés

8 Siegert, *Passagiere und Papiere.*

9 See the account of Cortés in Tabasco, where "Cortés encomendándoles la santa imagen y santascruces, y que las tuviesen muy limpias y barridas, y enramado y que las reverenciaseny hallarían salud y buenas sementeras." SeeDíaz del Castillo, chs 36–37; Donahue-Wallace, "Picturing Prints," pp. 328–331.

10 Armella de Aspe, *Testimonios artísticos*, with further literature, also Edgerton, *Theaters of Conversion.*

11 Mochizuki, "Sacred Art," p. 138.

12 MacGregor, "The Authority of Prints."

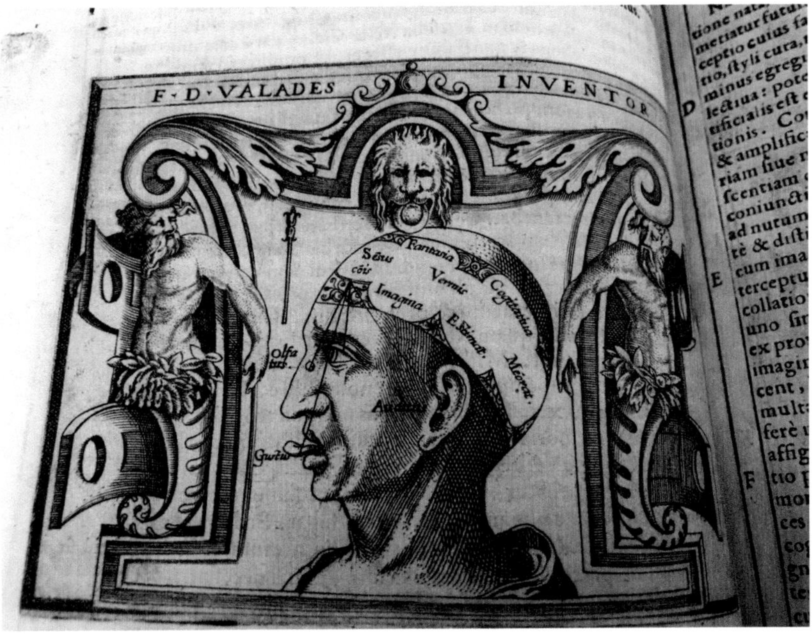

Figure 12.2 Diego Valadés (?), *Senses and the Operations of the Mind*, 1579, engraving from Diego Valadés, *Rhetorica Christiana* (Perugia: Pietro Giacopo Petrucci, 1579).

repeatedly emphasizes that these printed illustrations and the images used in Mexico were intended "to better impress or imprint information into the memory."[13]

One of the prints signed by Valadés in the text shows a cross section of the human mind as the repository of the senses and other mental faculties (Figure 12.2). Above this head, a curious instrument – a pointer whose rectangular edge recalls that of a burin – reiterates the connection between engraving, memory, picturing, and the process of conversion. This visualization of memory in Valadés is heavily indebted to an anonymous Roman work, the *Rhetorica ad Herennium*, and metaphors of cognition and impression stretching back to Galen. However, the text and images of the *Rhetorica* suggest that engravings were not just the bearer of cultural content, but in fact brought that culture into being, first within the mind, then the world. The ornamental borders surrounding the cranial cross-section and its interior divisions recall the printed borders of the book's other illustrations. Valadés connects the physical appearance of printed images to their mental

13 To cite two examples: "Per imagines itaque locis impressas in locatorum devenimus cognitionem: quibus adiuta memoria rite exoluit officium." Valadés, *Rhetorica*, p. 95 ; and "[...] uteturqúe frequentibus imaginibus ad rem imprimendam." Valadés, *Rhetorica*, p. 278.

location, visualizing the ontological function of prints, bringing concepts into being.

The abstract relationship between the senses and internal workings of the mind are internally and externally framed by grotesque decorative flourishes. This exemplar illustrates Michael Baxandall's concept of the "period eye" to articulate the ways in which both artists and viewers attended to and thought about the visual world, as shaped by shared cultural practices.[14] Valadés's illustration reflects and contributes to a period eye that equated incision and engraving with memory and knowledge on the one hand, the decorative embellishments of the frame with the synthesis of antique and modern forms of knowing, on the other. As much as his more famous descriptions of pre-Hispanic rituals and missionary practices in the same volume, the image testifies to how European print was used to actively shape the vision of colonial subjects, replacing previously held beliefs and political systems, cultural practices, and ways of life. The Franciscans at San José de los Naturales used prints and other visual material to restructure local conceptions of divinity and territorial rule, and to enforce a change in how indigenous actors saw the physical and spiritual worlds. Artworks helped Spanish colonizers construct and enforce a new period eye. Via the pedagogical programs of missionary orders and the political authority of colonial powers, printed images proposed for their subjects a new way of seeing, in order to impose a new way of being.[15]

Seeing Differently

Studying the early modern movement of prints outside of Europe, their reception, rejection, use and re-use, presents art historians with the opportunity to recover some of early modern prints' epistemological and ontological operations. Beyond replicating and reproducing knowledge, styles, and iconographies, prints produced new subjects and visualities. In order to acknowledge the latter, it is necessary to *forget how art historians typically see*. Put differently, one must divorce oneself from the interests of the artistic author and initial or intended viewership, lingering instead on instances where a print's operative authority has misfired or been transformed in ways

14 Baxandall, *Painting and Experience*, pp. 29–57.

15 Bernhard Siegert and Horst Bredekamp (from quite different perspectives) have both argued for images' active role, beyond representation, but in actively constituting reality. Both, however, have not explicitly addressed how European artworks were brought to bear on a non-European viewer in shaping globalization. See Siegert, "The map is the territory"; Bredekamp, *Theorie des Bildakts*.

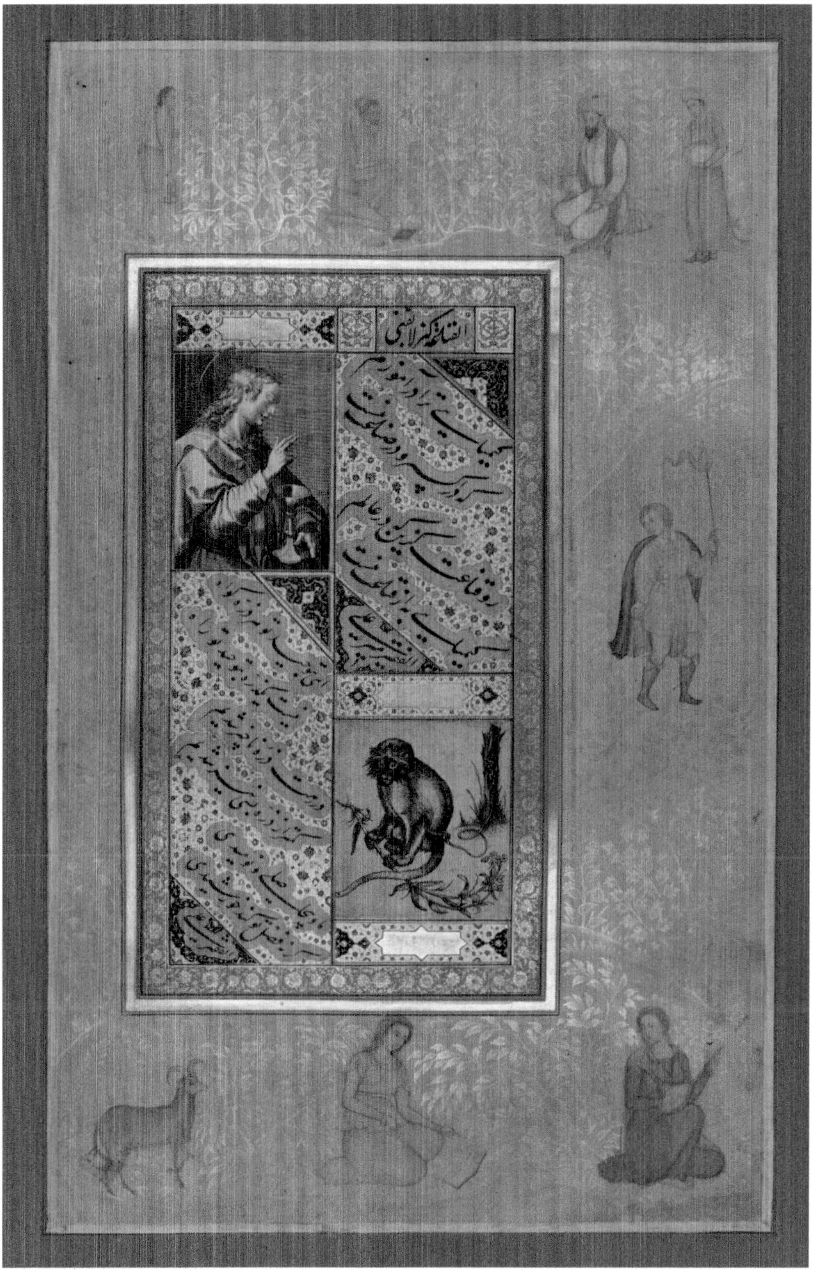

Figure 12.3 Mir 'Ali, Persian calligraphies with illumination, and pasted impression of Franz Brun, *Chained Monkey*, c. 1596, engraving, from the Berlin Album, assembled 1608–1618, fos 9r. Staatsbibliothek, Berlin (A.117)

their authors could not have anticipated, or where a print seems to have been seen differently, not as an iconographic totality but as a system of lines.[16]

Consider for example, the engraved monkey on this album page, a connoisseurial quotation from a famous engraving by Albrecht Dürer, an animal excerpted from one of the most copied compositions in early modern print (Figure 12.3).[17] Rather than focusing on how this particular impression arrived in this album, or on Franz Brun's intended audience for this curious print, what would it mean to see the engraving as an expression of calligraphic line? As Yael Rice has compellingly argued, this was precisely how this print was viewed at the Mughal court of Jahangir, where the engraving was included in an album alongside examples of calligraphic specimens and paintings.[18] Jahangir's father, Jalal al-Din Muhammad Akbar, had received gifts from a 1580 Jesuit delegation, including seven of the eight volumes of the monumental Polyglot Bible, a copy of the first European atlas (Abraham Ortelius's *Theatrum Orbis Terrarum*), engravings by Philipp Galle and other Northern European printmakers, and several large oil paintings and statues. European prints in this context were intended as lures to Catholic conversion, exemplars of European technological and cartographic supremacy. But the composition of the Jahangir albums reveals a wholly different reception for at least some of these engravings, which were seen primarily as aesthetic exemplars of line.

The pasted prints in the Jahangir albums illuminate what Whitney Davis calls the bivisibility of an artwork, the fact that pictorial spaces constitute one of the primary sites of interaction between visualities.[19] Extending Davis's observation, here I would suggest that early modern prints were a key space for this kind of negotiation of competing or intersecting visualities. Prints were the most mobile of early modern art objects, most likely to move far beyond their authors' ambit, to interact with competing aesthetic principles and modes of seeing.

Thus, while Jahangir's albums reveal a Mughal appreciation for the swelling and tapering lines of engraving, this same graphic language made the engravings of the illustrated Jesuit pedagogical title *Evangelicae historiae imagines* aesthetically limited as a tool for conversion in Southern China (Figure 12.1). In 1598, the Jesuit Niccolò Longobardo specifically requested

16 See Peters, "Systems and Swells."

17 The monkey can be found in the foreground of Albrecht Dürer's circa 1498 engraving, *The Virgin and Child with a Monkey*, Schoch, Mende, & Scherbaum, no. 20.

18 Rice, "Lines of Perception."

19 Davis, *Visuality and Virtuality*, p. 110.

the title from Rome; a 1605 letter from Matteo Ricci reports there was a copy of the *Imagines* available to missionaries in Southern China and requested another volume for use in the North.[20] Three woodcut versions of the *Imagines* were published in China in the first half of the seventeenth century. Between 1619 and 1623, the Portuguese Jesuits Gaspar Ferreira and João da Rocha supervised the production of the *Song nianzhu guicheng* (誦念珠規程 / Method of Saying the Rosary) in Nanjing (Figure 12.4).[21] Little more than a decade later, in Fuzhou, along China's southern coast, Giulio Aleni published the *Tianzhu jiangsheng chuxiang jingjie* (天主降生出像經解 / Illustrated Explanation of the Incarnation of the Lord of Heaven).[22] Finally, in 1640, scenes from the *Imagines* were reprised for the *Jincheng shuxiang* (進呈書像 / Images [from a] Book Presented to the Emperor) printed in Beijing.[23] The publication of these volumes suggests the continued Jesuit belief in the power of printed images to assist in the education and conversion of the Chinese. Instead of attempting to import their own press, as the Order did in Japan and elsewhere, the Jesuits supported the creation of local woodblock versions, embracing the dominant print technology in China.

The dramatic alterations to the printed model in the resultant woodcuts call into question the supposed universality of the *Imagines*' engravings. The Chinese prints eliminate the engraving's chiaroscuro and cross hatching, and introduce Ming spatial and iconographic conventions (Figure 12.4). The subsidiary scene of the Crucifixion (see detail at left Figure 12.1) is deleted entirely. These alterations represent the limits of the *Imagines*' legibility in China, making visible those moments where pictorial decorum was threatened (such as the Chinese distaste for images of the Crucifixion) or where local stylistic idiom was seen as preferable (the chiaroscuro of Wierix's clouds exchanged for swirling lines). The woodcut is the site of negotiation between monocular perspective associated with European artworks and local conceptions of spatial order.[24] Although they retained the combination of text and image of the earlier Antwerp engravings, and their pedagogical and devotional function, these changes in the *Song nianzhu guicheng* grant insight into the epistemic limits of European engravings, signaling where the

20 "[...] più utile è anco quell libro che questo della Bibbia per adesso, poichè con quello dichiariamo, anzi poniamo Avanti agli occhi quello che alle volte con parole non possiamo dichiarare." Ricci, *Lettere*, no. 43, p. 406.

21 Yi, *Song Nianzhu Guicheng*, pp. 195–290.

22 Shin, "The Reception of *Evangelicae Historiae Imagines*," pp. 303–333.

23 Standaert, *An Illustrated Life of Christ*.

24 Purtle, "*Double Take*," pp. 89–96.

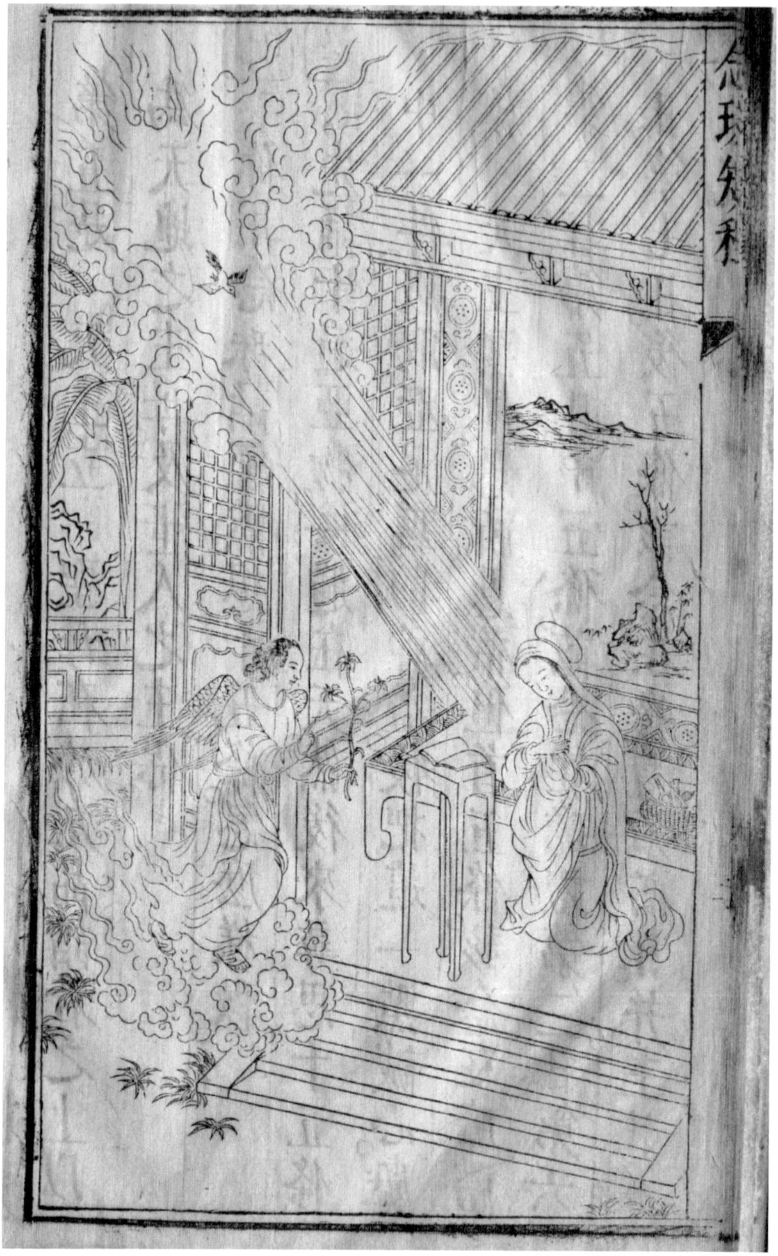

Figure 12.4 *Annunciation*, from Gaspar Ferreira (Fei Qikuei), but usually attributed to João da Rocha, SJ (Luo Ruwang), *Song nianzhu guicheng* (Rules for reciting the rosary), woodcut blockbook Nanjing, 1619–1623.

specific aesthetic and representational claims of the print had the potential to break down, or to meet resistance from the viewer.

Jesuit concessions to Chinese viewership were critiqued by rival orders as symptomatic of the irreconcilable distance between Christian values and Chinese traditions. The Chinese Rites controversy, the centuries' long debate about the compatibility of Chinese ancestral rites with Catholic belief, pitted the Jesuits against other Catholic missionary orders and the Pope himself during the seventeenth and eighteenth centuries.[25] The debate centered on invisible questions of belief, whether the continuation of Confucian ancestor rites and imperial veneration was antithetical to true Catholic conversion. The Jesuits did not see the persistence of these practices as impinging on the adoption of the true faith; but their critics stressed their fundamental incompatibility.

Across the globe, the early modern Catholic missionary project was freighted with anxiety as to how converts could learn to discern and discriminate between modes of Catholic veneration and pre-Christian customs. So the Third catechism of Lima, published in 1585, admonished the converted indigenous believer:

> When you say that Christians would adore the pictures that are painted or made from wood or metal, and kiss them and kneel down before them and begin to talk to them, and ask, 'aren't those *guacas* much like ours?,' [...] if Christians revere images [...] it is because of what the images represent, and not because of what they are. Just as the corregidor kisses the provision and royal seal, and brings it to his head, not because of that wax or the paper, but because it is the king's *quillca*.[26]

Just like the Chinese missionaries, the Spanish colonizers of the viceroyalty of Peru worried about the superficiality of Catholic conversion masking the continuation of Pre-Hispanic traditions. They feared Catholic altars would be seen as like the *guaca* ·(*huaca/wa'ka*), the Quechua/Aymara term for a sacred place/object/person or animal imbued with spiritual power, enabling

25 Mungello, *The Chinese Rites Controversy*; Standaert, *Chinese Voices in the Rites Controversy*.

26 "Padre, ¿cómo nos decís que no adoremos ídolos ni guacas, pues los cristianos adoran las imágenes que están pintadas y hechas de palo o metal, y las besan y se hincan de rodillas delante de ellas, y se dan en los pechos y hablan con ellas? ¿Estas no son guacas también como las nuestras?... Y si reverencian las imágenes, y las besan y se descubren delante de ellas, e hincan las rodillas y hieren los pechos, es por lo que aquellas imágenes representan, y no por lo que en sí son. Como el corregidor besa la provisión y sello real, y lo pone sobre su cabeza, no por aquella cera ni el papel, sino porque es quillca del rey." *Tercero catecismo*, sermón XIX, p. 576.

the transfer of this local mode of reverence to a nominally Catholic practice. Further complicating matters was the absence of a word for "representation" in the dominant local idioms. In Quechua and Aymara, representation and embodiment were conceptually joined, leading to the continual frustration of Catholic missionaries who sought to extirpate idolatry.[27] In explaining representation, the author of the 1585 catechism makes an analogy between religious images and the royal seal, distinguishing between the object's material (wax, paint, paper) and its referential authority, denoted by the Quechua word *quillca*, which conveyed both alphabetic and pictorial mark-making.[28]

In viceregal Peru, distrust of indigenous vision and the assumed continuation of pre-Hispanic beliefs encouraged colonial patrons to enforce the copying of European printed or sculpted models in contracts with local artists, often stipulating that the artist had to follow a given print or other model.[29] In addition to ensuring doctrinal compliance, the provision of a model print was also intended to ontologically model Catholic representation, the difference between veneration and idolatry. The 1585 catechism stressed the material qualities of devotional art, citing images made of paper, wood, paint or metal. Surviving artists' contracts from the viceregal Americas often require that commissions "conform" to a printed model, but also stress the resultant objects' material dissimilarity from these printed models by stipulating the size and materials to be used for the finished product.[30] These patrons saw the linear and monochrome print as clearly un-mimetic, incapable of being confused with the holy figures depicted; the distance between printed model and the resultant artwork mirrored the metaphoric operation of Catholic devotional art, pointing to the divine.

Yet, as Aaron Hyman has shown, this practice of producing conforming copies helped to reify particular local images of divinity, forming networks of

27 See Mills, *Idolatry and Its Enemies*.

28 See Rappaport and Cummins, *Beyond the Lettered City*, pp. 191–218.

29 On viceregal Peruvian contracts stipulating that the artist copy a specific print, see Cornejo Bouroncle, *Derroteros de arte cuzqueño*; and Ojeda di Ninno, "El grabado como fuente."

30 See for example: "*En la ciudad del Cuzco, a cinco dias del mes de diciembre de mil quinientos ochenta y tres años, ante mi el Escribano Público y testigos, Pedro de Santangel, se obligó de hacer un retablo de Nuestra Señora de la Soledad, conforme a una estampa, que tiene en su poder* [...] *el cual retablo ha de ir pintado al oleo y la guarnición y las dos moldurillas doradas y la cenefa del medio color, todo a costa del dicho pedro de Santangel* [...]" quoted by Cornejo Bouroncle, *Derroteros de arte cuzqueño*, 166; for an example from New Spain, see the 1675 contract from Puebla commissioning Baltazar de Echave y Rioja for two paintings for the sacristry; "ha de tener, puestos y entregados [...], costeando todo lo necesario de lienzos y pintura en todo arte y perfección, conforme a dos estampas, que [...] se le han entregado [...], dentro de cuatro meses," quoted in Pérez de Salazar, *Historia de la Pintura*, pp. 178–179.

related images and concomitant communities of faith.[31] Images copied from prints could work miracles, bridging supposed gaps between the representation and enactment of divine power. Indigenous viewers recognized the agency of prints. In 1661, for example, an indigenous man in the parish of Ambar, along the Huacho coast north of Lima, performed an unsanctioned form of the Eucharist before an altar adorned with *estampas*, small printed images of saints.[32] The *hechizero* of Ambar appeared to use prints as Catholics did, to adorn a ritual space. However, this similarity in practice did not signal epistemological or ontological continuity with the Catholic faith. In fact, for the inquisitorial investigator, the *hechizero's* prints inverted the Church's Post-Tridentine stance on the role of art as a mediator, actualizing the Protestant critiques of Catholic art as idolatrous. The *hechizero*, to the Church official, appeared to take the epistemological claims of devotional prints as an ontological operation, whereby the print, instead of representing the divine, was mobilized as an agent of spiritual power.

III. Training Sets and Taxonomies

The emperor Jahangir, the Chinese copyist, the *hechizero* in Ambar all saw European prints differently. How do we reconstruct and analyze these different visualities? As an art historian interested in digital methods, here I would introduce an unexpected analogue to the global movement and reception of early modern prints: the present-day world of computer vision. I would suggest that both the technologies of early modern print and of computer vision are modes of estranged vision, where one must break down and reconstitute how one sees, where one is forced to contemplate the visibility or invisibility of ontological difference, as well as the possibility of misinterpretation. I introduce computer vision here not only to open up the study of early modern print in a global context, but also to suggest ways in which art historical ways of seeing may inform the development of computer vision, particularly as institutions of art history seek to develop, critique and use these technologies.

Ontology, in computing, refers to the formal naming and defining of categories and their properties. In order for computers to learn to see, complex neural network-based algorithms have to be trained to build

31 Hyman, *Rubens in Repeat*.
32 Archivo del Arzobispado de Lima, Sección hechizerías y idolatrías, II–A, 12, Ambar December 20 1661, fos. 29v–33, cited in Brosseder, *The Power of Huacas*, p. 221.

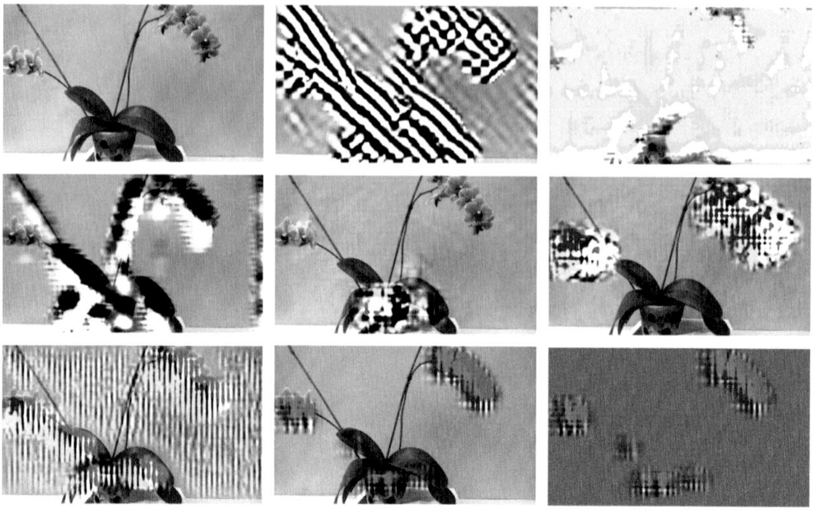

Figure 12.5 Screenshots of the *20/X* computer vision system looking at an orchid using image overlays backprojected from different neural network depths and feature maps, June 2015. Image: Shannon McMullen and Fabian Winkler.

these ontologies. Some training tasks, because they have been monetized and supported by corporate ventures, are well defined and developed: for example, as self-driving car technology advances, computers are becoming very adept at recognizing cars and street signs. Yet, other tasks remain surprisingly difficult. Investment from agribusiness means that AI can recognize cows well, as long as they are in familiar settings. However, if one shows such a neural network trained in these contexts the image of a cow on a beach, the computer will only be able to recognize the animal as "non-human." Computer vision works by first breaking the image into pixels, defining edges and combinations of edges, then identifying key features and their combination. Therefore, it is hard for a computer to see a cow if it is an unfamiliar textural environment (like the beach), when it has been trained to see cows against grass, other cows, or livestock pens.[33]

The 20/X project by artists Shannon McMullen and Fabian Winkler allows viewers to see and intervene in the computer vision process via an interactive interface (Figure 12.5). The viewer/user can experience and interfere with the levels of abstraction employed by computer vision algorithms, moving from coarse and geometry-driven in the beginning, to more specific and detail-oriented in the end, at which point distinctive patterns, areas and

33 For a recent discussion of these issues with computer vision, see Rosenfeld, Zemel, and Tsotsos, "The Elephant in the Room" ; Hart, "Machine Learning."

objects that 'excite' the computer vision system can be recognized.[34] These are processes that typically take place outside the experiential realm of human sense perception. 20/X exposes usually infinitesimally quick and invisible computer vision to the human gaze. By slowing down the algorithm, the project allows the viewer to experience the unfamiliarity of computer vision and the cascade of decisions and judgments that underpin these algorithms.

In training a computer to recognize certain things, the biases and limitations of the training set are replicated and reinforced in how the neural network operates, that is, how the computer sees. The well-studied failures of facial recognition software to "recognize" individuals with darker complexions is a clear example of this problem.[35] Trevor Paglen and Kate Crawford's recent project *Training Humans* exposed these issues via their scripted viral phenomenon *ImageNetRoulette*.[36] *ImageNetRoulette* used an open source Caffe deep learning framework, first developed at UC Berkeley, to classify images of individual faces uploaded to the website. The application would then label the image, drawing on the labels from faces the computer saw as similar, drawing in turn from faces/labels from the "person" categories held in ImageNet, a large visual database of over 14 million annotated images designed for use in visual object recognition software research. But *ImageNetRoulette* revealed that the ImageNet database, one of the dominant training sets for computer vision in the United States, is not a neutral repository. Labels for ImageNet are assigned manually, often via crowdsourcing, which injects and perpetuates conscious and unconscious bias into the system.

As Crawford and Paglen's project made clear, ImageNet was full of racist, misogynist and other problematic labels, including: jezebel, streetwalker, spinster, stud, debtor, Slav, mulatto, etc. BIPOC individuals who uploaded images of their faces to *ImageNetRoulette* publicized their offensive tags via social media; in response to the ensuing public outcry, the researchers behind ImageNet announced that they would scrub more than half of the 1.2 million pictures in the dataset's "person" category. But as Paglen and Crawford point out, "the whole endeavor of collecting images, categorizing

34 McMullen and Fabian Winkler, "20/X: Visuality, Representation And Epistemology."

35 As outlined in several studies, including the 2019 US National Institute of Standards and Technology; citing these concerns the European Commission considered a ban on the use of facial recognition software in public spaces. See Harwell, "Federal Study" ; Zhou, *et al.*, "Neural Dynamics."

36 The website is no longer live, but Kate Crawford and Trevor Paglen's statement on the *ImageNetRoulette* project is available at: (https://www.excavating.ai/) (accessed 3 July 2022).

them, and labeling them is itself a form of politics, filled with questions about who gets to decide what images mean and what kinds of social and political work those representations perform."[37] That computer vision is not possible without training is not simply a problem of problematic labels, but an indictment of how computer vision functions, via training sets and taxonomies. Computers can only see how we teach them.

What would computer vision look like if art historians taught computers how to see? The limitations of computer vision, the particular problems of training a computer to see and to describe images has been made clear to me when working as a consultant researcher on the Getty Research Institute's (GRI) Phototech project.[38] This project will digitize approximately 700,000 photos of paintings, drawings, and prints from the GRI's Photo Archive, as part of a consortium of photo archives (PHAROS) working toward digitizing their massive archives.[39] Artworks – specifically, photos of artworks – pose particular problems for computer vision. Take, for example, this photo of a daguerreotype of Louis-Jacques-Mandé *Daguerre* (Figure 12.6). The computer recognizes it is a person, but the presence of the framing lines and the black and white tone have led the computer to assign tags of "vintage," "car," "white goods," "train," and, most delightfully, "black-and-white photo of a microwave."[40] The problem is that computers, thus far, have been trained to see within very particular parameters that by and large do not include artworks. Computers can "see" resemblance, but they do so in very different ways than art historians do. Most of the work of recognition is done with image training sets in which the evaluation of new images is done on the basis of what previous images the neural network has been shown. So, Daguerre, to this computer, looked like a microwave because the bounding lines of the photograph have been read as edges, and this particular neural network had previously primarily trained on images of cars. Compounding these issues is the fact that the GRI collection comprises largely pictures of paintings. Uneven photographic captures (with mounts or without), blurry photographs, residual moiré patterns, and other variables mean that images of the very same painting can register very differently. As a result, computer vision constructs likeness very differently to an art historian.

37 Crawford and Paglen, "Epistemics of Training Sets," (www.Excavating.ai) (accessed 3 July 2022).

38 On PhotoTech, see (https://www.getty.edu/research/scholars/research_projects/phototech.html) (accessed 3 July 2022).

39 http://pharosartresearch.org/ (accessed 3 July 2022).

40 My thanks to Tracy Stuber for this fantastic example, which was generated using Google Cloud's Vision API in September 2019.

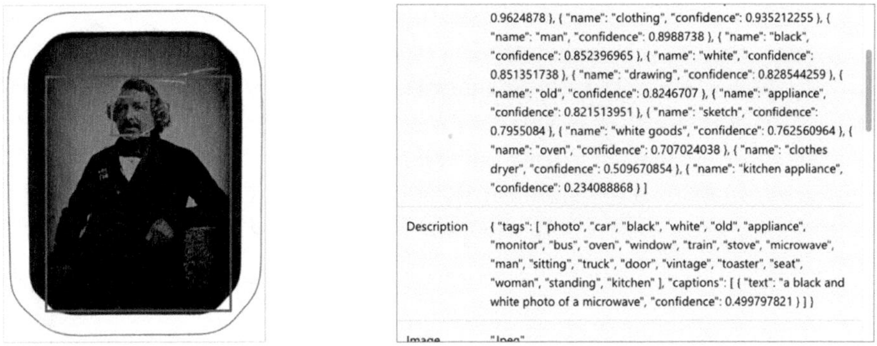

Figure 12.6 PhotoTech test of computer vision on a photograph of a daguerreotype of Daguerre.

The GRI's idea is to use computer vision to capture metadata from individual photos – information about the artist, creation date, medium, stamps and annotations made by scholars and librarians along the mounts – in order to ask art historical questions at scale about images. The GRI's Photo Archive is a fascinating art historical document of the discipline from c. 1950 to 2000, a corpus organized by regional or national school, then by artist name, then iconography. By digitizing these images, the GRI wants to build new access points into the archive, and to escape the formative physical taxonomies (Netherlandish paintings in these boxes, Italian ones several aisles away) by allowing for searches across different fields, ideally using text and visual search features. But as ImageNetRoulette has shown us, when scraping user-generated data, unforeseen issues arise. Historical categorical descriptions subsuming problematic terms and user biases continue to structure how computer vision works. Right now, there is no way to teach computers to see without the use of training sets and taxonomies.

However, computer vision offers more than just metadata capture; it offers an opportunity for methodological reflection on art historical praxis. Teaching computers to see estranges the sense of vision and forces a reassessment of how one understands likeness. I believe some computer vision terms offer art historians new ways to approach the global study of print. What would it mean to describe Antwerp prints, the most dominant form of imagery sent to colonial contexts from the Arctic to the South China Sea, as a kind of training set for the hegemonic imposition of European early modern vision? How did Antwerp prints structure vision, label objects and create ontologies for Chinese ivory carvers in Manila, indigenous artists and Catholic converts in the Americas, all of whom copied prints to make new artworks? Similarly, in what fundamental ways did extra-European actors of the past "mis-see" prints, or see them quite differently (and productively) than their makers anticipated?

Ralph DeKoninck, Oscar Flores Flores, and Ligia Fernández Flores have used the term *koiné* to describe both Antwerp engraving and its particular stylistic impact on European and colonial devotional art.[41] However, I think the linguistic metaphor insufficiently describes a much larger manifold process, the ways in which engraving functions as a media technology or a cultural technique. Media studies scholars after Friedrich Kittler have used these terms to described how changes in media process and/or blur distinctions basic to the sense production of a given culture.[42] Jussi Parikka has suggested the potentiality for investigating historic media technologies in this way, and Bernard Siegert has described Renaissance perspective, the grid of the Mercator projection as a "cultural technique of designing, uniting the will to create with the will to conquer."[43] More than simply an artistic technique or mathematical principle, in this accounting, perspective helped to shape a system of knowledge and a mode of vision central to the colonial project.

Yet just how early modern printed images operated as a media technology and cultural technique entrenched within colonial projects necessitates more robust analysis. The foundational studies of early modern print by Eisenstein, Chartier and Giesecke were primarily concerned with the printed words' social and epistemological impact on European viewers and readers.[44] The use of printed images outside of Europe requires renewed and sustained theoretical engagement. The examples sketched in this essay suggest potential avenues for further consideration.

IV. Conclusion

The close study of printed images and their variant copies has the potential to reveal the particular role of print as a medium in instrumentalizing vision in colonial contexts. Just as the artists of the 20/x project expose the functional operations of computer vision to the contemporary viewer, art historians need to work to recover how prints instantiated particular

41 Flores Flores and Fernández Flores. "En torno a la *koineización* pictórica"; and Dekoninck, "A Graphic Koiné."

42 Siegert, "The Map is the Territory."

43 *Idem*, *Cultural Techniques*, pp. 121–136; Parikka, *A Geology of Media*.

44 Eisenstein, *The Printing Press*; Chartier, *Usages de l'imprimé*, Giesecke, *Der Buchdruck*.

modes of vision—not just for particular communities of European readers and viewers (be they Franciscans or anatomists or armchair travelers), but for missionary subjects, courts and intellectuals far beyond the European continent – and to try to grasp how such prints were apprehended by these viewers. While it is not possible to truly forget how to see, I think there is value in deaccelerating, querying and resisting the ways in which art historians tend to see prints particularly, but not only, within colonial contexts. What kinds of prints were seen as suitable for what kinds of audiences, and how were the training sets for specific intercultural contexts curated and adapted? What enabled or foreclosed on a print's mobility and legibility? What can copies tell us about the relative 'stickiness' of certain visualities, iconographies, motifs? How did printed images operate as a media technology shaping ways of knowing as well as ways of seeing? The fantasy of a universalizing vision underpins both today's experiments in computer vision and the sixteenth century's explosion of printed imagery sent across oceans and continents. Thus in both in the past and in the present moment, forgetting and relearning how to see can be an act of control or resistance, and the gaze simultaneously a space of conformity, consideration and contestation.

Bibliography

Armella de Aspe, Virginia. *Testimonios artísticos de la evangelización*. México City: Grupo Gutsa, 1995.

Baxandall, Michael. *Painting and Experience in Fifteenth Century Italy: A Primer in the Social History of Pictorial Style*. Oxford: Oxford University Press, 1988.

Bredekamp, Horst. *Theorie des Bildakts*. Berlin: Suhrkamp Verlag, 2010.

Brosseder, Claudia. *The Power of Huacas: Change and Resistance in the Andean World of Colonial Peru*. Austin, TX: University of Texas Press, 2014.

Chartier, Rogier, ed, *Usages de l'Imprimé. XVe–XIXe Siècle*. Paris: Fayard, 1987.

Cornejo Bouroncle, Jorge. *Derroteros de Arte Cuzqueño Datos para una Historia del Arte en el Perú*. Cuzco: Ediciones Inca, 1960.

Crawford, Kate and Trevor Paglen, https://www.excavating.ai/ (accessed 3 July 2022).

Cummins, Thomas B. F. "The Indulgent Image: Prints in the New World." In *Contested Visions in the Spanish Colonial World*, edited by Ilona Katzew, William B. Taylor, and Luisa Elena Alcalá, pp. 203–225. New Haven, CT: Yale University Press, 2011.

Dekoninck, Ralph. "A Graphic Koiné for a New Religious Value. The Visual Translatability of the *Evangelicae Historiae Imagines*." *Nederlands Kunsthistorisch Jaarboek* 64 (2014): 272–297.

Díaz del Castillo, Bernal. *Historia Verdadera de la Conquista dela Nueva España. 1585*, edited by Joaquín Ramírez Cabañas. Mexico City: Editorial Porrua, 1976.

Davis, Whitney. *Visuality and Virtuality: Images and Pictures from Prehistory to Perspective.* Princeton, NJ: Princeton University Press, 2017.

Donahue-Wallace, Kelly. "Picturing Prints in Early Modern New Spain." *The Americas* 64, no. 3 (2008): 325–349.

Durham Peters, John. *The Marvelous Clouds: Toward a Philosophy of Elemental Media.* Chicago, IL: University of Chicago Press, 2015.

Edgerton, Samuel. *Theaters of Conversion: Religious Architecture and Indian Artisans in Colonial Mexico.* Albuquerque, NM: University of New Mexico Press, 2001.

Eisenstein, Elizabeth L. *The Printing Press as an Agent of Change: Communications and Cultural Transformations in Early-Modern Europe*, 2 vols. Cambridge: Cambridge University Press, 1979.

Flores Flores, Oscar and Ligia Fernández Flores. "En torno a la *koineización* pictórica en los reinos de la monarquía hispánica. Identidad y variadades dialectales." In *Pinturas de los reinos: Identidades compartidas en el mundo hispánico*, edited by Juana Gutiérrez-Haces, pp. 187–332. Mexico City: Banamex, 2008–2009.

Giesecke, Michael. *Der Buchdruck in der frühen Neuzeit: eine historische Fallstudie über die Durchsetzung neuer Informations- und Kommunications-technologien.* Frankfurt: Suhrkamp, 1991.

Giffin, Erin. "Replicas of the Santa Casa: Networks of Geographic Translation," https://storymaps.arcgis.com/stories/01dbd5ff3a5d4037b8567c7b6b760fa6 (accessed 3 July 2022).

Hart, Kevin. "Machine Learning Confronts the Elephant in the Room," *Quanta*, 20 September 2018: https://www.quantamagazine.org/machine-learning-confronts-the-elephant-in-the-room-20180920/ (accessed 3 July 2022).

Heuer, Christopher. *The City Rehearsed: Object, Architecture, and Print in the Worlds of Hans Vredeman de Vries.* New York: Routledge, 2009.

Hyman, Aaron. *Rubens in Repeat: The Logic of the Copy in Colonial Latin America.* Los Angeles, CA: J. Paul Getty Press, 2021.

MacGregor, William B. "The Authority of Prints: An Early Modern Perspective." *Art History* 22 (1999): 389–420.

McMullen, Shannon C. and Fabian Winkler. "20/X: Visuality, Representation and Epistemology in the Age of Intelligent Seeing Machines." *Leonardo* 51 no. 1 (2018): 71–74.

Mills, Kenneth. *Idolatry and Its Enemies: Colonial Andean Religion and Extirpation, 1640–1750.* Princeton, NJ: Princeton University Press, 1997.

Mochizuki, Mia. "Sacred Art in an Age of Mechanical Reproduction: The Salus Populi Romani Madonna in the World." In *Sacred and Profane in Early Modern Art*, edited by Kayo Hirakawa, pp. 129–144. Kyoto: Kyoto University, 2016.

Mungello, D.E., ed. *The Chinese Rites Controversy: Its History and Meaning*. Abingdon: Routledge, 2018.

Ojeda di Ninno, Almerindo. "El Grabado como Fuente del Arte Colonial: Estado de la Cuestión." In *De Amberes al Cuzco*, edited by Cécile Michaud and José Torres della Pina, pp. 10–21. Lima: Impulso Empresa de Servicios, 2009.

Osorio, Alejandra. "The King in Lima, Simulacra, Ritual, and Rule in Seventeenth-Century Peru." *Hispanic American Historical Review* 84, no. 3 (2004): 447–474.

Parikka, Jussi. *A Geology of Media*. Minneapolis, MN: University of Minnesota Press, 2015.

Pérez de *Salazar, Francisco. Historia de la Pintura en Puebla. Mexico City: Universidad Nacional Autónoma* de *México, 1963*.

Peters, Emily. "Systems and Swells: The Collective Lineage of Engraved Lines 1480–1650." In *The Brilliant Line: Following the Early Modern Engraver, ca. 1480–1650*, edited by Emily J. Peters, pp. 13–48. Providence, RI: Rhode Island School of Design Museum, 2009.

Pon, Lisa. *Raphael, Dürer, and Marcantonio Raimondi: Copying and the Italian Renaissance*. New Haven, CT: Yale University Press, 2004.

Purtle, Jennifer. "*Double Take: Chinese Optics and their Media in Postglobal Perspective.*" *Ars Orientalis* 48 (2018): 89–96.

Rice, Yael. "Lines of Perception: European Prints and the Mughal Kitābkhāna." In *Prints in Translation, 1450–1750: Image, Materiality, Space*, edited by Suzanne Karr Schmidt and Edward H. Wouk, pp. 202–223. London: Routledge, 2017.

Ricci, Matteo. *Lettere: 1580–1609*, edited by Francesco d'Arelli. Macerata: Quodlibet, 2001.

Rosenfeld, Amir, Richard Zemel, and John K. Tsotsos. "The Elephant in the Room." arXiv preprint arXiv:1808.03305 (2018), https://arxiv.org/abs/1808.03305 (accessed 3 July 2022).

Schoch, Rainer, Matthias Mende, and Anna Scherbaum, eds. *Albrecht Dürer. Das Druckgraphische Werk*. Vol. 1. Munich: Prestel, 2001.

Shin, Junhyoung Michael. "The Reception of *Evangelicae Historiae Imagines* in Late Ming China: Visualizing Holy Topography in Jesuit Spirituality and Pure Land Buddhism." *The Sixteenth Century Journal* 40, no. 2 (2009): 303–333.

Siegert, Bernhard. *Passagiere und Papiere. Schreibakte auf der Schwelle zwischen Spanien und Amerika*. Zürich: Neue Zürcher Zeitung, 2006.

Siegert, Bernhard. "The Map is the Territory." *Radical History* 169 (2011): 13–16.

Siegert, Bernhard. *Cultural Techniques: Grids, Filters, Doors and Other Articulations of the Real*. New York: Fordham University Press, 2015.

Standaert, Nicolas. *An Illustrated Life of Christ Presented to the Chinese Emperor: The History of Jincheng shuxiang (1640)*. Sankt Augustin: Steyler Verlag, 2007.

Standaert, Nicolas, ed. *Chinese Voices in the Rites Controversy: Travelling Books, Community Networks, Intercultural Arguments*. Rome: Institutum Historicum Societatis Iesu, 2012.

Tercero Cathecismo y Exposicion de la Doctrina Christiana, por Sermones. Para qve los Cvras y Otros Ministros Prediquen y Enseñen a los Yndios y a las Demas Personas, Conforme a lo qve en el Sancto Concilio Prouincial de Lima se Proueyo. Lima: Antonio Ricardo, 1585.

Valadés, Diego. *Retorica Cristiana*. Perugia: Petrus Jacobus Petrutius, 1579.

Webster, Susan Verdi. "The Secret Lives of Buildings in Colonial Quito." *Construction History* 28, no. 3 (2013): 21–46.

Yi, Qu and 藝 曲. Song Nianzhu Guicheng (Die Anweisung zur Rezitation des Rosenkranzes," *Monumenta Serica* 60, no. 1 (2012): 195–290.

Zhou, Yuqing, *et al*. "Neural Dynamics of Racial Categorization Predicts Racial Bias in Face Recognition and Altruism." *Nature Human Behaviour* 4, no. 1 (2020): 69–87.

About the author

Stephanie Porras is associate professor of Art History at Tulane University. The author of *Pieter Bruegel's Historical Imagination* (2016), and *Art of the Northern Renaissance* (2018), her forthcoming book, *The First Viral Images*, considers the role artworks played in early modern globalization.

Bibliography

A Matre Dei, Sebastianus. *Firmamentum Symbolicum*. Lublin [=Antwerp]: Georg Forster, 1652.

Ackerman, James. *Distance Points: Essays in Theory and Renaissance Art and Architecture*. Cambridge, MA: MIT Press, 1991.

Afdrukken Houtblokken Museum Plantin-Moretus. 6 Volumes. Antwerp: Museum Plantin-Moretus, 1960.

Agata, Mari. "Stop-Press Variants in the Gutenberg Bible: The First Report of the Collation." *Papers of the Bibliographical Society of America* 97 (2003): 139–165.

Aguilonius, Franciscus. *Opticorum Libri Sex Philosophis Iuxtà ac Mathematicis Utiles*. Antwerp: Ex Officina Plantiniana, apud Viduam et Filios Jo. Moreti, 1613.

Ambrose, Charles T. "The Conquest of Pus: A History of Bitumen, Creosote and Carbolic Acid." *Journal of Infectious Disease and Preventive Medicine* 6, no. 2 (2018): 1–8.

Americas 64, no. 3 (2008): 325–349.

Andretta, Elisa. *Roma Medica. Anatomie d'un Système Médical au XVIe Siècle*. Rome: École Française de Rome, 2011.

Apian, Peter. *Cosmographicus Liber Petri Apiani Mathematici Studiose Collectus*. Landshutae: Impensis P. Apiani, 1524.

Areford, David S. *The Viewer and the Printed Image in Late Medieval Europe*. Farnham: Ashgate, 2010.

Armella de Aspe, Virginia. *Testimonios Artísticos de la Evangelización*. México City: Grupo Gutsa, 1995.

Baglione, Giovanni. *Le Vite de' Pittori Scultori et Architetti*. Rome: Nella Stamperia d'Andrea Fei, 1642.

Baigrie, Brian Scott. *Picturing Knowledge: Historical and Philosophical Problems Concerning the Use of Art in Science*. Toronto: University of Toronto Press, 1996.

Baird, Davis. *Thing Knowledge : A Philosophy of Scientific Instruments*. Berkeley, CA: University of California Press, 2004.

Baker, Tawrin. "Dissection, Instruction, and Debate: Visual Theory at the Anatomy Theatre in the Sixteenth Century." In *Perspective as Practice: Renaissance Cultures of Optics*, edited by Sven Dupré, pp. 123–147. Turnhout, Belgium: Brepols, 2019.

Baker, Cathleen A. "The Wove Paper in John Baskerville's Virgil (1757): Made on a Cloth-Covered Laid Mould." In *Papermaker's Tears: Essays on the Art and Craft of Paper*, edited by Tatiana Ginsberg, pp. 2–44. Ann Arbor: The Legacy Press, 2019.

Bakker, Boudewijn. "The Amsterdam 'Paradise' by Herri met de Bles and the Fountain of Life." *The Rijksmuseum Bulletin* 63, no. 2 (2015): 114–133.

Baldriga, Irène. "'La fatiga di pigliar i disegni dalle piante'. Federico Cesi, la pittura filosofica e la riproduzione del mondo vegetale." In *Federico Cesi. Un principe*

naturalista. Acquasparta, 29 e 30 settembre 2003, edited by Antonio Graniti, pp. 503–523. Rome: Bardi, 2006.
Baldriga, Irène. "Reading The Universal Book of Nature: The Accademia dei Lincei in Rome (1603–1630)." In *The Reach of the Republic of Letters: Literary and Learned Societies in Late Medieval and Early Modern Europe*, edited by Arjan van Dixhoorn and Susie Speakman Sutch, pp. 353–388. Leiden: Brill, 2008.
Baldriga, Irène. *L'occhio della lince. I primi Lincei tra arte, scienza e collezionismo: 1603–1630*. Rome: Accademia Nazionale dei Lincei, 2002.
Balfe, Thomas, and Joanna Woodall. "Introduction: From Living Presence to Lively Likeness: The Lives of Ad Vivum." In *Ad Vivum? Visual Materials and the Vocabulary of Life-Likeness in Europe before 1800*, edited by Thomas Balfe, Joanna Woodall, and Claus Zittel, pp. 1–43. Leiden: Brill, 2019.
Bartholin, Caspar. *Institutiones Anatomicae*. Leiden: Franciscus Hackius, 1641.
Bartsch, Adam von. *The Illustrated Bartsch*, 50 vols. Edited by Walter L. Strauss. New York: Abaris Books, 1978.
Baskins, Cristelle and Lisa Rosenthal, eds. *Early Modern Visual Allegory: Embodying Meaning*. Aldershot: 2007.
Baxandall, Michael. *Giotto and the Orators: Humanist Observers of Painting in Italy and the Discovery of Pictorial Composition.* Oxford: Clarendon, 1971.
Baxandall, Michael. *Painting and Experience in Fifteenth Century Italy: A Primer in the Social History of Pictorial Style*. Oxford: Oxford University Press, 1988.
Bedos-Rezak, Brigitte Miriam. "Replica: Images of Identity and the Identity of Images in Prescholastic France." In *The Mind's Eye: Art and Theological Argument in the Middle Ages*, edited by Jeffrey Hamburger and Anne-Marie Bouché, pp. 46–64. Princeton, NJ: Princeton University Press, 2006.
Behrmann, Carolin. "Metrics of Justice: A Sundial's Nomological Figuration." *Nuncius.* 30, no.1 (2015): 161–194.
Bell, Peter Jonathan. "Johann Friedrich Greuter" and "Matthaus Greuter," *Allgemeines Kunstlerlexikon: Die bildenden Kunstler aller Zeiten* 61: 500–505. Munich: De Gruyter, 2009.
Bennett, Jim. "Early Modern Mathematical Instruments." *ISIS* 102, no.4 (December 2011): 697–705.
Bennett, Jim. "Shopping for Instruments in Paris and London." In *Merchants and Marvels: Commerce, Science and Art in Early Modern Europe*, edited by Paula Findeln and Pamela H. Smith, pp. 370–395. New York: Routledge, 2001.
Berger, Susanna. *The Art of Philosophy: Visual Thinking in Europe from the Renaissance to the Early Enlightenment.* Princeton, NJ: Princeton University Press, 2017.
Bertoloni Meli, Domenico. *Mechanism. A Visual, Lexical, and Conceptual History.* Pittsburgh, PA: Pittsburgh University Press, 2019.

Biagioli, Mario. “From Print to Patents: Living on Instruments in Early Modern Europe.” *History of Science* 44, no. 2 (2006): 139–186.

Biagioli, Mario. “Picturing Objects in the Making: Scheiner, Galileo and the Discovery of Sunspots.” In *Wissensideale und Wissenskulturen in der Frühen Neuzeit: Ideals and Cultures of Knowledge in Early Modern Europe*, edited by Wolfgang Detel and Claus Zittel, pp. 39–96. Berlin: Akademie Verlag, 2002.

Bilancia, Fernando. “Le lastre e le stampe di Matthäus Greuter in una lista di beni assegnati in dote alla figlia.” In *Roma nel primo Seicento. Una città moderna nella veduta di Matthäus Greuter*, edited by Augusto Roca De Amicis, pp. 347–353. Rome: Editoriale Artemide, 2018.

Blair, Ann M. “The Rise of Note-Taking in Early Modern Europe.” *Intellectual History Review* 20, no. 3 (2010): 303–316.

Blair, Ann. *Too Much to Know: Managing Scholarly Information Before the Modern Age*. New Haven, CT: Yale University Press, 2010.

Blanchard, Jean-Vincent, *L'Optique du Discours au XVIIe Siècle. De la Rhétorique des Jésuites au Style de la Raison Moderne (Descartes, Pascal)*. Sainte Foy: Presses de l'Université Laval, 2005.

Blumenberg, Hans. *Die Lesbarkeit der Welt.* Frankfurt am Main: Surhkamp, 1981.

Blumenberg, Hans. *Paradigms for a Metaphorology*, transl. Robert Savage. Ithaca, NY: Cornell University Press, 2010.

Bobinger, Maximilian. *Christoph Schissler der Ältere und der Jüngere.* Augsburg: Verlag Die Brigg, 1954.

Booth, Sara Elizabeth, and Albert van Helden. “The Virgin and the Telescope: The Moons of Cigoli and Galileo.” *Science in Context* 13, no. 3–4 (2000): 463–486.

Borrelli, Arianna. “Optical Diagrams as ‘Paper Tools’: Della Porta's Analysis of Biconvex Lenses from De Refractione to De Telescopio.” In *The Optics of Giambattista Della Porta (ca. 1535–1615): A Reassessment*, edited by Arianna Borrelli, Giora Hon, and Yaakov Zik, pp. 57–96. Cham: Springer, 2017.

Borrelli, Arianna, Michael Korey, and Volker R. Remmert, eds. *Iconography on Early Modern Scientific Instruments*, special issue of *Nuncius. Journal of the Material and Visual History of Science* 30, no. 1 (2015).

Bosse, Abraham. *Traicté des Manières de Graver en Taille Douce sur l'Airin.* Paris: Chez Bosse, 1645.

Bousquet-Bressolier, Catherine. “Pédagogie de l'Image Jésuite: De l'Image Emblématique Spirituelle aux *Emblemata* Mathématiques.” In *François de Dainville S.J. (1909–1971). Pionnier de l'Histoire de la Cartographie et de l'Éducation*, edited by Catherine Bousquet-Bressolier, pp. 143–166. Paris/Geneva: Publications de l'École Nationale des Chartes, 2004.

Bowen, Karen L., and Dirk Imhof. “18,257 Impressions from a Plate.” *Print Quarterly* 22 (2005): 265–279.

Bowen, Karen Lee, and Dirk Imhof. *Christopher Plantin and Engraved Book Illustrations in Sixteenth-Century Europe.* Cambridge: Cambridge University Press, 2008.

Brahe, Tycho. *Astronomiae Instauratae Mechanica.* Hamburg: Philippi de Ohr chalcographi, 1598.

Brahe, Tycho. *Instruments of the Renewed Astronomy*, edited by Alena Hadravová, Petr Hadrava, and Jole R. Shackelford. Prague: Koniasch Latin Press, 1996.

Bredekamp, Horst. "Gazing Hands and Blind Spots: Galileo as Draftsman." *Science in Context* 13, no. 3–4 (2000): 423–462.

Bredekamp, Horst. *Galilei der Künstler. Der Mond, die Sonne, die Hand*. Berlin: Akademie Verlag, 2009.

Bredekamp, Horst. *Theorie des Bildakts.* Berlin: Suhrkamp Verlag, 2010.

Bredekamp, Horst, Vera Dünkel, and Birgit Schneider, eds. *The Technical Image: A History of Styles in Scientific Imagery.* Chicago, IL: University of Chicago Press, 2015.

Brenni, Paolo. "Art and Science: Design of Scientific Instruments." In *Creativity in the Arts and Science*, edited by William R. Shea and Antonio Spadafora, pp. 121–141. Canton: Science History Publications, 1990.

Brevaglieri, Sabina. "Science, Books and Censorship in the Academy of the Lincei. Johannes Faber as Cultural Mediator." In *Conflicting Duties. Science, Medicine and Religion in Rome (1550–1750)*, edited by Maria Pia Donato and Jill Kraye, pp. 109–133. London: Warburg Institute Colloquia 15, 2009.

Brevaglieri, Sabina. *Natural Desiderio di Sapere. Roma Barocca fra Vecchi e Nuovi Mondi.* Rome: Viella, 2019.

Brian, Peter. *Galen on Bloodletting: A Study of the Origins, Development and Validity of his Opinions, with a Translation of the Three Works.* Cambridge: Cambridge University Press, 1986.

Brosseder, Claudia. *The Power of Huacas: Change and Resistance in the Andean World of Colonial Peru.* Austin, TX: University of Texas Press, 2014.

Bucciantini, Massimo. "Teologia e Nuova Filosofia. Galileo, Federico Cesi, Giovambattista Agucchi e la Discussione sulla Fluidità e Corruttibilità del Cielo." In *Sciences et religions de Copernic à Galilée*, pp. 411–441. Rome: École Française de Rome, 1999.

Burke, Peter. "The Renaissance Translator as Go-Between." In *Renaissance Go-Betweens: Cultural Exchange in Early Modern Europe*, edited by Andreas Höfele and Werner Von Koppenfels, pp. 17–31. Berlin: De Gruyter, Inc., 2005.

Bury, Michael. "Beatrizet and the 'Reproduction' of Antique Relief Sculpture." *Print Quarterly* 13, no. 2 (1996): 111–126.

Bury, Michael. *The Print in Italy, 1550–1620.* London: British Museum Press, 2001.

Campbell, Angela, and Andrew Raftery. "Remaking Dürer: Investigating the Mater Engravings by Masterful Engraving." *Art in Print* 2, no. 4 (2012): 15–21.

Campbell, Angela. "Finding Folds: Albrecht Dürer's *Meisterstiche* Papers." *Print Quarterly* 24, no. 4 (2012): 405–411.

Carlino, Andrea. "The Book, the Body, the Scalpel: Six Engraved Title Pages for Anatomical Treatises of the First Half of the Sixteenth Century." *RES: Anthropology and Aesthetics* 16 (1988): 33–50.

Carlino, Andrea. *Books of the Body: Anatomical Ritual and Renaissance Learning.* Transl. John Tedeschi and Anne Tedeschi. Chicago, IL: University of Chicago Press, 1999.

Carlino, Andrea. "Tre Piste per l'Anatomia di Juan de Valverde. Logiche d'Edizione, Solidarietà Nazionali e Cultura Artistica a Roma nel Rinascimento." *Mélanges de l'Ecole française de Rome. Italie et Méditerranée* 114, no. 2 (2002): 513–541.

Carlino, Andrea. ""Vesalio e la Cultura Visiva delle Anatomie a Stampa del Rinascimento." In *Rappresentare il Corpo. Arte e Anatomia da Leonardo all'Illuminismo*, edited by Giuseppe Olmi, pp. 75–91. Bologna: Bononia University Press, 2004.

Carruthers, Mary. "Rhetorical *Ductus*, or, Moving through a Composition." In *Acting on the Past: Historical Performances Across the Disciplines*, edited by Mark Franko and Annette Richards, pp. 99–125. Hanover, NH: University Press of New England, Wesleyan University, 2000.

Cavallo, Sandra. *Artisans of the Body in Early Modern Italy: Identities, Families and Masculinities.* Manchester: Manchester University Press, 2010.

Chappell, Miles. "Lodovico Cigoli, Matthäus Greuter and Santa Maria del Fiore in Florence." *Paragone/Arte* 3, 58.75/76 (2007): 96–112.

Charreaux, Stéphanie, and Jérôme van Wijland. "Recensement et Description des Exemplaires de la Première Édition du De Fabrica 1543 Conservés en France dans les Bibliothèques Publiques." In *La Fabrique du Vésale. La Mémoire d'un Livre*, edited by Jacqueline Vons, 253–312. Paris: BIUS, 2014.

Chartier, Rogier, ed, *Usages de l'Imprimé. XVe–XIXe Siècle*. Paris: Fayard, 1987.

Châtelain, Jean-Marc. *Livres d'Emblèmes et de Devises. Une Anthologie (1531–1735)*. Paris: Klincksieck, 1993.

Chevallier, Jacques, and Jean-Christophe Neidhardt. "La Destinée des Bois de la *Fabrique* de 1543." *Histoire des Sciences Médicales* 48 (2014): 485–494.

Christianson, John Robert. *On Tycho's Island: Tycho Brahe and His Assistants, 1570–1601.* New York: Cambridge University Press, 2000.

Cocles, Bartolommeo della Rocca. *Phisionomi Vnd || Chiromanci.|| Eyn Newß Complexion Büch=||lein/ Der Menschen Geburt/ Sitten/ Geberden Vnd Neyg=||ligkeyten/ Auß Der Phisionomi/ Chiromanci/ Den Siben || Planeten/ Zwölff Zeichen/ Vnnd Den XXXvj. Bildern || Deß Himels/ Auch Nach Den Zwölff Monaten/ Leichtlich || Vnd Grundtlich Zulernen/ Auß Platone/ Aristotele/ Pto||lomeo/ Hali/ Albumasar/ Vnnd Johanne Künigsper=||ger #c. in Fünff Büchlin Kurtzlich Gezogen/ Vnd Im Sech||sten Werden Erzelt Wunder Gestalt Lewt Mit Jren Siten.*||. Strasbourg: Jakob Cammerlander, 1540.

Cole, Michael. "Cellini's Blood." *Art Bulletin* 81, no. 2 (1999): 215–235.

Coleman, Dorothy Gabe. *An Illustrated Love "Canzoniere". The* Délie *of Maurice Scève*. Geneva: Slatkine, 1981.
Conley, Tom. *The Self-Made Map: Cartographic Writing in Early Modern France*. Minneapolis, MN: University of Minnesota Press, 1996.
Conway, Sir William Martin. *Literary Remains of Albrecht Dürer*. Cambridge: Cambridge University Press, 1889.
Copper Through the Ages. London: Copper Development Association, 1955.
Cornejo Bouroncle, Jorge. *Derroteros de Arte Cuzqueño Datos para una Historia del Arte en el Perú*. Cuzco: Ediciones Inca, 1960.
Crawford, Kate and Trevor Paglen, https://www.excavating.ai/ (accessed 3 July 2022).
Crooke, Helkiah. *Mikrokosmographia: A Description of the Body of Man*. London: Thomas and Richard Cotes, 1631.
Cropper, Elizabeth. "On Beautiful Women, Parmigianino, Petrarchismo, and the Vernacular Style." *The Art Bulletin* 58, no. 3 (1976): 374–394.
Cummins, Thomas B. F. "The Indulgent Image: Prints in the New World." In *Contested Visions in the Spanish Colonial World*, edited by Ilona Katzew, William B. Taylor, and Luisa Elena Alcalá, pp. 203–225. New Haven, CT: Yale University Press, 2011.
Cuneo, Pia F. "Just a Bit of Control: The Historical Significance of Sixteenth- and Seventeenth-Century German Bit-Books." In *The Culture of the Horse: Status, Discipline, and Identity in the Early Modern World*, edited by Karen Raber, Treva J. Tucker, and Pia F. Cuneo, pp. 141–173. New York: Palgrave Macmillan, 2005.
Dackerman, Susan, ed. *Prints and the Pursuit of Knowledge in Early Modern Europe*. Cambridge: Harvard Art Museum, 2011.
Daly, Peter M., and G. Richard Dimler, eds. *The Jesuit Series. Part Five (P–Z)*. Toronto: University of Toronto Press, 2007.
Daly, Peter. "Emblematic Publications by the Jesuits of the Flanders Belgium Province to the Year 1700." In *The Jesuits and the Emblem Tradition*, edited by John Manning and Marc Van Vaeck, pp. 249–278. Turnhout: Brepols, 1999.
Daston, Lorraine. "Science Studies and the History of Science." *Critical Inquiry* 35, no. 4 (2009): 798–813.
Daston, Lorraine. "Epistemic Images." In *Vision and Its Instruments: Art, Science, and Technology in Early Modern Europe*, edited by Alina Payne, pp. 13–35. University Park, PA: Pennsylvania State University Press, 2015.
Davis, Whitney. *Visuality and Virtuality: Images and Pictures from Prehistory to Perspective*. Princeton, NJ: Princeton University Press, 2017.
Daxecker, Franz. *The Physicist and Astronomer Christoph Scheiner: Biography, Letters, Works*. Innsbruck: Leopold-Franzens-University of Innsbruck, 2004.
De Brézé, Nathalie. "Picturing the Soul, Living and Departed." In *The Anthropomorphic Lens: Anthropomorphism, Microcosmism, and Analogy in Early Modern*

Thought and Visual Arts, edited by Walter Melion, Bret Rothstein, and Michel Weemans, pp. 119–144. Leiden: Brill, 2015.

De Caro, Mario. "On Galileo's Platonism, Again." In *Hypotheses and Perspectives in the History and Philosophy of Science: Homage to Alexandre Koyré 1892–1964*, edited by Raffaele Pisano, Joseph Agassi, and Daria Drozdova, pp. 85–104. Dordrecht: Springer, 2018.

De Keyser, Marten. *Fons Vitae, ex quo Scaturiunt Suauissimae Co[n]solationes, Adflictis Me[n]tibus in Primis Necessariae, ne in Aduersitate & Dolore Protinus Animum Despondeant.* Antwerp: apud Martinum Caesarem, 1533.

De Mello, Francisco. *Argumentum Salutationis Gratulatoriae Dicatae Excellentissimo Domino D. Francisco de Mello….* Brussels: Jan Mommaert, 1642.

De Mûelenaere, Gwendoline. *Early Modern Thesis Prints in the Southern Netherlands: An Iconological Analysis of the Relationships between Art, Science and Power.* Leiden: Brill, 2022.

De Mûelenaere, Gwendoline. "The Art of Learning: Illustrated Lecture Notebooks at the Old University of Louvain." In *The Epistemic Function of Vision in Science*, edited by Giulia Giannini, Enrico Giannetto, and Matteo Valleriani. Cham: Springer, forthcoming.

De Ridder-Symoens, Hilde. "De Praktijk van Kennisoverdracht aan de Europese Universiteiten voor 1800." In *Ex Cathedra. Leuvense Collegedictaten van de 16de tot de 18de Eeuw*, edited by Geert Vanpaemel *et al.*, pp. 7–21. Louvain: Universiteitsbibliotheek, 2012.

De Schepper, Marcus. "Amblemata voor de Uldinge. Een Zestigtal 'Onbekende' Zuid-Nederlandse Embleemdrukken in de Brusselse Koninklijke Bibliotheek." In *The Stone of Alciato. Literature and Visual Culture in the Low Countries. Essays in Honour of Karel Porteman*, edited by Marc van Vaeck, Hugo Brems and Geert H.M. Claassens, pp. 1085–1118. Leuven: Peeters, 2003.

De Tollenaer, Jean, Jean Bolland, Sidronius de Hossche, and Godefroid Henschen. *Imago Primi Saeculi Societatis Iesu.* Antwerp: B. Moretus, 1640.

De Trauttmansdorff, Ferdinandus Ernestus. *Laudatio Funebris Ferdinandi III Austriaci….* S.l.: s.n., 1657.

Dear, Peter. *Discipline and Experience: The Mathematical Way in the Scientific Revolution.* Chicago, IL: University of Chicago Press, 1995.

Dear, Peter. "The Meanings of Experience." In *The Cambridge History of Science*, edited by Katharine Park and Lorraine Daston, pp. 106–131. Cambridge: Cambridge University Press, 2006.

Dekoninck, Ralph. "*Formatur unicus una, non alter.* Enjeux théologiques, Rhétoriques et Artistiques du Prototype dans le Monde de la Gravure Durant la Première Modernité." *Degrés: Revue de Synthèse à Orientation Sémiologique* 146 (2011): 1–16.

Dekoninck, Ralph. "A Graphic Koiné for a New Religious Value. The Visual Translatability of the *Evangelicae Historiae Imagines*." *Nederlands Kunsthistorisch Jaarboek* 64 (2014): 272–297.
Dekoninck, Ralph. "Le Double Mouvement de l'Âme vers l'Image. Une Théorie Aristotélico-Thomiste au Cœur des Débats du Milieu du XVIe Siècle sur le Juste Rapport à l'Image Religieuse." *Revue de Théologie et de Philosophie*, 154 (2022), 277-296.
Dekoninck, Ralph, and Agnès Guiderdoni. "Knowledge in Transition: a Case of 'Scientific Emblematics' (Ciermans and Vaenius) at the Turn of the Seventeenth Century." In *Embattled Territory: The Circulation of Knowledge in the Spanish Netherlands*, edited by Sven Dupré *et al.*, pp. 279–297. Ghent: Academia Press, 2015.
Delen, A.J.J. "Les Artistes Collaborateurs de Christophe Plantin." In *Sept Études Publiées à l'Occasion du Quatrième Centenaire du Célèbre Imprimeur Anversois Christophe Plantin, 1520–1920*, edited by Musée Du Livre Bruxelles, pp. 85–123. Brussels: J.E. Buschmann, 1920.
Delen, A.J.J. "Les Illustrateurs Français de Christophe Plantin." *Le Bibliophile* 2 (1932): 63–74 and 175–183.
Depauw, Carl, and Dirk Imhof. *De Boekillustratie ten Tijde van de Moretussen.* Antwerp: Stad Antwerpen, 1996.
Descartes, René. *La Dioptrique* [1637], in *Œuvres et lettres*, edited by André Bridoux. Paris: Bibliothèque de la Pléiade, 1953.
Descartes, René. *The Philosophical Writings of Descartes*, transl. John Cottingham, Robert Stoothoff, and Dugald Murdoch. 3 vols. Cambridge: Cambridge University Press, 1985.
Díaz del Castillo, Bernal. *Historia Verdadera de la Conquista dela Nueva España. 1585*, edited by Joaquín Ramírez Cabañas. Mexico City: Editorial Porrua, 1976.
Didi-Huberman, Georges. *La Ressemblance par Contact. Archéologie, Anachronisme et Modernité de l'Empreinte*. Paris: Éditions de Minuit, 2008.
Diefenbacher, Jörg, and Ekhard Leuschner, eds. *The New Hollstein German Engravings, Etchings and Woodcuts 1400–1700, vol. 10, The Greuter Family, pts. 1–2, Matthäus Greuter*. 2 vols. Ouderkerk aan den IJssel: Sound & Vision Interactive, 2016.
Dietz, Georg Josef. "Case Studies in Art and Science: Analysing Dürer's Early Pen-and-Ink Drawings." In *The Young Durer: Drawing the Figure*, pp. 96–97. London: The Courtauld Gallery, 2013.
Dillon, Anne. *Michelangelo and the English Martyrs*. Burlington, VT: Ashgate, 2012.
Doherty, Meghan. "Discovering the 'True Form:' Hooke's *Micrographia* and the Visual Vocabulary of Engraved Portraits." *Notes and Records of the Royal Society* 66 (2012): 211–234.

Donahue-Wallace, Kelly. "Picturing Prints in Early Modern New Spain." *The Americas* 64, no. 3 (2008): 325–349.

Drake, Stillman. "Galileo's Discovery of the Law of Free Fall." *Scientific American* 228 (1973): 84–92.

Drake, Stillman. "Galileo's Experimental Confirmation of Horizontal Inertia: Unpublished Manuscripts." *Isis* 64, no. 3 (1973): 291–305.

Dumont, Pascal. *Descartes et l'Esthétique. L'Art d'Émerveiller.* Paris: Presses Universitaires de France, 1997.

Dupré, Sven. "Ausonio's Mirrors and Galileo's Lenses: The Telescope and Sixteenth-Century Practical Optical Knowledge." *Galilaeana* 2 (2005): 145–180.

Dürer, Albrecht. *Hierin Sind Begriffen Vier Bücher von Menschlicher Proportion durch Albrechten Dürer von Nürnberg Erfunden und Beschriben….* Nuremberg: Hieronymus Andreae, 1528.

Durham Peters, John. *The Marvelous Clouds: Toward a Philosophy of Elemental Media*. Chicago, IL: University of Chicago Press, 2015.

Durling, Richard J. "A Chronological Census of Renaissance Editions and Translations of Galen." *Journal of the Warburg and Courtauld Institutes* 24, no. 3/4 (1961): 230–305.

Eamon, William. *Science and the Secrets of Nature: Books of Secrets in Medieval and Early Modern Culture*. Princeton, NJ: Princeton University Press, 1996.

Edgerton, Samuel. *Theaters of Conversion: Religious Architecture and Indian Artisans in Colonial Mexico.* Albuquerque, NM: University of New Mexico Press, 2001.

Edgerton, Samuel Y. "The Renaissance Artist as Quantifier." In *The Perception of Pictures*, edited by Margaret A. Hagen, 2 vols., I: 179–211. New York: Academic Press, 1980.

Edgerton, Samuel Y. "Galileo, Florentine 'Disegno,' and the 'Strange Spottednesse' of the Moon." *Art Journal* 44, no. 3 (1984): 225–232.

Egmond, Florike. *Eye for Detail: Images of Plants and Animals in Art and Science, 1500–1630.* London: Reaktion, 2017.

Egmond, Florike. "The Ad Vivum Conundrum: Eyewitnessing and the Artful Representation of Naturalia in Sixteenth-Century Natural Science." In *Zeigen – Überzeugen – Beweisen: Methoden der Wissensproduktion in Kunstliteratur, Kennerschaft und Sammlungspraxis der Frühen Neuzeit*, edited by Elisabeth Oy-Marra and Irina Schmiedel, pp. 33–62. Heidelberg: arthistoricum.net, 2020.

Eisenstein, Elizabeth L. *The Printing Press as an Agent of Change: Communications and Cultural Transformations in Early-Modern Europe*, 2 vols. Cambridge: Cambridge University Press, 1979.

Elen, Albert J. "Paper Analysis in Italian Drawing-Books of the 15th and 16th Centuries." In *Le Papier au Moyen Âge: historie et Techniques*, edited by Monique Zerdoun Bat-Yehouda, pp. 193–202. Turnhout: Brepols, 1999.

Elkins, James. "Art History and Images That Are Not Art." *The Art Bulletin* 77, no. 4 (1995): 553–571.

Ellinghaus, Julia. "Bilderwelten auf wissenschaftlichen Instrumenten der Frühen Neuzeit und ihre graphischen Vorlagen." *Imprimatur. Ein Jahrbuch für Bücherfreunde* 27 (2021): 107–130.

Ellinghaus, Julia, and Volker Remmert, eds. *Manipulating the Sun: Picturing Astronomical Miracles from the Bible in the Early Modern Era*. Leiden: Brill, forthcoming.

Emison, Patricia. "The Simple Art and Certain Complexities in Trying to Understand It." In *The Simple Art: Printed Images in an Age of Magnificance*, edited by Patricia Emison, 1–19. Durham, NC: The Art Gallery, University of New Hampshire, 2006.

Ems, Grégory. "Variété des 'Mises en Scène' Emblématiques dans la Province Jésuite Flandro-Belge au XVIIe Siècle. " *Dix-Septième Siècle* 269, no. 4 (2015): 705–734.

Ems, Grégory. *L'Emblématique au Service du Pouvoir. La Symbolique du Prince Chrétien dans les Expositions Emblématiques du Collège des Jésuites de Bruxelles sous le Gouvernorat de Léopold-Guillaume (1647–1656)*, 2 vols. Louvain-la-Neuve: Presses Universitaires de Louvain, 2016.

Engelgrave, Hendrik. *Lux Evangelica sub Velum Sacrorum Emblematum Recondita in Anni Dominicas: Selecta Historia et Morali Doctrina*. Antwerp: Widow of Jan Cnobbaert, 1648.

Erasmus, Desiderius. *Dulce Bellum Inexpertis*. Text edited and translated by Yvonne Remy and René Dunil-Marquebreucq. Brussels: Latomus, 1953.

Fabricius ab Aquapendente, Hieronymus. *De visione, Voce, Auditu*. Venice: Franciscus Bolzetta, 1600.

Favaro, Antonio. "Giovanni Battista Agucchi." In *Amici e corrispondenti di Galileo Galilei,* 3 vols. Florence: Libr. editrice Salimbeni, 1983, II: 373–395.

Featherstone, Rupert. "Bitumen." *Grove Art Online* 2003, https://www.oxfordartonline.com/groveart/view/10.1093/gao/9781884446054.001.0001/oao-9781884446054-e-7000009066 (accessed 5 March 2021).

Feldhay, Riva. "Producing Sunspots on an Iron Pan: Galileo's Scientific Discourse." In *Science, Reason, and Rhetoric,* edited by Henry Krips *et al.*, pp. 119–144. Pittsburgh, PA: University of Pittsburgh Press, 1995.

Fénelon, François de Pons de Salignac de La Mothe. *De l'éducation des filles* [1687], in *Œuvres*, edited by Jacques Le Brun. Paris: Bibliothèque de la Pléiade, 1983.

Ferrari, Michele C. *Le Monde en Images. Une Encyclopédie Bavaroise au Début du XVIe Siècle*. Dijon: Éditions Faton, 2021.

Ferrell, Lori Anne. "Page *Techne*: Interpreting Diagrams in Early Modern English 'How-To' Books." In *Printed Images in Early Modern Britain: Essays in Interpretation,* edited by Michael Hunter, pp. 113–126. Farnham: Ashgate, 2010.

Field, Richard. *Albrecht Dürer 1471–1528 : A Study Exhibition of Print Connoisseurship.* Philadelphia, PA: Philadelphia Museum of Art, 1970.

Finocchiaro, Maurice A. *The Galileo Affair: A Documentary History.* Berkeley, CA: University of California Press, 1989.

Finocchiaro, Maurice A. *Retrying Galileo, 1633–1992.* Berkeley, CA: University of California Press, 2005.

Finocchiaro, Maurice A. *The Trial of Galileo: Essential Documents.* Indianapolis, IN: Hackett Publishing, 2014.

Fissell, Mary Elizabeth. *Vernacular Bodies: The Politics of Reproduction in Early Modern England.* Oxford: Oxford University Press, 2004.

Fletcher, Shelley. "A Re-Evaluation of Two Mantegna Prints." *Print Quarterly* 14, no. 1 (1997): 67–77.

Flores Flores, Oscar, and Ligia Fernández Flores. "En torno a la *koineización* pictórica en los reinos de la monarquía hispánica. Identidad y variadades dialectales." In *Pinturas de los reinos. Identidades compartidas en el mundo hispánico,* edited by Juana Gutiérrez-Haces, pp. 187–332. Mexico City: Banamex, 2008–2009.

Forbes, Robert James. *Bitumen and Petroleum in Antiquity.* Leiden: Brill, 1936.

Fowler, Caroline. "Res Papirea: Mantegna's Paper Things." *The Art Bulletin* 99, no. 1 (2017): 8–35.

Franklin, James. "Diagrammatic Reasoning and Modeling in the Imagination: The Secret Weapons of the Scientific Revolution." In *1543 and All That: Image and Word, Change and Continuity in the Proto-Scientific Revolution*, edited by Guy Freeland and Anthony Corones, pp. 53–115. Dordrecht: Kluwer Academic Publishers, 2010.

Fransen, Sietske. "Latin in a Time of Change: The Choice of Language as Signifier of a New Science?" *Isis* 108, no. 3 (2017): 629–635.

Fransen, Sietske, and Katherine M. Reinhart, "The Practice of Copying in Making Knowledge in Early Modern Europe: An Introduction." *Word & Image* 35, no. 3 (2019): 211–222.

Freedberg, David. *The Eye of the Lynx: Art, Science and Nature in the Age of Galileo.* Chicago, IL: University of Chicago Press, 2002.

Furetière, Antoine. *Dictionaire Universel.* The Hague & Rotterdam: Arnout & Renier Leers, 1690.

Gabrieli, Giuseppe, ed. *Il Carteggio Linceo della Vecchia Accademia di Federico Cesi (1603–1630).* "Atti della Reale Accademia Nazionale dei Lincei," ser. 6, vol. 7. Rome: Dott. Giovanni Bardi, Tipografo Della R. Accademia Nazionale Dei Lincei, 1938.

Gaida, Margaret. "Reading Cosmographia: Peter Apian's Book-Instrument Hybrid and the Rise of the Mathematical Amateur in the Sixteenth Century." *Early Science and Medicine* 21, no. 4 (2016): 277–302.

Galen, Claudius. *Methodus Medendi.* Commentary by T. L. Croanglo. Paris, 1526.

Galen, Claudius. *De Usu Partium Corporis Humani Libri 17*. Basel: Andreas Cratander, 1533.

Galen, Claudius. *Claudii Galeni Pergameni de Hippocratis et Platonis Placitis Opus Eruditum, et Philosophis & Medicis Utilissimum....* Paris: Apud Simonem Colinaeum, 1534.

Galen, Claudius. "Galeni de Naturalibus Facultatibus Libri Tres, Thoma Linacro Anglo interprete." In *Galeni Omnia qual Extant Opera in Latinum Sermonem Conuersa*, pp. 290v–308r. Venice: apud haeredes Lucaeantonij Iuntae, 1565.

Galen, Claudius, and Johann Guenther. "Secundus". In *Libri tres*. Paris: Apud Simonem Colinaeum, 1528.

Galen, Claudius, and Leonhart Fuchs. *Claudii Galeni Pergameni*. Paris: Apud Iacobum Dupuys, 1549.

Galen, Claudius, and Theodor Zwinger. *In Artem Medicinalem Galeni, Tabulae & Commentarii....* Basil: apud Iacobum Parcum, expensis Ionnis Oporini, 1561.

Galilei, Galileo. *Istoria e dimostrazioni intorno alle macchie solari e loro accidenti. Comprese in tre lettere scritte all'illustrissimo Signor Marco Velseri*. Rome: Appresso Giacomo Mascardi, 1613.

Galilei, Galileo. *Le Opere di Galileo Galilei* (hereafter: *OG* and volume number), 21 vols, ed. Antonio Favaro. Florence: G. Barbèra, 1929–1939.

Galilei, Galileo, and Christoph Scheiner. *Galileo and Scheiner on Sunspots 1611–1613*, ed. and transl. Eileen Reeves and Albert Van Helden. Chicago, IL: University of Chicago Press, 2010.

Galluzzi, Paolo. *The Lynx and the Telescope: The Parallel Worlds of Cesi and Galileo*. Leiden: Brill, 2017.

Gänger, Stefanie. "World Trade in Medicinal Plants from Spanish America, 1717–1815." *Medical history* 59, no. 1 (2015): 44–62.

Gascoigne, Bamber. *How to Identify Prints*. Second edition. New York: Thames and Hudson, 2004.

Gaskell, Philip. *A New Introduction to Bibliography*. London: Oak Knoll Press, 1972.

Gaulke, Karsten. "Perfect in Every Sense: Scientific Iconography on an Equation Clock by Jost Bürgi and the Self-Understanding of the Astronomers at the Kassel Court in the Late 1580s." *Nuncius* 30, no. 1 (2015): 37–74.

Geminus, Thomas. *Compendiosa Totius Anatomie Delineatio*. London: John Herford, 1545.

Gerritsen, Johann. "Printing at Froben's: An Eye-Witness Account." *Studies in Bibliography* 44 (1991): 144–163.

Gersdorff, Hans von. *Feldbůch der Wund-Artzney, sampt vilen Instrumenten der Chirurgey*. Strasbourg: Johann Schott, 1517.

Giesecke, Michael. *Der Buchdruck in der frühen Neuzeit. Eine historische Fallstudie über die Durchsetzung neuer Informations- und Kommunications-technologien*. Frankfurt: Suhrkamp, 1991.

Giffin, Erin. "Replicas of the Santa Casa: Networks of Geographic Translation," https://storymaps.arcgis.com/stories/01dbd5ff3a5d4037b8567c7b6b760fa6 (accessed 3 July 2022).

Gilias, Guy, C. R. Van Tilburg, and Vincent Van Roy. *Rembert Dodoens. Een Zestiende-eeuwse Kruidenwetenschapper, Zijn Tijd- en Vakgenoten en Zijn Betekenis.* Antwerp-Apeldoorn: Garant, 2016.

Gill, Christopher. *Naturalistic Psychology in Galen and Stoicism.* Oxford: Oxford University Press, 2010.

Gingerich, Owen. "Astronomical Paper Instruments with Moving Parts." In *Making Instruments Count: Essays on Historical Scientific Instruments Presented to Gerard L'Estrange Turner,* edited by R. G. W. Anderson, J. A. Bennett, and W. F. Ryan, pp. 63–74. Aldershot: Variorum, 1993.

Goree, John. "The Woodblocks of Vesalius and the Printings: From the Renaissance to the Modern Era." *The Free Library* (2014): https://www.thefreelibrary.com/The+woodblocks+of+Vesalius+and+the+printings%3a+from+the+renaissance+to...-a0407670283 (accessed 27 February 2021).

Gorman, Michael John. "The Elusive Origins of the Immutable Mobile." 2001, https://web.archive.org/web/20120628201904/http://www.stanford.edu/group/STS/immutablemobile.htm (accessed 5 March 2021).

Goyet, Francis. "De la Rhétorique à la Création: Hypotypose, Type, Pathos." In *La Rhétorique. Enjeux de ses Résurgences,* edited by Jean Gayon, Jacques Poirier and Jean-Claude Gens, pp. 46–67. Brussels: Ousia, 1998.

Grafton, Anthony. "The Importance of Being Printed." *Journal of Interdisciplinary History* 11 (1980): 265–286.

Grafton, Anthony. *The Culture of Correction in Renaissance Europe.* London: British Library, 2011.

Graniti, Antonio, ed. *Federico Cesi. Un Principe Naturalista.* Rome: Bardi, 2006.

Grimm, Heinrich. "Geadelte deutsche Buchdrucker-Familien im 16. und 17. Jahrhundert." *Gutenberg-Jahrbuch* 36 (1961): 257–271.

Groebner, Valentin. "Complexio/Complexion: Categorizing Individual Natures, 1250–1600." In *The Moral Authority of Nature,* edited by Lorraine Daston, and Fernando Vidal, pp. 361–383. Chicago, IL: University of Chicago Press, 2010.

Guerra, Francisco. "Juan de Valverde de Amusco." *Clio Medica* 2 (1967): 339–362.

Guerrieri Borsoi, Maria Barbara. "GREUTER, Matthäus." *Dizionario Biografico degli Italiani* 59 (2002), www.treccani.it/enciclopedia/matthaus-greuter_(Dizionario-Biografico) (accessed 4 March 2021).

Hájek z Hájku, Tadeáis. *Aphorismorum Metoposcopicorum Libellus Unus.* Frankfurt am Main: Heirs of Andreas Wechel, 1584.

Hall, Bert S. "The Didactic and the Elegant: Some Thoughts on Scientific and Technological Illustrations in the Middle Ages and the Renaissance." In *Picturing*

Knowledge: Historical and Philosophical Problems Concerning the Use of Art in Science, edited by Brian Scott Baigrie, pp. 3–39. Toronto: University of Toronto Press, 1996.

Hall, James. *Dictionary of Subjects and Symbols in Art*. London: John Murray, 1996.

Hallyn, Fernand. "Rhétorique de la lunette." *Littératures classiques* 11 (1989): 13–23.

Hamm, Patricia. "A History of the Manufacture of Printing Ink From 1500–1900 With Notes for the Conservator." In *Conference Papers, Manchester 1992*, edited by Sheila Fairbrass, pp. 30–34. Institute of Paper Conservation, 1992.

Hankinson, R. J. "Galen's Anatomy of the Soul." *Phronesis* 36, no. 2 (1991): 197–233.

Harcourt, Glenn. "Andreas Vesalius and the Anatomy of Antique Sculpture." *Representations* 17 (1987): 28–61.

Harrison, Charles. "The Printed Picture in the Renaissance." In *Making Renaissance Art, Vol. 1*, edited by Kim W. Woods, pp. 211–247. New Haven, CT: Yale University Press, 2007.

Hart, Kevin. "Machine Learning Confronts the Elephant in the Room," *Quanta*, 20 September 2018: https://www.quantamagazine.org/machine-learning-confronts-the-elephant-in-the-room-20180920/ (accessed 3 July 2022).

Harvard, Stephen. *Ornamental Initials: The Woodcut Initials of Christopher Plantin. A Complete Catalogue*. New York: American Friends of the Plantin-Moretus Museum, 1974.

Hauschke, Sven. "Goldschmiede als Hersteller wissenschaftlicher Instrumente und Geräte." In *Nürnberger Goldschmiedekunst 1541–1868, vol. 2, Goldglanz und Silberstrahl*, edited by Karin Tebbe, Ursula Timann, and Thomas Eser, 216–332. Nuremburg: Verlag des Germanischen Nationalmuseums, 2007.

Havelange, Carl. *De L'œil et du Monde. Une Histoire du Regard au Seuil de la Modernité*. Paris: Fayard, 1998.

Hedges, S. Blair. "A Method for Dating Early Books and Prints Using Image Analysis." *Proceedings of the Royal Society* 462 (2006): 3555–3574.

Henderson, Felicity. "Images." In *A Handbook of Editing Early Modern Texts*, edited by Claire Loffman and Harriet Philipps, pp. 70–73. New York: Routledge, 2018.

Hentschel, Klaus. *Visual Cultures in Science and Technology: A Comparative History*. Oxford: Oxford University Press, 2014.

Heseler, Baldasar, and Ruben Eriksson. *Andreas Vesalius' First Public Anatomy at Bologna: 1540*. Uppsala: Almqvist & Wiksell, 1959.

Heuer, Christopher. *The City Rehearsed: Object, Architecture, and Print in the Worlds of Hans Vredeman de Vries*. New York: Routledge, 2009.

Hier Inn Begriffne Pis, Zaigen Clärlich an, Wie Ain Yedess Roß, Iüng Und Alt Soll Gezämbt, Dadurch Ime, Angenomene Pöse Gewonhait Ubelstandt, ... Benomen ... Werden Mag. VD 16 G 592. [Augsburg?], 1530.

Higgins, B. D. R. "Printing the First Folio." In: *The Cambridge Companion to Shakespeare's First Folio*, edited by Emma Smith, pp. 30–47. Cambridge: Cambridge University Press, 2016.

Hollstein, F. W. H. *German Engravings, Etchings, and Woodcuts ca. 1400–1700*. Amsterdam: M. Hertzberger, 1954.

Höltgen, Karl Josef. "Richard Haydocke: Translator, Engraver, Physician." *The Library* 33 (1978): 15–32.

Horovitz, Isabel. "The Materials and Techniques of European Paintings on Copper Supports." In *Copper as Canvas: Two Centuries of Masterpiece Paintings on Copper 1575–1775*, pp. 63–92. New York: Oxford University Press, 1998.

Horowitz, Michael, and Jack Collins. "A Census of Copies of the First Edition of Andreas Vesalius' *De humani corporis fabrica* (1543), with a Note on the Recently Discovered Variant Issue." *Journal of the History of Medicine and Allied Sciences* 39 (1984): 198–221.

Houtzager, Hans. "Editions, Copies and Replicas of Fabrica and Epitome, Published Between 1543 and 1725." In *Art of Vesalius*, edited by Robrecht van Hee, pp. 191–222. Antwerp: Garant, 2014.

Houwaert, Johan Baptist. *Sommare Beschrijvinghe vande Triumphelijcke Incomst vanden ... Aerts-Hertoge Matthias....* Antwerpen: Christoffel Plantijn, 1579.

Hunter, Dard. *Papermaking*. New York: Dover Press, 1978.

Hunter, Matthew C. *Wicked Intelligence: Visual Art and the Science of Experiment in Restoration London*. Chicago, IL: University of Chicago Press, 2013.

Hyman, Aaron. *Rubens in Repeat: The Logic of the Copy in Colonial Latin America*. Los Angeles, CA: J. Paul Getty Press, 2021.

Imhof, Dirk. *Jan Moretus and the Continuation of the Plantin Press: A Bibliography of the Works Published and Printed by Jan Moretus I in Antwerp (1589–1610)*. Leiden: Brill, 2014.

Ivins, William M. *On the Rationalization of Sight, with an Examination of Three Renaissance Texts on Perspective*. New York: Metropolitan Museum of Art, 1938.

Ivins, William M. *Prints and Visual Communication*. Cambridge, MA: MIT Press, 1969.

Ivins, William M. *How Prints Look: Photographs with Commentary*, rev. and exp. by Marjorie B. Cohn. Boston, MA: Beacon, 1987.

Jaanus, H. J. *Een Remonstrantse Voorzaat: Henricus Meilingius*. The Hague: Central Bureau of Genealogy, s.d. (mid-twentieth century).

Jaeger, Mary. *Archimedes and the Roman Imagination*. Ann Arbor, MI: University of Michigan Press, 2008.

Jaffé, David. "Mellan and Peiresc." *Print Quarterly* 7 (1990), 168–175.

Jenkins, Catherine, Nadine Orenstein, and Freyda Spira, eds. *The Renaissance of Etching*. New York: The Metropolitan Museum of Art, 2020.

Joffe, Stephen N., and Veronica Buchanan. "The Andreas Vesalius Woodblocks: A Four Hundred Year Journey from Creation to Destruction." *Acta medico-historica Adriatica* 14 (2016): 347–372.

Johns, Adrian. "Ink." In *Materials and Expertise in Early Modern Europe: Between Market and Laboratory*, edited by Ursula Klein and E. C. Spary, pp. 101–124. Chicago, IL: University of Chicago Press, 2010.

Johns, Adrian. *The Nature of the Book: Print and Knowledge in the Making.* Chicago: University of Chicago Press, 1998.

Johnson, C. Richard Jr., *et al.* "Chain Line Pattern Matching in Rembrandt's Prints." In *Rembrandt and His Circle: Insights and Discoveries*, edited by S. Dickey, 319–334. Amsterdam: Amsterdam University Press, 2017.

Johnston, Stephen. "Mathematical Practitioners and Instruments in Elizabethan England." *Annals of Science* 48, no. 4 (1991): 319–344.

Jordanova, Ludmilla. *The Look of the Past: Visual and Material Evidence in Historical Practice.* Cambridge: Cambridge University Press, 2012.

Jorink, Eric. "Beyond the Lines of Apelles: Johannes Swammerdam, Dutch Scientific Culture, and the Representation of Insect Anatomy." In *Art and Science in the Early Modern Netherlands*, edited by Eric Jorink and Bart Ramakers, pp. 148–184. Zwolle: WBooks, 2011.

Jurkowlaniec, Grażyna. "The Crucified Christ as the Source of the Seven Sacraments: Patters of Reception of a Sixteenth-Century Image on Both Sides of the Alps and on Both Sides of the Atlantic Ocean." Transl. by Zuzanna Sarnecka. In *Artistic Translations Between Fourteenth and Sixteenth Centuries: International Seminar for Young Researchers. Proceedings*, edited by Zuzanna Sarnecka and Aleksandra Fedorowicz-Jackowska, pp. 187–209. Warsaw: University of Warsaw, 2013.

Karpenko, Vladimir. "Some Notes on the Early History of Nitric Acid: 1300–1700." *Bulletin of the History of Chemistry* 34, no. 2 (2009): 105–116.

Karr Schmidt, Suzanne. *Altered and Adorned. Using Renaissance Prints in Daily Life.* New Haven, CT: Yale University Press, 2011.

Karr Schmidt, Suzanne. *Interactive and Sculptural Printmaking in the Renaissance.* Leiden: Brill, 2018.

Karr Schmidt, Suzanne, and Edward H. Wouk. *Prints in Translation, 1450–1750: Image, Materiality, Space.* Abingdon: Routledge, 2017.

Katzenellenbogen, Adolf. *Allegories of the Virtues and Vices in Mediaeval Art from Early Christian Times to the Thirteenth Century.* London: The Warburg Institute, 1939.

Keizer, Joost M., and Todd M. Richardson, eds. *The Transformation of Vernacular Expression in Early Modern Arts.* Leiden: Brill, 2012.

Kemp, Martin. "Temples of the Body and Temples of the Cosmos: Vision and Visualization in the Vesalian and Copernican Revolutions." In *Picturing Knowledge:*

Historical and Philosophical Problems Concerning the Use of Art in Science, pp. 40–85. Toronto: University of Toronto Press, 1996.

Kepler, Johannes. *Ad Vitellionem Paralipomena, quibus Astronomiae Pars Optica Traditur.* Frankfurt: Marnius and heirs of Aubrius, 1604.

Kessler, Herbert L, and Gerhard Wolf, eds. *The Holy Face and the Paradox of Representation: Papers from a Colloquium Held at the Bibliotheca Hertziana, Rome and the Villa Spelman, Florence, 1996*. Bologna: Nuova Alfa Ed., 1998.

King, David A. *Astrolabes and Angels, Epigrams and Enigmas: From Regiomontanus' Acrostic for Cardinal Bessarion to Piero della Francesca's Flagellation of Christ.* Stuttgart: Steiner Verlag, 2007.

Kirschbaum, Engelbert, ed. *Lexikon der christlichen Ikonographie.* Rome: Herder, 1970.

Klapisch-Zuber, Christiane. *L'Ombre des Ancêtres. Essai sur l'Imaginaire Médiéval de la Parenté.* Paris: Fayard, 2000.

Klein, Ursula. *Experiments, Models, Paper Tools: Cultures of Organic Chemistry in the Nineteenth Century.* Stanford, CA: Stanford University Press, 2003.

Klestinec, Cynthia Jennifer. "Juan Valverde de (H)Amusco and Print Culture: the Editorial Apparatus in Vernacular Anatomy Texts." *Zeitsprünge: Forschungen zur Frühen Neuzeit* 9 (2005): 78–94.

Knipping, John B. *Iconography of the Counter Reformation in the Netherlands: Heaven on Earth, Vol. II.* Nieuwkoop: B. de Graaf; Leiden: A. W. Sijthoff, 1974.

Kockelbergh, Iris. "The Rich Collection of Woodblocks of the Museum Plantin-Moretus and Its Use in the Officina Plantiniana." In *Ulisse Aldrovandi. Libri e Imagini di Storia Naturale nella prima Età Moderna,* edited by Guiseppe Olmi and Fulvio Simoni, pp. 109–118. Bologna: Bononia University Press, 2018.

Kodera, Sergius. "The Art of the Distillation of 'Spirits' as a Technological Model for Human Physiology: The Cases of Marsilio Ficino, Joseph Duchesne and Francis Bacon." In *Blood, Sweat and Tears: The Changing Concepts of Physiology from Antiquity into Early Modern Europe*, edited by Manfred Horstmanshoff, Helen King, and Claus Zittel, pp. 139–170. Leiden: Brill, 2012.

Koehler, Sylvester Rosa. *A Chronological Catalogue of the Engravings, Dry-Points and Etchings of Albert Dürer.* New York: The Grolier Club, 1897.

Korey, Michael. *Die Geometrie der Macht. Mathematische Instrumente und fürstliche Mechanik um 1600.* München: Deutscher Kunstverlag, 2007.

Kornell, Monique. "Jan Steven van Calcar, c. 1515–c. 1546, Vesalius' Illustrator." In *Andreas Vesalius and the* Fabrica *in the Age of Printing; Art, Anatomy, and Printing in the Italian Renaissance,* edited by Rinaldo Fernando Canalis and Massimio Ciavolella, pp. 99–130. Turnhout: Brepols, 2018.

Krebber, G. B. and Kotting, K. "Jean Bellegambe en zijn 'Mystiek Bad' voor Anchin." *Oud Holland* 104, no. 3/4 (1990): 123–139.

Kremer, Richard. "Experimenting with Paper Instruments in Fifteenth- and Sixteenth-Century Astronomy: Computing Syzygies with Isotemporal Lines and Salt Dishes." *Journal for the History of Astronomy* 42, no. 2 (2011): 223–258.

Krohn, Deborah L. *Food and Knowledge in Renaissance Italy: Bartolomeo Scappi's Paper Kitchens.* New York: Routledge, Taylor & Francis Group, 2018.

Kruse, Britta-Juliane. "Fachliteratur–Fachprosa, X.1 Medizinische Sammelhandschriften/ Kräuterbücher, X.2 Wahrsagebücher." In *Aderlass und Seelentrost. Die Überlieferung deutscher Texte im Spiegel Berliner Handschriften und Inkunabeln*, edited by Peter Jörg Becker and Eef Overgaauw, pp. 340–366. Mainz: P. von Zabern, 2003.

Kusukawa, Sachiko. "The Uses of Pictures in the Formation of Learned Knowledge: The Cases of Leonhard Fuchs and Andreas Vesalius." In *Transmitting Knowledge: Words, Images, and Instruments in Early Modern Europe*, edited by Sachiko Kusukawa and Ian Maclean, pp. 73–96. Oxford: Oxford University Press, 2006.

Kusukawa, Sachiko. "The Role of Images in the Development of Renaissance Natural History." *Archives of Natural History* 38 (2011): 189–213.

Kusukawa, Sachiko. *Picturing the Book of Nature: Image, Text, and Argument in Sixteenth-Century Human Anatomy and Medical Botany.* Chicago, IL: University of Chicago Press, 2012.

L'E Turner, Gerard. *Elizabethan Instrument Makers: The Origins of the London Trade in Precision Instrument Making.* Oxford: Oxford University Press, 2000.

Lamprey, John. *Hartmann's Practika: A Manual for making Sundials and Astrolabes with the Compass and Rule.* Bellevue: J. Lamprey, 2002.

Landau, David. "Printmaking in the Age of Lorenzo." In *Florentine Drawing at the Time of Lorenzo the Magnificent: Papers from a Colloquium Held at the Villa Spelman, Florence, 1992*, edited by Elizabeth Cropper, pp. 175–180. Baltimore, MD: Johns Hopkins University Press, 1992.

Larkey, Sanford V. "The Vesalian Compendium of Geminus and Nicholas Udall's Translation: Their Relation to Vesalius, Caius, Vicary and de Mondeville." *The Library* 13, no. 4 (1933): 367–394.

Larkin, Graham, and Lisa Pon. "Introduction: the Materiality of Printed Words and Images." *Word & Image* 17, no. 1–2 (2001): 1–6.

Latour, Bruno. "Visualisation and Cognition: Drawing Things Together." Edited by H. Kuklick. *Knowledge and Society Studies in the Sociology of Culture Past and Present* 6 (1986): 1–40.

Latour, Bruno, ed. *Science in Action.* Cambridge, MA: Harvard University Press, 1987.

Latour, Bruno. "Visualisation and Cognition: Drawing Things Together." In *Knowledge and Society: Studies in the Sociology of Culture Past and Present* 6, edited by H. Kuklick, pp. 1–40. Greenwich: Jai, 1990.

Laurenza, Domenico. *Art and Anatomy in Renaissance Italy: Images from a Scientific Revolution.* New York: Metropolitan Museum of Art, 2012.

Lavin, Irving. "Il Volto Santo de Claude Mellan: *ostendatque etiam quae occultet.*" In *République des Lettres, République des Arts. Mélanges offerts à Marc Fumaroli*, edited by Christian Mouchel and Colette Nativel, pp. 385–414. Geneva: Librairie Droz, 2008.

Lawson, Ian. "Crafting the Microworld: How Robert Hooke Constructed Knowledge About Small Things." *Notes and Records of the Royal Society of London*, 70 (2016): 23–44.

Le Boeuffle, André. *Astronomie. Astrologie. Lexique latin.* Paris: Picard, 1987.

Lecercle, François. "De la Relique à l'Image: La Promotion du Suaire de Turin." In *Symboles de la Renaissance III*, edited by Daniel Arasse, Maurice Brock, and Georges Didi-Huberman, pp. 95–112. Paris: Éditions Rue d'Ulm, 1990.

Lecercle, François. "Le Signe et la Relique. Les Théologies de l'Image à la Renaissance." PhD Dissertation, 2 vols. Montpellier: Université de Montpellier, 1987.

Legros, Philippe. *François de Sales, une Poétique de l'Imaginaire. Étude des Représentations Visuelles dans l'*Introduction à la Vie Dévote *et le* Traité de l'Amour de Dieu. Tübingen: Gunter Narr Verlag, 2004.

Leitch, Stephanie. "Visual Acuity and the Physiognomer's Art of Observation." *Oxford Art Journal* 38, no. 2 (2015): 187–206.

Leitch, Stephanie. "Dürer's Rhinoceros Underway: The Epistemology of the Copy in the Early Modern Print." In *The Primacy of the Image in Northern European Art, 1400–1700: Essays in Honor of Larry Silver*, edited by D. Cashion *et al.*, pp. 241–255. Boston: Brill, 2017.

Leitch, Stephanie. "Getting to How-To: Chiromancy, Physiognomy, Metoscopy and Prints in Secrets' Service." In *Quid Est Secretum? Visual Representation of Secrets in Early Modern Europe, 1500–1700*, edited by Walter Melion, Ralph Dekoninck, and Agnès Guiderdoni, pp. 635–659. Leiden: Brill, 2020.

Lemli, Jozef. "De Officinae Plantinianae te Antwerpen en te Leiden en hun Botanische Edities van 1589 tot 1647." In *De Botanica in de Zuidelijke Nederlanden (Einde 15de Eeuw–ca. 1650)*, edited by Dirk Imhof and Francine de Nave, pp. 57–59. Antwerp: Museum Plantin-Moretus, 1993.

Leuschner, Eckhard. "The Papal Printing Privilege." *Print Quarterly* 15, no. 4 (1998): 359–370.

Lincoln, Evelyn. *Brilliant Discourse: Pictures and Readers in Early Modern Rome.* New Haven, CT: Yale University Press, 2014.

Loach, Judy. "The Teaching of Emblematics and Other Symbolic Imagery by Jesuits within Town Colleges in Seventeenth and Eighteenth-Century France." In *The Jesuits and the Emblem Tradition*, edited by John Manning and Marc van Vaeck, pp. 161–186. Turnhout: Brepols, 1999.

Lobelius, Matthias. *Icones Stirpium, seu Plantarum tam Exoticarum, quam Indigenarum....* Antwerp: ex Officina Plantiniana, 1591.

Lobelius, Matthias. *Kruydtboeck oft Beschrijvinghe van allerleye Ghewassen, Kruyderen, Hesteren, ende Gheboomten.* Antwerp: Christoffel Plantyn, 1581.

López Piñero, José María. "La Disección y el Saber Anatómico en la España de la Primera Mitad del Siglo XVI." *Cuadernos de Historia de la Medicina Española* 13 (1974): 51–110.

Lüthy, Christoph, and Alexis Smets. "Words, Lines, Diagrams, Images: Towards a History of Scientific Imagery." *Early Science and Medicine* 14 (2009): 398–439.

Macgregor, Neil, and Olivier Bonfait, eds. *Il Dio Nascosto. I Grandi Maestri del Seicento e l'Immagine di Dio.* Rome: Edizioni De Luca, 2000.

MacGregor, William B. "The Authority of Prints: An Early Modern Perspective." *Art History* 22 (1999): 389–420.

MacLachlan, James. "Galileo's Experiments with Pendulums: Real and Imaginary." *Annals of Science* 33, no. 2 (1976): 173–185.

MacLachlan, James. "Experimenting in the History of Science." *Isis* 89, no. 1 (1998): 90–92.

Maclean, Ian. "Diagrams in the Defense of Galen: Medical Uses of Tables, Squares, Dichotomies, Wheels, and Latitudes, 1480–1574." In *Transmitting Knowledge: Words, Images, and Instruments in Early Modern Europe*, edited by Sachiko Kusukawa and Ian Maclean, pp. 135–164. Oxford: Oxford University Press, 2006.

Mahon, Denis. *Studies in Seicento Art and Theory.* London: The Warburg Institute University of London, 1947.

Malebranche, Nicolas de. *De la Recherche de la Vérité* [1674–1675], edited by Geneviève Rodis-Lewis. Paris: Bibliothèque de la Pléiade, 1979.

Manick, Annette. "A Note on Printing Inks." In *Italian Etchers of the Renaissance and Baroque*, edited by Sue Welsh Reed and Richard W. Wallace, pp. xliv–xlvii. Boston, MA: Museum of Fine Arts, 1989.

Manning, John. "Towards an Emblematic Rhetoric." In *The Emblem*, pp. 90–101. London: Reaktion, 2002.

Margócsy, Dániel. "From Vesalius through Ivins to Latour: Imitation, Emulation and Exactly Repeatable Pictorial Statements in the *Fabrica*." *Word & Image* 35, no. 3 (2019): 315–333.

Margócsy, Dániel, Mark Somos, and Stephen N. Joffe. *The Fabrica of Andreas Vesalius: A Worldwide Descriptive Census, Ownership, and Annotations of the 1543 and 1555 Editions.* Leiden: Brill, 2018.

Marías, Fernando. *El Largo Siglo XVI. Los Usos Artísticos del Renacimiento Español.* Madrid: Taurus, 1989.

Marion, Jean-Luc. *Sur la Théologie Blanche de Descartes. Analogie, Création des Vérités Éternelles et Fondement.* Paris: Presses Universitaires de France, 1981.

Marolles, Michel de. *Les Mémoires de Michel de Marolles*. Paris: Antoine de Sommaville, 1656.

Marr, Alexander. "Knowing Images." *Renaissance Quarterly* 69, no. 3 (2016): 1000–1013.

Marr, Alexander. "Ingenuity in Nuremberg: Dürer and Stabius's Instrument Prints." *The Art Bulletin* 100, no. 3 (2018): 48–79.

Marr, Alexander, and Christopher P. Heuer. "Introduction: The Uncertainty of Epistemic Images." *21: Inquiries into Art, History, and the Visual* 2 (2020): DOI: https://doi.org/10.11588/xxi.2020.2.76226.

Martin Mulsow, "History of Knowledge." In *Debating New Approaches to History*, edited by Marek Tamm and Peter Burke, pp. 159–188. London: Bloomsbury, 2019.

Mayer, Thomas F. "The Censoring of Galileo's Sunspot Letters and the First Phase of his Trial." *Studies in History and Philosophy of Science* 42 (2011): 1–10.

Mayer, Thomas F. "An Interim Report on a Census of Galileo's Sunspot Letters." *History of Science* 50, no. 2 (2012): 155–196.

Mayer, Thomas F. *The Trial of Galileo, 1612–1633*. Toronto: University of Toronto Press, 2012.

McDonald, Mark P. *The Print Collection of Ferdinand Columbus (1488–1539): A Renaissance Collector in Seville*, 3 vols. London: British Museum Press, 2004.

McKenzie, D. F. "Printers of the Mind: Some Notes on Bibliographical Theories and Printing-House Practices." *Studies in Bibliography* 22 (1969): 1–75.

McKitterick, David. *Print, Manuscript and the Search for Order*. Cambridge: Cambridge University Press, 2003.

McLuhan, Marshall. *Understanding Media: The Extensions of Man*. New York: McGraw-Hill, 1964.

McMullin, Ernan. "The Galileo Affair: Two Decisions." *Journal for the History of Astronomy* 40, no. 2 (2009): 191–212.

McMullen, Shannon C., and Fabian Winkler. "20/X: Visuality, Representation and Epistemology in the Age of Intelligent Seeing Machines." *Leonardo* 51 no. 1 (2018): 71–74.

Meder, Joseph. *Durer-Katalog: Ein Handbuch uber Albrecht Durers Stiche, Radierungen, Holzschnitte, Deren Zustande, Ausgaben und Wasserzeichen*. Vienna: Verlag Gilhofer & Ranschburg, 1932.

Meilingius, Henricus Jacobi. *Disputatio Medica Inauguralis*. Trajecti ad Rhenum, 1660.

Meilingius, Henricus Jacobi. "Henricus Meilingius (Borculo, NL) to Hugo de Groot (Paris, France), 1 October 1629." In *Briefwisseling van Hugo Grotius. Deel 4*, edited by B.L. Meulenbroek, pp. 107–108. The Hague: Martinus Nijhoff, 1964.

Melion, Walter S. "Prayerful Artifice: The Fine Style as Marian Devotion in Hieronymus Wierix's Maria of ca. 1611." In *The Authority of the Word: Reflecting on*

Image and Text in Northern Europe, 1400–1700, edited by Celeste Brusati, Karl A. E. Enenkel, and Walter S. Melion, pp. 589–637. Leiden: Brill, 2011.

Melion, Walter S., and Bart Ramakers, eds. *Personification. Embodying Meaning and Emotion*. Leiden: Brill, 2016.

Ménestrier, Claude-François. *L'Art des Emblèmes où s'Enseigne la Morale par les Figures de la Fable, de l'Histoire, & de la Nature*. Lyon: Benoît Coral, 1662.

Ménestrier, Claude-François. *Le Veritable Art du Blason ou l'Usage des Armoiries*. Paris: Estienne Michallet, 1673.

Mersmann, Jasmin. "Moving Shadows, Moving Sun. Early Modern Sundials Restaging Miracles." *Nuncius* 30, no. 1 (2015): 96–123.

Meyer, A. W., and Sheldon K. Wirt. "The Amuscan Illustrations." *Bulletin of the History of Medicine* 14, no. 5 (1943): 667–687.

Miller, George. "The Magical Number Seven, Plus or Minus Two: Some Limits on Our Capacity for Processing Information." *The Psychological Review* 63 (1956): 81–97.

Mills, Kenneth. *Idolatry and Its Enemies: Colonial Andean Religion and Extirpation, 1640–1750*. Princeton, NJ: Princeton University Press, 1997.

Mochizuki, Mia. "Sacred Art in an Age of Mechanical Reproduction: The Salus Populi Romani Madonna in the World." In *Sacred and Profane in Early Modern Art*, edited by Kayo Hirakawa, pp. 129–144. Kyoto: Kyoto University, 2016.

Moore, Rosemary. "Paper Cuts. The Early Modern Fugitive Print." *Object* 17 (2015): 54–76.

Morrall, Andrew. "Entrepreneurial Craftsmen in Late Sixteenth-Century Augsburg." In *Mapping Markets for Paintings in Europe 1450–1750*, edited by Neil de Marchi and Hans J. van Miegroet, pp. 211–236. Turnhout: Brepols, 2006.

Muccilo, Maria. "Fabrici d'Acquapendente, Girolamo." In *Dizionario Biographico Degli Italiani*, Vol. 43, 1993. https://www.treccani.it/enciclopedia/girolamo-fabrici-d-acquapendente_%28Dizionario-Biografico%29/ (accessed 18 February 2021).

Mungello, D.E. ed. *The Chinese Rites Controversy: Its History and Meaning*. Abingdon: Routledge, 2018.

Nauta, D., *et al.*, eds. *Biografisch Lexicon voor de Geschiedenis van het Nederlandse Protestantisme, Deel 3*. Kampen: Uitgeversmaatschappij J.H. Kok, 1988.

Needham, Paul. "Copy-Specifics in the Printing Shop." In *Early Printed Books as Material Objects*, edited by Bettina Wagner and Marcia Reed, pp. 9–20. Berlin: De Gruyter, 2010.

Nowotny, Helga. *Es ist So. Es Könnte auch Anders Sein*. Frankfurt a. M.: Suhrkamp Verlag, 1999.

Noyes, Ruth Sargent. "*Aut Numquid Post Annos Mille Quingentos Docenda est Ecclesia Catholica quomodo Sacrae Imagines Pingantur?* Post-Tridentine Image Reform and the Myth of Paleotti." *Catholic Historical Review* 99, no. 2 (2013): 239–261.

Noyes, Ruth Sargent. "Mattheus Greuter's Sunspot Etchings for Galileo Galilei's *Macchie Solari* (1613)." *The Art Bulletin* 98, no. 4 (2016): 464–485.

Noyes, Ruth Sargent. "'One of Those Lutherans We Used to Burn in Campo de' Fiori.' Engraving Sublimated Suffering in Counter-Reformation Rome." In *Visualizing Sensuous Suffering and Affective Pain in Early Modern Europe and the Spanish Americas*, edited by Heather Graham and Lauren Kilroy-Ewbank, pp. 116–165. Leiden: Brill, 2018.

Nutton, Vivian. "Representation and Memory in Renaissance Anatomical Illustration." In *Immagini per Conoscere. Dal Rinascimento alla Rivoluzione Scientifica*, edited by Fabrizio Meroi and Claudio Pogliano, pp. 61–80. Florence: Olschki, 2001.

Nutton, Vivian. "Vesalius Revised: His Annotations to the 1555 *Fabrica*." *Medical History* 56 (2012): 415–443.

Nutton, Vivian. "Historical Introduction." In *On the Fabric of the Human Body: An Annotated Translation of the 1543 and 1555 Editions of Andreas Vesalius' De Humani Corporis Fabrica*, edited by Daniel Garrison and Malcolm Hast, pp. lxxv–ciii. Basel: Karger, 2014.

Nutton, Vivian. "More Vesalian Second Thoughts: The Annotations to the *Institutiones anatomicae secundum Galeni sententiam*." *Gesnerus* 11 (2015): 94–116.

Ogilvie, Brian W. *The Science of Describing: Natural History in Renaissance Europe*. Chicago, IL: University of Chicago Press, 2006.

Ojeda di Ninno, Almerindo. "El Grabado como Fuente del Arte Colonial: Estado de la Cuestión." In *De Amberes al Cuzco*, edited by Cécile Michaud and José Torres della Pina, pp. 10–21. Lima: Impulso Empresa de Servicios, 2009.

O'Malley, Charles Donald. *Andreas Vesalius of Brussels, 1514–1564*. Berkeley, CA: University of California Press, 1964.

Osorio, Alejandra. "The King in Lima, Simulacra, Ritual, and Rule in Seventeenth-Century Peru." *Hispanic American Historical Review* 84, no. 3 (2004): 447–474.

Ostrow, Steven F. "Cigoli's Immacolata and Galileo's Moon: Astronomy and the Virgin in Early Seicento Rome." *Art Bulletin*, 78, no. 2 (1996): 218–235.

Panofsky, Erwin. "Die Perspektive Als 'Symbolische Form.'" In *Vorträge Der Bibliothek Warburg 1924/25*, 258–330. Liepzig/Berlin: Teubner, 1927.

Panofsky, Erwin. *Artist, Scientist, Genius: Notes on the 'Renaissance-Dämmerung'*. New York: Metropolitan Museum of Art, 1953.

Panofsky, Erwin. *Galileo as a Critic of the Arts*. The Hague: Martinus Nijhoff, 1954.

Panofsky, Erwin. "Galileo as a Critic of the Arts: Aesthetic Attitude and Scientific Thought." *Isis* 47, no. 1 (1956): 3–15.

Panofsky, Erwin. *Perspective as Symbolic Form*. Transl. by Christopher S. Wood. New York: Zone Books, 1997.

Panse, Melanie. *Hans von Gersdorffs Feldbuch der Wundarznei. Produktion, Präsentation und Rezeption von Wissen*. Wiesbaden: Reichert, 2012.

Pantin, Isabelle. "Analogy and Difference: A Comparative Study of Medical and Astronomical Images in Books, 1470–1550." *Early Science and Medicine* 18, no. 1–2 (2013): 9–44.

Paradin, Claude. *Symbola Heroica M. Claudii Paradini Belliiocensi Canonici, et D. Gabrielis Symeonis Multo quam Antea Fidelius de Gallica Lingua in Latinam Conuersa*. Antwerp: ex officina Christophori Plantini, 1583.

Paradin, Claude, and Gabriele Simeoni. *Les Devises Heroiques*. Antwerp: De l'imprimerie de Christophle Plantin, 1561.

Parikka, Jussi. *A Geology of Media*. Minneapolis. MN: University of Minnesota Press, 2015.

Park, Katharine. "Psychology: The Organic Soul." In *The Cambridge History of Renaissance Philosophy*, edited by Charles B. Schmitt, Quentin Skinner, Eckhard Kessler, and Jill Kraye, pp. 464–484. Cambridge: Cambridge University Press, 1988.

Park, Katharine. *Secrets of Women: Gender, Generation, and the Origins of Human Dissection*. New York: Zone Books, 2006.

Park, Katharine. "Allegories of Knowledge." In *Prints and the Pursuit of Knowledge in Early Modern Europe*, edited by Susan Dackerman, pp. 358–409. Cambridge, MA: Harvard Art Museum, 2011.

Parshall, Peter. "*Imago contrafacta*: Images and Facts in the Northern Renaissance." *Art History* 16 (1993): 554–579.

Parshall, Peter. "Graphic Knowledge: Albrecht Dürer and the Imagination." *The Art Bulletin* 95, no. 3 (2013): 393–410.

Pechstein, Klaus. "Hans Epischofer. Der Monogrammist HsE." In *Anzeiger des Germanischen Nationalmuseums 1970*, pp. 96–102. Nuremburg: Verlag des Germanischen Nationalmuseums, 1970.

Peckham, John. *Perspectiva communis*. [Milan]: Petrus de Corneno [Bonus Accursius?], 1482.

Peckham, John. *John Pecham and the Science of Optics: Perspectiva Communis*. Edited by David C. Lindberg. Madison, WI: University of Wisconsin Press, 1970.

Pérez de *Salazar, Francisco: Historia de la Pintura en Puebla. Mexico City. Universidad Nacional Autónoma de México, 1963*.

Perkins, Emma. "Instruments of Authority: Tycho Brahe's Technological Illustrations." *History and Technology* 34, no. 3–4 (2018): 259–272.

Peters, Emily. "Systems and Swells: The Collective Lineage of Engraved Lines 1480–1650." In *The Brilliant Line: Following the Early Modern Engraver, ca. 1480–1650*, edited by Emily J. Peters, 13–48. Providence, RI: Rhode Island School of Design Museum, 2009.

Pettegree, Andrew. "The Renaissance Library and the Challenge of Print." In *The Meaning of the Library: A Cultural History*, edited by Alice Crawford, pp. 72–102. Princeton, NJ: Princeton University Press, 2015.

Pigler, Andor. *Barockthemen Vol. 1*. Budapest: Akadémia Kiadó, 1974.
Pinder, Ulrich. *Epiphanie Medicorum*. Nuremburg: [Friedrich Peypus], 1506.
Plaßmeyer, Peter. "Christoph Schissler: The Elector's Dealer." In *European Collections of Scientific Instruments, 1550–1750*, edited by Giorgio Strano, Stephen Johnston, Mara Miniate, and Alison Morrison-Low, pp. 15–25. Leiden: Brill, 2009.
Pon, Lisa. *Raphael, Dürer, and Marcantonio Raimondi: Copying and the Italian Renaissance*. New Haven, CT: Yale University Press, 2004.
Popper, Nicholas. "'Abraham, Planter of Mathematics': Histories of Mathematics and Astrology in Early Modern Europe." *Journal of the History of Ideas* 67, no. 1 (2006): 87–106.
Popplow, Marcus. "Why Draw Pictures of Machines?: The Social Contexts of Early Modern Machine Drawings." In *Picturing Machines, 1400–1700*, edited by Edited by Wolfgang Lefèvre, pp. 17–48. Cambridge, MA: MIT Press, 2004.
Porteman, Karel. *Emblematic Exhibitions (Affixiones) at the Brussels Jesuit College (1630–1685). A Study of the Commemorative Manuscripts (Royal Library, Brussels)*. Turnhout: Brepols, 1996.
Porteman, Karel. "The Use of the Visual in Classical Jesuit Teaching and Education." *Paedagogica Historica* 36, 1 (2000): 179–196.
Préaud, Maxime, and Barbara Brejon de Lavergnée, eds. *L'Œil d'Or: Claude Mellan, 1598–1688*. Paris: Bibliothèque nationale, 1988.
Purtle, Jennifer. "*Double Take: Chinese Optics and their Media in Postglobal Perspective.*" *Ars Orientalis* 48 (2018): 89–96.
Raber, Karen, Treva J. Tucker, and Pia F. Cuneo, eds. *The Culture of the Horse: Status, Discipline, and Identity in the Early Modern World*. New York: Palgrave Macmillan, 2005.
Rangel, Luz M., and Aureli Alabert. "Two Corrections to the Recent Census of the Gutenberg Bible." Unpublished Manuscript, 2019. DOI: 10.31219/osf.io/mg59f.
Rasmussen, Eric. "The Date of Q4 'Hamlet.'" *Papers of the Bibliographical Society of America* 95 (2001): 21–29.
Raynaud, Dominique. *Eye Representation and Ocular Terminology from Antiquity to Helmholtz*. Amsterdam: Wayenborgh Publications, 2020.
Redín Michaus, Gonzalo. *Pedro Rubiales, Gaspar Becerra y los Pintores Españoles en Roma, 1527–1600*. Madrid: Departamento de Historia del Arte, Instituto de Historia, Consejo Superior de Investigaciones Científicas, 2007.
Reeves, Eileen. *Painting the Heavens: Art and Astronomy in the Age of Galileo*. Princeton, NJ: Princeton University Press, 1997.
Reeves, Eileen. "Mere projections: Sunspots and the Camera Obscura." *Galilaeana* 4 (2007): 47–77.
Reeves, Eileen. "From Dante's Moonspots to Galileo's Sunspots." *MLN* 124, no. 5 (2009): 190–209.

Reeves, Eileen, and Albert Van Helden. "Idiom and Image: Translating the Letters on Sunspots." *Isis* 109, no. 4 (2018): 767–773.

Remmert, Volker R. "On Picturing the Past: Arithmetic and Geometry as Wings of the Mind." *The Mathematical Intelligencer* 31, no. 3 (2009): 42–47.

Remmert, Volker R. *Picturing the Scientific Revolution: Title Engravings in Early Modern Scientific Publications.* Philadelphia, PA: St. Joseph's University Press, 2011.Remmert, Volker R. "Inventing Tradition in 16th- and 17th-Century Mathematical Sciences: Abraham as Teacher of Arithmetic and Astronomy." *The Mathematical Intelligencer* 37, no. 3 (2015): 55–59.

Remmert, Volker R. "Analyse des Imageries des Instruments Scientifiques de l'Époque Moderne (XVIe–XVIIe Siècle)." *Images des Mathématiques* (2019): https://images.math.cnrs.fr/Analyse-des-imageries-des-instruments-scientifiques-de-l-epoque-moderne-XVIe.html?lang=fr (accessed 3 July 2022).

Rheinberger, Hans-Jörg. *An Epistemology of the Concrete: Twentieth-Century Histories of Life*. Durham, NC: Duke University Press, 2010.

Ricci, Matteo. *Lettere: 1580–1609*, edited by Francesco d'Arelli. Macerata: Quodlibet, 2001.

Rice, Yael. "Lines of Perception: European Prints and the Mughal Kitābkhāna." In *Prints in Translation, 1450–1750: Image, Materiality, Space*, edited by Suzanne Karr Schmidt and Edward H. Wouk, pp. 202–223. London: Routledge, 2017.

Richeome, Louis. *Trois Discours pour la Religion Catholique. Des miracles, des Saincts et des Images.* Bordeaux: Simon Millange, 1598.

Ripa, Cesare. *Della Novissima Iconologia di Cesare Ripa Perugino Cavalier de SS. Mauritio & Lazzaro...* Padua: Per Pietro Paolo Tozzi, 1625.

Ripa, Cesare. *Iconologie où les Principales Choses qui Peuvent Tomber dans la Pensée Touchant les Vices sont Représentées*, edited by Jean Baudouin. Paris: s.n., 1643.

Roberts, K. B. and J. D. W. Tomlinson. *The Fabric of the Body: European Traditions of Anatomical Illustrations.* Oxford: Clarendon Press, 1992.

Roberts, Sean. "Francesco Rosselli and Berlinghieri's Geographia Re-Examined," *Print Quarterly* 28 (2011): 4–17.

Roberts, Sean. "Tricks of the Trade. The Technical Secrets of Early Engraving." In *Visual Cultures of Secrecy in Early Modern Europe*, edited by Timothy McCall, Sean Roberts, and Giancarlo Fiorenza, pp. 182–207. Kirksville, MI: Truman State University Press, 2013.

Roberts, Sean. "Inventing Engraving in Vasari's Florence." *Intellectual History Review* 24, no. 3 (2014): 367–388.

Rocca, Julius. "Galen and Greek Neuroscience (Notes Towards a Preliminary Survey)." *Early Science and Medicine* 3, no. 3 (1998): 216–240.

Rosenfeld, Amir, Richard Zemel, and John K. Tsotsos. "The Elephant in the Room." arXiv

Rosenkranzes," *Monumenta Serica* 60, no. 1 (2012): 195–290.

Röttel, Karl. *Peter Apian: Astronomie, Kosmographie und Mathematik am Beginn der Neuzeit. Mit Ausstellungskatalog*. Buxheim: Polygon-Verlag, 1995.

Rudolph, Pia. "Hausbücher. Bayerische Bild-Enzyklopädien." In *Katalog der deutschsprachigen illustrierten Handschriften des Mittelalters* (*KdiH*), vol. 6., edited by Ulrike Bodemann *et al.*, pp. 466–473. Munich: C.H. Beck, 2015.

Rudy, Kathryn. *Image, Knife and Gluepot: Early Assemblage in Manuscript and Print*. Cambridge: Open Book Publishers, 2019.

Russel, Daniel. "The Emblem and Authority." *Word & Image* 4, 1 (1988): 81–87.

Russel, Daniel. *Emblematic Structures in Renaissance French Culture*. Toronto: University of Toronto Press, 1995.

Sales, François de. *Traicté de l'Amour de Dieu*. Lyon: Pierre Rigaud, 1626.

Salmon, William. *Polygraphice, or, The Arts of Drawing, Engraving, Etching, Limning, Painting, Washing, Varnishing, Gilding, Colouring, Dying, Beautifying and Perfuming*. London: R. Jones, 1685.

San Juan, Rose Marie. "Restoration and Translation in Juan de Valverde's *Historia de la composision del cuerpo humano*." In *The Virtual Tourist in Renaissance Rome: Printing and Collecting the Speculum Romanae Magnificentiae*, edited by Rebecca Zorach, pp. 53–61. Chicago, IL: University of Chicago Library, 2008.

Sánchez-Molero, José Luis Gonzalo. "Antonio de Salamanca y los Libros Españoles en la Roma del Siglo XVI." In *Roma y España un Crisol de la Cultura Europea en la Edad Moderna*, edited by Carlos José Hernando Sánchez, pp. 335–365, vol. I. Madrid: Sociedad Estatal para la Acción Cultural Exterior, 2007.

Saunders, Alison. "Make Pupils Do Themselves: Emblems, Plays and Performances in Jesuit Colleges Seventeenth Century." In *The Jesuits and the Emblem Tradition*, edited by John Manning and Marc van Vaeck, pp. 187–206. Turnhout: Brepols, 1999.

Saunders, J. B. de C M., and Charles Donald O'Malley. *The Illustrations from the Works of Andreas Vesalius of Brussels with Annotations and Translations, a Discussion of the Plates and Their Background, Authorship and Influence, and a Biographical Sketch of Vesalius*. Cleveland, OH: World Publishing Company, 1950.

Savoi, Denis. "Le Cadran Solaire de Hauteur de Wenzel Jamnitzer de l'Oberservatoire de Paris." *Commission des Cadrans Solaires – Société Astronomique de France, Cadran Info* 36 (2017): 136–151.

Saxl, F. "A Spiritual Encyclopaedia of the Later Middle Ages." *Journal of the Warburg and Courtauld Institutes* 5 (1942): 82–142.

Sayre, Eleanor A. "Introduction." In *Albrecht Dürer Master Printmaker*. Boston, MA: Museum of Fine Arts, 1971.

Schadt, Hermann. *Die Darstellungen der Arbore Consanguinitatis und der Arbore Affinitatis Bildschemata in Juristischen Handschriften*. Tübingen: Wasmuth, 1982.

Schaller, Andrea. *Prunkuhren der Renaissance. Württembergisches Landesmuseum Stuttgart.* Stuttgart: Württembergisches Landesmuseum, 2001.

Schenck, Kimberly. "Goltzius's Use of Grey Ink." *Print Quarterly* 15, no. 2 (1998): 186–190.

Schiefsky, Mark. "Galen and the Tripartite Soul." In *Plato and the Divided Self,* edited by Rachel Barney, Tad Brennan, and Charles Brittain, pp. 331–349. Cambridge: Cambridge University Press, 2012.

Schoch, Rainer, Matthias Mende, and Anna Scherbaum, eds. *Albrecht Dürer. Das Druckgraphische Werk.* Vol. 1. Munich: Prestel, 2001.

Schwenke, Paul, and Karl Preisendanz. "Zwei neue Exemplare der Ablassbriefe von 1455." *Zentralblatt für Bibliothekswesen* 36 (1919): 175–177.

Selleslach, Kristof. "Discipline en Solidariteit in de Officina Plantiniana." Lecture, Friends of the Plantin-Moretus Museum, Antwerp, 11 September 2019.

Serjeantson, R. W. "Proof and Persuasion." In *The Cambridge History of Science,* edited by Katharine Park and Lorraine Daston, pp. 132–176. Cambridge: Cambridge University Press, 2006.

Sethares, William A., Margaret Holben Ellis, and C. Richard Johnson, Jr. "Computational Watermark Enhancement in Leonardo's *Codex Leicester.*" *Journal of the American Institute for Conservation* 59, no. 2 (2020): 87–96.

Settle, Thomas B. "An Experiment in the History of Science." *Science* 133, no. 3445 (1961): 19–23.

Sgard, Jean. *La Sainte Face de Claude Mellan. Étude des Bases Géométriques du Dessin.* Abbeville: Société d'Émulation Historique et Littéraire d'Abbeville, 1957.

Shea, William R., and Mariano Artigas. *Galileo in Rome: The Rise and Fall of a Troublesome Genius.* Oxford: Oxford University Press, 2003.

Sherman, William H. "On the Threshold: Architecture, Paratext, and Early Print Culture." In *Agent of Change: Print Culture Studies after Elizabeth L. Eisenstein,* edited by Sabrina A. Baron, Eric N. Lindquist, and Eleanor F. Shevlin, pp. 67–81. Amherst, MA: University of Massachusetts Press, 2007.

Shin, Junhyoung Michael. "The Reception of *Evangelicae Historiae Imagines* in Late Ming China: Visualizing Holy Topography in Jesuit Spirituality and Pure Land Buddhism." *The Sixteenth Century Journal* 40, no. 2 (2009): 303–333.

Shotwell, R. Allen. "Animals, Pictures, and Skeletons: Andreas Vesalius's Reinvention of the Public Anatomy Lesson." *Journal of the History of Medicine and Allied Sciences* 71, no. 1 (2016): 1–18.

Siegert, Bernhard. *Passagiere und Papiere. Schreibakte auf der Schwelle zwischen Spanien und Amerika.* Zürich: Neue Zürcher Zeitung, 2006.

Siegert, Bernhard. "The Map is the Territory." *Radical History* 169 (2011): 13–16.

Siegert, Bernhard. *Cultural Techniques: Grids, Filters, Doors and Other Articulations of the Real.* New York: Fordham University Press, 2015.

Skaarup, Bjørn Okholm. *Anatomy and Anatomists in Early Modern Spain*. Farnham: Ashgate, 2015.

Skaarup, Bjørn Okholm. "The Unexpected Success of a Spanish Anatomy Book: Juan Valverde de Amusco's Historia de la Composicion del Cuerpo Humano (Rome, 1556), and Its Many Later Editions." In *Specialist Markets in the Early Modern Book World*, edited by Richard Kirwan and Sophie Mullins, pp. 123–141. Leiden: Brill, 2015.

Smeyers, Maurits. "Een Collegeschrift van de Oude Leuvense Universiteit (1481–1482). Een Codicologisch en Iconografisch Onderzoek. Bijdrage tot de Studie van het Universitaire Onderricht Tjdens de Middeleeuwen." *Arca Lovaniensis* 4 (1975): 243–303.

Smith, Pamela H. *The Body of the Artisan: Art and Experience in the Scientific Revolution*. Chicago, IL: University of Chicago Press, 2004.

Smith, Steven Escar. "'Armadillos of Invention': A Census of Mechanical Collators." *Studies in Bibliography* 55 (2002): 133–170.

Sommervogel, Charles. *Bibliothèque des Écrivains de la Compagnie de Jésus*. Vol. 7 (Roeder-Thonhauser). Brussels/Paris: Schepens/Picard, 1896.

Spica, Anne-Élisabeth. "Les Jésuites et l'Emblématique." *Dix-Septième Siècle* 237, no. 4 (2007): 633–651.

Spiegel, Gabrielle. "Forging the Past: The Language of Historical Truth in the Middle Ages." *History Teacher* 17 (1984): 267–288.

Stabile, Giorgio. "Linguaggio della Natura e Linguaggio della Scrittura in Galilei. Dalla 'Istoria' Sulle Macchie Solari alle Lettere Copernicane." *Nuncius* 9 (1994): 37–64.

Standaert, Nicolas. *An Illustrated Life of Christ Presented to the Chinese Emperor: The History of Jincheng Shuxiang (1640)*. Sankt Augustin: Steyler Verlag, 2007.

Standaert, Nicolas, ed. *Chinese Voices in the Rites Controversy: Travelling Books, Community Networks, Intercultural Arguments*. Rome: Institutum Historicum Societatis Iesu, 2012.

Stijnman, Ad. *Engraving and Etching, 1400–2000: A History of the Development of Manual Intaglio Printmaking Processes*. London: Archetype, 2012.

Sugg, Richard. *The Smoke of the Soul: Medicine, Physiology and Religion in Early Modern England*. Basingstoke: Palgrave Macmillan, 2013.

Sullivan, Erin, and Andrew Wear. "Materiality, Nature and the Body." In *The Routledge Handbook of Material Culture in Early Modern Europe*, edited by Catherine Richardson, Tara Hamling and David Gaimster, pp. 141–157. New York: Routledge 2017.

Swan, Claudia. "Illustrated Natural History." In *Prints and the Pursuit of Knowledge in Early Modern Europe*, edited by Susan Dackerman, pp. 186–191. Cambridge, MA: Harvard Art Museums, 2011.

Taub, Liba. “On Scientific Instruments.” *Studies in History and Philosophy of Science* 40, no. 4 (December 2009): 337–438.

Taub, Liba. “Introduction: Reengaging with Instruments.” *ISIS* 102, no.4 (2011): 689–696.

Temkin, Owsei. “On Galen’s Pneumatology.” *Gesnerus: Swiss Journal of the History of Medicine and Sciences* 8, no. 1–2 (1951): 180–189.

Tercero Cathecismo y Exposicion de la Doctrina Christiana, por Sermones. Para qve los Cvras y Otros Ministros Prediquen y Enseñen a los Yndios y a las Demas Personas, Conforme a lo qve en el Sancto Concilio Prouincial de Lima se Proueyo. Lima: Antonio Ricardo, 1585.

Toesca, Ilaria, and Roberto Zapperi. “AGUCCHI, Giovanni Battista.” *Dizionario Biografico degli Italiani* 1 (1960), https://www.treccani.it/enciclopedia/giovanni-battista-agucchi_(Dizionario-Biografico)/ (accessed 27 February 2021).

Tomai, Camillo. *De Febribus Commentarius.* Venice: Andrea Arrivabene, 1555.

Torrentius, Laevinus. *Laeuini Torrentii in C. Suetonii Tranquilli 12. Caesares Commentarii.* Antwerp: ex officina Christophori Plantini, 1578.

Torrini, Maurizio. “Galileo Copernicano.” *Giornale Critico della Filosofia Italiana* 72, no. 1 (1993): 26–42.

Trémolières, François. “L’Art Sacré au Prisme de Fénelon.” *Dix-Septième Siècle* 230, no. 1 (2006): 49–69.

Turner, Anthony. “From Mathematical Practice to the History of Science: The Pattern of Collecting Scientific Instruments.” *Journal of the History of Collections* 7, no. 2 (1995): 135–150.

Tylecote, R.F. *A History of Metallurgy.* London: The Metals Society, 1976.

Underwood, Paul A. “The Fountain of Life in Manuscripts of the Gospels.” *Dumbarton Oaks Papers* 5 (1950): 41–148.

Valadés, Diego. *Retorica Cristiana.* Perugia: Petrus Jacobus Petrutius, 1579.

Valleriani, Matteo, ed. *The Structures of Practical Knowledge.* Dordrecht: Springer, 2017.

Valverde de Amusco, Juan. *Historia de la Composición del Cuerpo Humano.* Rome: Impressa por Antonio Salamanca, y Antonio Lafrerij, 1556.

Valverde, Juan. *Anatomia del corpo humano.* Rome: Salamanca et Lafreri, 1559.

Valverde de Amusco, Juan. *La Anatomia del Corpo Umano Composta da m. Giouanni Valuerde.* Venice: Giunti, 1586.

Valverde, Juan. *Anatome Corporis Humani Auctore Joanne Valverdo Nunc Primum a Michaele Columbo Latine Reddita et Additis Novis Aliquot Taulis Exornata.* Venice: Giunta, 1607.

Van den Wijngaert, Frank. “De late Moretussen en de Boekillustratie.” *De Gulden Passer* 25 (1947): 186–240.

Van der Aa, A. J. *Biographisch Woordenboek der Nederlanden*. Haarlem: J. J. van Brederode, 1869.

Van der Stock, Jan. *Printing Images in Antwerp: The Introduction of Printmaking in a City; Fifteenth Century to 1585*. Rotterdam: Sound and Vision Publishing, 1998.

Van Even, Edward. "Blendeff (Lambert)." In *Biographie Nationale vol. 2*, col. 470–471. Brussels: Thiry, 1868.

Van Helden, Albert. "Galileo and Scheiner on Sunspots: A Case Study in the Visual Language of Astronomy." *Proceedings of the American Philosophical Society* 140, no. 3 (1996): 358–396.

Van Helden, Albert, and Thomas L. Hankins. "Introduction: Instruments in the History of Science." *Osiris* 9 (1994): 1–6.

Van Vaeck, Marc en Johan Verberkmoes. "Humor in Collegedictaten." In *Ex Cathedra. Leuvense Collegedictaten van de 16de tot de 18de Eeuw*, edited by. Geert Vanpaemel *et al.*, pp. 187–121. Louvain: Universiteitsbibliotheek, 2012.

Van Vaeck, Marc. "Printjes [...] tot Ciraet van de Dictata Reeksen van Embleemgravures in de Leuvense Collegedictaten." In *Ex Cathedra. Leuvense Collegedictaten van de 16de tot de 18de Eeuw*, edited by Geert Vanpaemel *et al.*, pp. 151–168. Louvain: Universiteitsbibliotheek, 2012.

Vanpaemel, Geert. "The Louvain Printers and the Establishment of the Cartesian Curriculum." *Studium* 4, no. 4 (2011): 241–254.

Veldman, Ijla M. "Maarten van Heemskercks Visie op het Geloof." *Bulletin van het Rijksmuseum* 35, no. 3 (1987): 193–210.

Venema, Cornelis P. "The Doctrine of the Lord's Supper in the Reformed Confessions." *Mid-America Journal of Theology* 12 (2001): 135–199.

Vermeir, René. *In Staat van Oorlog. Filips IV en de Zuidelijke Nederlanden, 1629–1648*. Maastricht: Shaker Publishing, 2001.

Vesalius, Andreas. *Andreae Vesalii Bruxellensis, Scholae Medicorum Patauinae Professoris, Suorum de Humani Corporis Fabrica Librorum Epitome*. Basel: ex officina Ioannis Oporini, anno 1543 mense iunio.

Vesalius, Andreas. *De Humani Corporis Fabrica Libri Septem*. Basel: Ex Officina Ioannis Oporini, 1543.

Vesalius, Andreas. *De Humani Corporis Fabrica Libri Septem*. Basel: Ex Officina Ioannis Oporini, 1555.

Vesalius, Andreas. *Andreae Vesalii Anatomia*. Venice: Apvd J. A. et J. de Franciscis, 1604.

Vesalius, Andreas. *Icones anatomicae*. Munich: Bremer Presse, 1934.

Vesalius, Andreas. *The Fabric of the Human Body: An Annotated Translation of the 1543 and 1555 Editions of "De Humani Corporis Fabrica Libri Septem."* Transl. by Daniel H. Garrison and Malcolm H. Hast. 2 vols. Basel: Karger, 2014.

Vidal, Fernando. *The Sciences of the Soul: The Early Modern Origins of Psychology.* Transl. by Saskia Brown. Chicago, IL: University of Chicago Press, 2011.

Voet, Leon. *The Golden Compasses. The History of the House of Plantin-Moretus.* Amsterdam: Vangendt & Co, 1969.

Voet, Leon. *The Plantin Press, 1555–1589: A Bibliography of the Works Printed and Published by Christopher Plantin at Antwerp and Leiden.* Amsterdam: Van Hoeve, 1982.

Vons, Jacqueline. "Books for Learning about the Body." In *Art of Vesalius*, edited by R. van Hee, pp. 51–70. Antwerp: Garant, 2014.

Wadell, Maj-Brit. *Fons Pietatis. Eine Ikonographische Studie.* Translated by Dieter Rosenthal and Anne-Marie Thiberg. Göteborg: Elanders Boktryckeri Aktiebolag, 1969.

Walkling, Andrew R. "Hinman, Redux." *The Collation Blog.* May 2018. https://collation.folger.edu/2018/05/hinman–redux/ (accessed 3 July 2022).

Webster, Susan Verdi. "The Secret Lives of Buildings in Colonial Quito." *Construction History* 28, no. 3 (2013): 21–46

Wei-Hsuan Chen, Jessie. "The Blocks, the Tools, and the Printed Plants: The Making and Printing of Botanical Woodblocks at the Early Modern *Officina Plantiniana.*" Master's thesis, Utrecht University, 2017–2018.

Wei-Hsuan Chen, Jessie. "A Woodblock's Career: Transferring Visual Botanical Knowledge in the Early Modern Low Countries." *Nuncius: Journal of the Material and Visual History of Science* 35, no. 1 (2020): 20–63.

Westermann, Ekkehard. "Copper Production, Trade, and Use in Europe from the End of the Fifteenth Century to the End of the Eighteenth Century." In *Copper as Canvas: Two Centuries of Masterpiece Paintings on Copper 1575–1775*, pp. 117–130. New York: Oxford University Press, 1998.

White, Raymond, and Jo Kirby. "Rembrandt and His Circle: Seventeenth-Century Dutch Paint Media Re-examined." *National Gallery Technical Bulletin* 15 (1994): 64–78.

Wilding, Nick. "Galileo and the Stain of Time." *California Italian Studies* 2, no. 1 (2011): https://escholarship.org/uc/item/7rh9r96p (accessed 5 March 2021).

Wouk, Edward H. "Toward an Anthropology of Prints." In *Prints in Translation, 1450–1750. Image, Materiality, Space*, edited by Suzanne Karr Schmidt and Edward H. Wouk, pp. 1–18. London/New York: Routledge, 2017.

Wouters, Jan. *Dat Epitome ofte Cort Begriip der Anatomien.* Bruges: P. de Clerck, 1569.

Wunderlich, Herbert. *Das Dresdner 'Quadratum geometricum' aus dem Jahre 1569 von Christoph Schißler d. A.* Berlin: VEB Deutscher Verlag der Wissenschaften, 1960.

Yi, Qu and 藝 曲. Song Nianzhu Guicheng (Die Anweisung zur Rezitation des Rosenkranzes," *Monumenta Serica* 60, no. 1 (2012): 195–290.

Zanen, Sylvia van. *Planten op Papier. Het Pionierswerk van Carolus Clusius (1526–1609)*. Zutphen: WalburgPers Algemeen, 2020.

Zhou, Yuqing, *et al.* "Neural Dynamics of Racial Categorization Predicts Racial Bias in Face Recognition and Altruism." *Nature Human Behaviour* 4, no. 1 (2020): 69–87.

Zorach, Rebecca. *Blood, Milk, Ink, Gold: Abundance and Excess in the French Renaissance*. Chicago, IL and London: The University of Chicago Press, 2005.

Index